Charles Cardale Babington

Flora of Cambridgeshire

A catalogue of plants found in the county of Cambridge

Charles Cardale Babington

Flora of Cambridgeshire
A catalogue of plants found in the county of Cambridge

ISBN/EAN: 9783337271695

Printed in Europe, USA, Canada, Australia, Japan

Cover: Foto ©berggeist007 / pixelio.de

More available books at **www.hansebooks.com**

Flora of Cambridgeshire:

OR

A CATALOGUE OF PLANTS

FOUND IN

The County of Cambridge,

WITH REFERENCES TO FORMER CATALOGUES, AND THE
LOCALITIES OF THE RARER SPECIES.

BY

CHARLES CARDALE BABINGTON, M.A.

F.R.S. F.L.S. &c.

"Turpe est in patria vivere et patriam ignorare." LINN.

LONDON:
JOHN VAN VOORST, PATERNOSTER ROW,
1860.

PREFACE.

AN interval of two hundred years has elapsed since the publication by the celebrated Ray of the first *Catalogue* of the plants found in the county of Cambridge, and forty since the third and last edition of Relhan's *Flora Cantabrigiensis* appeared. As even the latter of these works has now become nearly obsolete, owing to the great advances which have been made in Botany since Relhan wrote, a new Catalogue seems to be desirable. I therefore venture to place the present work before botanists notwithstanding its many imperfections.

The list contained in this book is the result of my own researches, extending over a long period, combined with those of other botanists, who have kindly informed me of their discoveries and usually directed me to the spots where the plants grew. When I have seen the plant in any locality, the station is usually recorded as resting upon my own authority; for I unfortunately many times neglected to make a note of the name of my original informant. Those to whom I am the most indebted in this respect are Professor Henslow, the Rev. W. H. Coleman (one of the authors of the *Flora Hertfordiensis*), the Rev. W. W. Newbould, Mr S. W. Wanton, M.A., Mr D. Britten and Mr H. Fordham, of Royston, Mr Algernon Peckover and family and Mr J. Balding, of Wisbech, Miss A. M. Barnard, formerly of Odsey, Mr W. Marshall, of Ely, and Mr G. S. Gibson, of Saffron Walden. Prof. Henslow permitted me to make use of his interleaved and annotated copy of Relhan's *Flora*. Mr Wanton, who most assiduously traced out and confirmed a great number of Relhan's localities of plants, allowed me to transcribe all his notes. Mr Coleman left an extensive list of the localities of plants in the hands of the Rev. F. W. Collison, late Fellow of St John's College, who transferred it to me. Mr Newbould has given me the greatest possible assistance by the communication of notes and specimens; indeed without his help I could not have now ventured to offer this book to botanists. The appearance of his initial upon nearly every page will shew the great extent and value of his contributions. Messrs Britten and Fordham have supplied very full lists of the plants found near Royston. Miss Barnard, grand-niece of the late Sir J. E. Smith, sent a catalogue of plants found about Odsey, at the extreme south-western corner of the county. Mr Peckover and members of his family added largely to the Flora of Wisbech, to which Mr Balding also made a valuable contribution. To Mr Marshall I am indebted for many localities of plants in the Isle of Ely. And to Mr G. S. Gibson for notes concerning the plants inhabiting the south-eastern part of the county.

Mr Relhan, son of the author of the *Flora*, placed in my hands a copy of that book containing a few additions made by his father. A copy of the second volume of the first edition of Berkenhout's *Outlines of the Natural History of Great Britain and Ireland* has fallen into my hands, which belonged to "J. Fisher, B.A. of Christ's College," shortly after its publication in 1770. Very many localities for Cambridgeshire plants, noticed apparently by him, are recorded in the margins of this book. He seems to have been the Dr John Fisher, LL.D. who subscribed for a copy of the first edition of the *Flora Cantabrigiensis* in 1785, and the stations were probably nearly all communicated to Relhan. It is a curious list, and shews the great extent of Dr Fisher's researches. His name deserves to be recorded as an active contributor to our knowledge of the local Flora.

A few plants from Cambridgeshire are preserved, with their localities, in Relhan's *Herbarium*, which now belongs to the Linnean Society of London. Unfortunately the great majority of his specimens have no notes of the places where they were found appended to them. The *Herbarium* is thus rendered of very little value.

There are also some curious entries made by Mr R. Jackson in a copy of Martyn's *Methodus* preserved in the library of Trinity College: occasionally they are dated 1730.

Mr Jackson is not mentioned in the list of those 118 men "who, though they never published anything upon the subject, have nevertheless contributed in some degree to improve this part of the natural history of their native country," which is given by Professor T. Martyn in his *Plantæ Cantabrigienses*. But the Rev. Richard Jackson, the founder of the Jacksonian Professorship, was a Fellow of Trinity College, and took his B.A. degree in the year 1727, his M.A. in 1731. That he took an interest in Botany is shewn by his founding a perpetual annuity to be paid to the "head or chief gardener of the University Physic Garden," and by several of the directions given for the guidance of his Professor. In all probability he was the R. Jackson to whom the *Methodus* belonged.

It is hoped, and earnestly requested, that those botanists who may use this *Catalogue* will communicate to me any additional localities that they may observe, or any confirmation of the older ones, and point out such improvements in the book as may occur to them.

St John's College, Cambridge.
26 *April*, 1860.

INTRODUCTION.

THOSE botanists who have resided in the University having for at least two hundred years made the Flora of Cambridgeshire a subject of study, many works relating especially to it have issued from the press. It will be well to give some account of them. They commence with the celebrated and singularly excellent *Catalogus Plantarum circa Cantabrigiam nascentium* of Ray, which was published in the year 1660. This forms a small 12mo volume of 182 pages; and, after deducting all the plants which were cultivated or otherwise do not come within the plan of the present Flora, it contains 671 plants found in this county by Ray. The names are arranged alphabetically, and this, in addition to the obscurity attendant upon the old nomenclature of plants, renders the book rather difficult to consult.

In 1663 Ray published an *Appendix* of 13 pages, in which 37 plants were added; and of this a second edition, consisting of 30 pages, was edited by Mr Peter Dent, an apothecary of Cambridge, in 1685. Mr Dent inserted in this edition 59 more plants unnoticed in the *Catalogus*. These additions are made almost wholly in the words of

Ray, as found in his *Catalogus Plantarum Angliæ*. Concerning this botanist Ray remarks, "D. Pet. Dent, Medicus pharmacopœus Cantabrigiensis, insignis botanicus et vetus amicus noster." *Hist. Plant.* ii. 856. These Appendices, especially the second edition, have long been of great rarity.

In 1670 Ray published his *Catalogus Plantarum Angliæ*, and in 1677 he issued a second edition of it. He states in the Preface that as all the copies of the *Cat. Pl. Cantabr.* were sold, he had contemplated a new edition of that book, but had ultimately determined to extend its range so as to include "totius Britanniæ stirpium;" but at the same time to render it convenient for use at the University by marking the plants of Cambridgeshire. Accordingly, if we take account solely of the plants to the names of which there is prefixed a "C," we have a second and a third edition of the *Catalogus Pl. Cantabr.* in these two editions of the *Cat. Pl. Angliæ*. The seventeen years intervening between the first and last of these publications did not add very much to the Flora of this county; but it must be remembered that Ray was deprived of his Fellowship by the Bartholomew Act, and ceased to reside at Cambridge in the autumn of 1662. In 1695 he contributed a list of the rarer plants of this county to Gibson's edition of *Camden's Britannia*, but I do not find any new information in it.

In 1727 Professor John Martyn arranged the plants of Ray's *Catalogus* and its *Appendices* according to the then most approved system of classification, in a little work entitled, *Methodus Plantarum circa Cantabrigiam nascentium*. Although no new plants are to be found in this book, it must have been most useful at the time of its compilation. It appears never to have been regularly published, and was prepared for the use of the students attending his first course of botanical lectures at Cambridge. In addition to the species derived from the *Catalogus*, it contains the generic characters taken from Ray's *Methodus emendata et aucta*, and other books, much improved by Martyn's own observations. Its value now consists in its containing the whole of Ray's and Dent's *Appendices*, although so intermixed with Martyn's own remarks as to render their separation a business of some difficulty. Professor J. Martyn commenced a second edition of the *Methodus*, as we learn from his son, the Rev. Professor T. Martyn; but of this, which I have not seen, and of which a copy probably does not exist, only a sheet and a half were printed. Prof. T. Martyn included the new plants noticed in that abortive edition and others recorded in manuscript by his father, to the amount of 150, in his *Plantæ Cantabrigienses*, published in 1763. Only the names of the plants are given in the *Pl. Cantabr.*, and they are arranged according to the Linnæan system and nomenclature. The chief value of this book is caused by the Professor having added references to Martyn's *Methodus* and Ray's *Catalogus;* thus rendering those works more accessible to the modern student.

The *Plantæ Cantabrigienses* records 829 "distinct sorts of plants" as inhabiting Cambridgeshire. Appended to it are *Herbationes Cantabrigienses*, in which many localities of plants are recorded.

In the same year, 1763, but three months later, Mr Israel Lyons published a *Fasciculus Plantarum circa Cantabrigiam nascentium, quæ post Rajam observatæ fuere*. It contains 105 species, arranged after the Linnæan Method. The specific characters are given, chiefly in the words of

Linnæus, and numerous synonyms are added. An account of Mr Lyons will be found in Cooper's *Annals of Cambridge,* iv. 381.

The next work relating to the plants of our county is the *Flora Cantabrigiensis* of Relhan. The first edition of this book appeared in 1785 and was quickly followed by three *Supplements,* published in 1786, 1788, and 1793 respectively. The second edition of the *Flora* issued from the press in 1802, and the third and last in 1820. Messrs Turner and Dillwyn inform us that Relhan had "himself seen every species in the station he has assigned to it in his *Flora;*" but I have some doubt concerning the correctness of this statement, because many of the localities seem to be only repetitions of those recorded by his predecessors. Mr H. C. Watson thinks that Relhan's *Flora* included "the borders of the counties adjoining Cambridgeshire," and extended "a considerable distance into Norfolk;" but that seems not to be the case. Relhan included the small piece of Suffolk adjoining the town of Newmarket (as also did Ray); but, as far as I have seen, gives no other locality which lies beyond the boundaries of the county, and omits several so situated which are recorded by Ray in the *Catalogus*. Probably Mr Watson was misled by the fact that there are places named Swaffham and Thetford in Cambridgeshire as well as in Norfolk.

Relhan's *Flora* contains a valuable account of the Mosses, Algæ, and Fungi found in the county. These orders are not included in the present Catalogue on account of their study having been too much neglected to allow of a proper list of them being now made.

The localities of the rarer species found in the county

are given in Turner and Dillwyn's *Botanist's Guide;* derived chiefly, if not wholly, from Relhan's *Flora*. It was published in 1805.

In 1829 Prof. Henslow published a *Catalogue of British Plants*, in which he marked, by printing their names in *Italics*, the species "not found in Cambridgeshire;" and in 1835 he issued a second edition of the *Catalogue*, where the letter "c" is appended to the names of all the plants "included in the Flora of Cambridgeshire."

In the first volume of the *New Botanist's Guide*, also dated 1835, Mr H. C. Watson gave a list of localities for plants found in this county, mostly taken from Relhan's *Flora*, but with a few additions. In the second volume of the *Guide*, dated 1837, he added a rather long supplement to this list, derived chiefly from notes supplied to him by the Rev. W. H. Coleman.

Since that year only a few localities of Cambridgeshire plants have been announced in the various periodical and other publications devoted to Botany. Amongst them must especially be noticed Mr G. S. Gibson's "Flora of the neighbourhood of Saffron Walden." It is to be found in the first volume of the *Phytologist* (pages 408, 838, and 1123), and contains many plants of Cambridgeshire.

It is proper in the study of a Flora which is so much mixed up with the History of Botany in England to endeavour to trace each plant to the person by whom it was first noticed or recorded. Accordingly a reference will be found under each species to the earliest writer who mentions it as a native of our county, and when possible the name of the original finder of it is stated. Also it seems highly desirable to record not merely the present state of our Flora but also its condition before the great alterations caused by modern enclosures and drainage: alterations still advancing so rapidly that probably many of the places in which I have myself gathered plants within the last few years do not now produce them. The localities given by the older Botanists, but which have not been confirmed by recent observers, are inserted on their authority and markedly separated from the rest by being printed in Italics, so as to point out their historical not modern character.

Several of the less accessible parts of the county have not been examined so fully as could be desired; and indeed the publication of this book has been delayed in the hope of removing that defect from it. As it does not seem probable, from what we know of them, that the distant parts of the Fens possess plants differing from those found in other portions of that district, the want of their more thorough examination does not warrant longer delay.

It is now generally considered that the introduction of descriptions of the plants, or even their generic and specific characters, into a local Flora, such as the present, only swells its bulk without adding proportionably to its usefulness. The time when such descriptive local Floras were useful seems to have long since passed. The only modern books of the kind which now attract attention are rendered valuable, not by their local peculiarities, but from their authors having inserted so much original matter as to render them fragments of a general Flora. In this *Catalogue* I have naturally adopted the names and the arrangement used in my own *Manual*, edition 4; but have always quoted the above mentioned works of Ray, the Martyns, Lyons, and Relhan, and occasionally referred to some other book in which a plant is noticed as a native of Cambridgeshire.

The ponderous and inconvenient character of the specific names used by Ray and J. Martyn cannot fail to attract the attention of those botanical students who may use this book, and will shew to them how much we are indebted to Linnæus for the introduction of the binomial nomenclature now adopted in Botany and Zoology. This which seems to us to be so simple and natural an arrangement will be seen not to have occurred even to so eminent a man as Ray, and ought to be regarded as perhaps the firmest support of the reputation of Linnæus. His classification may be given up; but the use of his system of nomenclature must continue. Let us therefore never forget what we owe to the great Linnæus.

TOPOGRAPHICAL REMARKS.

CAMBRIDGESHIRE may be described as flat and naked; nevertheless it is not so absolutely flat as is generally supposed. A range of conspicuous chalk-hills extends across the southern part of the county; and the south-eastern district, consisting also of chalk, is undulating and well furnished with wood. To the north of the chalk country a broad belt of level clayey land occurs, having much flint gravel distributed over its surface. Formerly each watercourse traversing this clayey district was bordered, more or less widely, by a morass, and some of its depressed parts formed tracts of fen.

Speaking generally, a line drawn from Biggleswade to Newmarket will have the chalk on its southern and the clay (Gault and Boulder clay) on its northern side. Another line connecting Huntingdon with Newmarket would separate the clay-district from the great fen-country, which occupies the whole of the northern half of the county, and is usually known as the Isle of Ely.

Chalk Country.—Until recently (within 60 years) most of the chalk district was open and covered with a beautiful coating of turf, profusely decorated with *Anemone Pulsatilla, Astragalus Hypoglottis,* and other interesting plants. It is now converted into arable land, and its peculiar plants mostly confined to small waste spots by road-sides, pits, and the very few banks which are too steep for the plough. Thus many species which were formerly abundant have become rare; so rare as to have caused an unjust suspicion of their not being really natives to arise in the minds of some modern botanists. Even the tumuli, entrenchments, and other interesting works of the ancient inhabitants have seldom escaped the rapacity of the modern agriculturist, who too frequently looks upon the native plants of the country as weeds, and its antiquities as deformities.

Clayey District.—Until within about sixty years of the present time the whole of the clay district was open, although cultivated. The homesteads were collected together so as to form villages, and each had one or two little paddocks attached to it; the remainder of the parish, the "field," being without fences, and divided by slender lines of ancient turf, denominated "balks," into long narrow strips, called "yard lands." With a very few slight exceptions all the "field" is now inclosed, and the "balks," with the various plants which grew upon them, destroyed by the plough. Thus the plants native to the clay have suffered nearly as much as those indigenous to the chalk. Where they were once abundant they are now rarely to be found.

Fens.—The Fens are usually supposed to be absolutely flat and depressed below the level of the sea. But such is not the fact; since they slope with tolerable regularity from their inner border to the coast so as once to have possessed a natural drainage. Dugdale states that there was in the seventeenth century a fall of 10 feet from the general level of the country to low water-mark at the junction of Salter's

Lode with the Ouse, which takes place at a little below Denver Sluice. There seems to be the strongest reason to believe that in the time of the Roman occupation of the country the greater part of the Fen was kept tolerably dry by the natural drainage; and that the defence given by the great sea-banks, which are still to be seen near Wisbech, was only required to keep out exceedingly high tides driven by a northerly wind into the estuary, which was of great extent. The country then formed a swampy plain interspersed with drier places and deep morasses. Land near Thorney, which is now thoroughly fenny, was, even as late as the reign of Henry II. (1154—89), covered with orchards and vineyards, aud quite a paradise ("paradisi simularum." *Will. of Malms. De gestis Pontif.*, in *Script. post Bedam*, ed. 1601, p. 294). The remains of many ligneous plants have been found at the bottom of the peat, rooted in the clay upon which they grew, such as the Hazel. Wells tells us that near Whittlesey, "in digging through the moor, at a depth of eight feet, the labourers came upon a perfect soil, and swarths of grass lying thereon, as if it had been newly mown." Dugdale states that oak and fir trees have been found with their roots in the firm earth below the peat; and that in Marshland, a part of the Fen lying between Wisbech and the sea, furze-bushes and nut-trees rooted in the solid earth were met with at 16 feet below the present surface. Indeed there can be no doubt that the trees, fir, oak, yew, &c. really grew on the soil which is now deeply buried beneath the peat.

In the course of time the outlets of the rivers became choked with the sediment brought from the upper country, and the water was driven back so as to flood the greater part of the district. A map of it in that condition will be found in Dugdale's *Imbanking and Draining*. The worst condition seems to have been attained at about the year 1600. This unfortunate state of things was attempted to be improved, with variable success, by extensive drainage works, most of which are now considered to have been devised on a wrong plan; and, although parts of the district were very much improved by them, it is only recently that the natural drainage has been to any considerable extent restored by the thorough clearing of the mouths of the rivers. Thus in large tracts near to the coast the pumping mills are dispensed with and the water escapes to the sea at nearly all times.

Small spots formed of mounds of Boulder Clay or Gravel, and the true Isle of Ely, which consists of an outlier of the Lower Green Sand, are slightly raised about the peaty flats, and are scattered like islands (which formerly they often were) over the Fens.

Botanically speaking the Fens have undergone an equally if not more destructive change than the Chalk district. The employment of steam has made the removal of the water so certain that nearly the whole level may be cited as a pattern in farming. With the water many of the most interesting and characteristic plants have disappeared, or are become so exceedingly rare that the discovery of single individuals of them is a subject for wonder and congratulation. There is scarcely a spot remaining (I only know of one, near Wicken) in which the ancient vegetation continues undisturbed and the land is sufficiently wet to allow of its coming to perfection. Owing to the necessary existence of numerous ditches, to divide the fields and collect the water, those plants which are absolutely aquatic have not suffered so greatly as the others; but they are fast decreasing, now that the steam-engine causes even many of the ditches to be dry in summer. As the character of the Fen district is very little known, it is well to remark that the peat is not formed of *Sphagnum*, like that of bogs, but consists chiefly of the decomposed remains of various aquatic herbaceous plants. At the bottom there is a layer formed mostly of the remains of the woody plants and trees which constituted the forest which formerly covered the country. The remains of oak, yew, hazel and willow are found abundantly in some parts of the Fens, and pine wood is plentiful in others. The wood of the larger trees is often well preserved and turned quite black, but a few inches of the surface have become soft and spongy. The latter is the condition of most of the smaller branches and the lesser ligneous plants. The Rev. Leonard Jenyns informs us that it is the opinion of the turf-cutters at Isleham that, before the present more perfect drainage of the Fen, the turf grew at the rate of about twenty inches in sixteen years. Now the want of sufficient water has put an end to this restoration of the turf in the places where it has been cut for fuel, and what little is obtained for that purpose is of very inferior quality to the former supply. The ditches in the Fen and the holes made by the turf-cutters are soon occupied by a few plants, such as *Chara hispida, Utricularia vulgaris, Callitriche,* several species of *Potamogeton, Sagittaria,* and *Alisma ranunculoides.* As soon as these have formed a tolerably firm mass by the decomposition of their lower parts, the *Junci, Carices, Cladium,* and similar plants, establish themselves upon it. It then ceases to increase much in height, but in the course of years becomes firm enough to bear the weight of a man, although long affording a very treacherous footing. (See "Sedgwick on the Geology of Cambridge," in *Rep. Brit. Assoc.* 1845, *Sections*, p. 40, and "Jenyns on the Turf of Cambridgeshire," in the same *Report*, p. 75.

Formerly the natural produce of the unreclaimed Fen land was cut, at intervals of two or three years, dried, formed into bundles, and sold in the neighbouring towns and villages for use in the lighting of fires, and for other purposes. Within the last twenty years large quantities were so used in our Colleges, but now its scarcity has caused other materials to be employed. Then whole gangs of barges loaded with sedge were often to be seen on the Cam, arriving at this town; now a single barge load is only to be observed occasionally.

In addition to these large districts of Chalk, Clay, and Fen, there are three small and peculiar tracts to notice. At the extreme west of the county the parish of Gamlingay stands chiefly upon the Lower-green-sand Formation, and included, until recently, a broad sandy heath, and extensive quaking bogs. The enclosure of the Heath, and especially the effectual drainage of the bogs, has been destructive to some plants, although most of those which inhabited the Heath may still be found.

The following list contains the names of some of the plants which are especially characteristic of the Gamlingay district.

Teesdalia nudicaulis.	Trifolium subterraneum.
Moenchia erecta.	Vicia sativa β angustifolia.
Sagina ciliata.	Ornithopus perpusillus.
Tilia parvifolia.	Peplis portula.
Genista anglica.	Filago apiculata.

Gnaphalium sylvaticum.
Senecio sylvaticus.
Arnoceris pusilla.
Hypochæris glabra (extinct).
Erica tetralix.
Erica cinerea.
Vaccinium oxycoccos (extinct).
Teucrium Scorodonia.
Centunculus minimus (extinct).
Littorella lacustris (extinct).
Malaxis paludosa (extinct).
Narthesium ossifragum.
Rhynchospora alba (extinct).
Carex dioica (extinct).
C. curta.
Aira flexuosa.
Nardus stricta.
Polystichum aculeatum (extinct).
Blechnum boreale.
Lycopodium clavatum.
L. inundatum (extinct).
Chara syncarpa.

A small portion of a botanically interesting sandy district occurs just within the eastern edge of the county. It extends far into the counties of Suffolk and Norfolk, Mildenhall, Brandon and Thetford being situated in it. The parish of Chippenham includes this small tract, which is almost entirely situated between the village of that name and the county of Suffolk. The soil consists chiefly of very loose sand and gravel overlaying the chalk formation. Although it is almost wholly under cultivation it produces some peculiar plants : the following are especially deserving of mention.

Silene anglica.
S. Otites.
Medicago falcata.
M. sylvestris.
M. minima.
Ornithopus perpusillus.
Galium anglicum.
Apera interrupta.

The third of these smaller districts is the most changed of them. It is situated at the extreme northern point of the county, both above and below the town of Wisbech, and also on both sides of the river at Foul Anchour, near Tydd. It was formerly a salt-marsh, but is now nearly, or, for the most part, quite dry. The soil is not peat, but a kind of silt deposited by the river in the ancient estuary. The actual banks of the river produce a few marine plants, but the extensive salt-marshes are nearly gone, and with them much of their peculiar vegetation.

The following list includes the names of the most characteristic plants which remain, together with those of a few which were gathered there formerly, but have not been seen by any botanist for many years.

Frankenia lævis.
Althæa officinalis.
Lepigonum marinum.
Aster Tripolium.
Bidens tripartita.
Glaux maritima.
Statice Limonium.
Statice caspia.
Armeria maritima.
Plantago maritima.
Sueda maritima.
Beta maritima (extinct).
Salicornia herbacea.
Atriplex littoralis.
A. Babingtonii.
Obione pedunculata (extinct).
Obione portulacoïdes.
Triglochin maritimum.
Ruppia (rostellata?)
Sclerochloa maritima.
S. distans.
S. loliacea.
Lepturus incurvatus.
Triticum pungens.
Hordeum maritimum.

Rivers.—Little need be said concerning the rivers of this county. With the exception of the Cam, they are wholly included in the Fens, are often raised high above the neighbouring country, and have lofty banks to confine their waters. Within these banks a space is left, on one or both sides of the usual channel, to convey the water brought down by floods, and prevent it from overflowing the adjoining country. This space, called the "wash," is often of considerable extent, (that between the great cuts, known as the Bedford Rivers, varies from half to three quarters of a mile in width,) it is constantly liable to be overflowed, and usually forms boggy pasture abounding in the plants which inhabit such places.

The Cam is formed by two, or perhaps we should say three, small rivers or brooks. The two larger of these rise in the chalk district and unite with one another and with the third stream within a very few miles of Cambridge. After leaving the chalk, which one of them does very near to its source, and the other at only a short distance above their junction, these streams are continuously bounded by a narrow belt of low marshy land until the joint-river enters the Fens at a short distance below Cambridge. From thence downwards it has the usual character of the fen-rivers. The former of these streams rises in a very powerful spring at Ashwell in Hertfordshire, and its whole course is nearly parallel to the range of chalk-hills which divide this county from those of Hertford and Essex; the second stream rises near Quendon in Essex, and passes rapidly over a gravelly and chalky soil until after entering our county. It then acquires the swampy borders and sluggish character common to our streams. The names of these streams are much confused. Either is called the Cam or Rhe, according to fancy; but, if any weight is to be given to the Celtic meaning of those names, the first described is the Cam (crooked or meandering), the latter the Rhe (swift). The third stream contributing to the formation of the Cam (or Grauut river, as it is properly called from Grantchester to Harrimere[1]) is only known as the Bourn Brook. It passes through a clay-country, and usually conveys a small volume of water, although it is sometimes much swollen by floods.

Districts.—How very difficult it is to divide this county into districts, otherwise than artificially, will be seen from the above remarks. The river-basins cannot be used, for we can hardly be said to have more than one. Indeed it is rather the elevations that point out a difference in the vegetation than the depressions. The fens also are so uniform in character throughout their whole extent that, but for that extent, they might well be considered as a single district. An endeavour has been made to give as natural a character as possible to the districts: to keep the chalk country separate from the gault: and that from the fen. To do this artificial boundaries are unavoidable; for it is requisite that they should be lines easily to be traced upon the map, and recognized in the country. Turnpike-roads and main water-courses are therefore used as boundaries. The county is thus divided into eight districts, of which Nos. 1 and 2 include the main mass of Chalk. No. 3 contains most of the Drift Clay, together with the small tract of Lower Green Sand at Gamlingay: 4 and 5 occupy the country bordering on the fen (most of which was formerly very swampy) and a small portion of the fen itself. For fen and "high land," as it is locally called, are so intermixed at their edges that no satisfactory line can be drawn to separate them. The small patch of Suffolk sand at Chippenham is included in No. 5. The remaining

[1] This name is now nearly obsolete. It was a morass or lake formed by the junction of the old channel of the Ouse with the Cam, at about five miles above the city of Ely.

three districts (6, 7, and 8) are wholly in the Fens: the only "high land" in them consists of the fen "islands," which are spots elevated sufficiently to escape being flooded in times when the fen was often overflowed. No. 8 extends to the ancient estuary of the river Nene, but did not quite reach the sea even at the time when the boundaries of the counties were fixed, for Norfolk and Lincolnshire join at the old mouth of the river.

An endeavour must now be made to define the limits of these districts which are severally named from some chief place included in them. Each district reaches to the boundary of this and the neighbouring counties.

No. 1. *Cambridge* is bounded to the south and east by the counties of Suffolk and Essex; to the north and west by the two turnpike-roads which lead from Newmarket and from Chesterford to Cambridge: except that when the former arrives at the place called the Paper Mills the boundary of my division descends the stream called the Stour to the river Cam, and ascending that river to the mouth of the stream called the Vicar's Brook, by the side of Cow Fen, ascends that brook until the road to Trumpington and Chesterford is attained, at one mile from Cambridge. This district, besides the chalk range, extending from the Gogmagog Hills to Linton in one direction and to Newmarket in another, contains also the only well-wooded tract that we have, which consists of the parishes intervening between those chalk-hills and the county of Suffolk.

No. 2. *Royston.* This is bounded to the south and west by the neighbouring counties of Essex, Hertford and Bedford, and to the east by the Cambridge and Chesterford road. The northern boundary follows the road from Trumpington to Royston as far as the cross-roads at Shepreth; then turning to the right traverses that leading to Meldreth as far as the brook in the centre of the latter village; descending that brook to the river, it ascends the Cam to Arrington Bridge, from whence it is drawn along the road leading to Potton until it reaches the boundary of the county. By far the greater part of this is a chalk-district.

No. 3, *Wimpole,* is bounded westwards by the edge of the county; southwards by No. 2; eastwards by No. 1; and northwards by the Cambridge and St Neots turnpike-road, except that at Eltisley the St Neots road is left and that leading to St Ives is used for a very short distance, in order to reach the edge of the county.

No. 4, *Cottenham.* The three internal boundaries of this district are the above-mentioned Cambridge and St Neots road, the river Cam from Cambridge until it joins the old Ouse, and then that old river up to Earith.

No. 5, *Burwell,* is bounded by districts 1 and 4; and by the road from Mildenhall by Soham to the river Ouse below Barraway. The small detached piece of Suffolk adjoining Newmarket is included in it.

No. 6, *Ely,* is bounded towards the south by Nos. 4 and 5; to the north-east by the edge of the county; and to the north-west by the New Bedford River. It includes the whole of the true island of Ely, but is otherwise a fen country.

No. 7, *Chatteris,* is a fen district, but includes one considerable island, namely, that of Doddington. It is bounded by the New Bedford River on the one side, and the Old River Nene (passing by March) and the cut called Bevill's River on the other; the two remaining boundaries being the adjoining counties of Norfolk and Huntingdon.

No. 8, *Wisbech*, includes all that part of the county which lies to the north of No. 7. This is the most continuously fenny tract included in our Flora: its fen-islands are small and few in number: but, as has been already stated, the silted-up estuary of the Ouse forms a remarkable feature at its northern extremity.

If now we reconsider these districts we may combine them so as to divide the county into three unequal parts, which may be called the Chalk, Clay, and Fen Tracts.

A, CHALK, is a tract formed of the districts 1 and 2, in which nearly all our chalky land is included. *B*, CLAY, includes 3, 4, and 5 which, if the border of the Fens is excepted, lie almost wholly on the Boulder Clay, Gault, or Kimmeredge Clay. *C*, FEN, consists of the remaining three districts 6, 7, and 8.

The Map in outline will illustrate these Districts, but they can only be traced exactly upon the Ordnance Map: nevertheless this small sketch will probably suffice in most cases.

The county of Cambridge is about 50 miles in length from north to south, and its greatest breadth is 25 miles. It is said to contain nearly 550,000 acres. It lies wholly between the 52 and 53 parallels of Latitude, and the town of Cambridge is situate 5 miles to the east of the meridian of Greenwich. We learn from the observations of the Rev. L. Jenyns that the mean temperature of the seasons is, Spring 47.18°, Summer 60.87°, Autumn 49.86°, and Winter 38.09°. The mean annual range of the Barometer is 1.890 inches. The mean fall of rain about 22.5 inches.

Deep drifting snow is not common, so that the roads are seldom obstructed by it. The prevalent winds are from the south-west and north-west, but cold north-easterly winds are very common in April and May. Those persons who desire more detailed information relative to our climate will obtain it by reference to Mr Jenyns's valuable work entitled *Observations in Meteorology*.

No part of the county is much elevated above the sea. The level of the river at Cambridge is $24\frac{1}{2}$ feet above low water in Lynn Deeps; but as the tide there rises 22 feet, it is only $2\frac{1}{2}$ feet above high water and $13\frac{1}{2}$ feet above mean tide. The latter therefore, viz. $13\frac{1}{2}$ feet, is our height above the sea according to the usual mode of calculation. As there is a descent of about 10 feet from the level of the fen at Denver Sluice to low water there, the fall of the country from Cambridge to Denver is not more, and probably much less, than 14 feet in a direct line of about 30 miles, or about $5\frac{1}{2}$ inches to the mile. I am informed that the fall between Ely and Denver is only at the rate of $1\frac{1}{2}$ inch per mile. The fall from Peterborough to the mouth of the river Nene below Wisbech is stated by Rennie to be $17\frac{1}{2}$ feet to low water mark, in 32 miles, or an average of $6\frac{1}{2}$ inches per mile. But if we reduce this to the mean level of the sea we find the fall to be only 5 feet, or hardly 2 inches in the mile. Before the formation of Denver Sluice the tide ascended the river Cam nearly to the mouth of Swaffham Lode. It rose 4 feet at Harrimere, the former confluence of the Cam and Ouse. If we now consider the higher elevations we find that the ground near the wind-mill by Madingley Wood is 174.3 feet above the mean level of the sea; the place where the turnpike-road crosses the Gogmagog Hills is 233.7 feet;

the top of the hill between Wimpole and Great Eversden 244.9; the Ordnance trigonometrical point near Newmarket 261.3; the base of Balsham Church-tower 378 feet; the base of the tower of Ely Cathedral 51.6.

I am indebted for these latter heights, to the kindness of Colonel Sir Henry James who had them extracted for me from the records of the Ordnance Survey.

A few of the names of places mentioned in this book require explanation.

Mare Way. There are two ancient roads thus named. One in District 3 runs along the top of the range of chalk-hills on the north-east side of Wimpole Park. The other in District 4, extends from Aldreth High Bridge, by Balsar's Hill, to a spot between Willingham and Rampton. It was apparently continued by Cuckoo Lane (between Westwick and Cottenham Field) to Histon. On the Ordnance Map it is called Ancient Bridle Way.

Hill of Health. After this name had been in use for about two centuries it totally disappeared upon the enclosure of St Giles's Parish. It is the moderately elevated ground lying to the south of the Huntingdon road just outside of the town of Cambridge. It is now partly planted and partly occupied by a gentleman's house.

Hinton Moor. This was once an exceedingly wet fen. It is now completely drained and cultivated. The footpath leading from Cambridge to the Church at Cherry Hinton and the carriage-road to the same village, cross it. It once extended from the Hills Road, near Red Cross, nearly to the Stone Bridge in Coldham's Lane.

Hall Wood near Wood Ditton. This wood has been completely removed and its site brought into cultivation.

Triplow Heath is now enclosed and cultivated.

TREATISES ON THE PLANTS OF CAMBRIDGESHIRE.

1. (R. C.)—Catalogus Plantarum circa Cantabrigiam nascentium: In quo exhibentur quotquot hactenus inventæ sunt, quæ vel sponte proveniunt, vel in agris seruntur; una cum synonymis electioribus, locis natalibus et observationibus quibusdam oppido raris. Adjiciuntur in gratiam tyronum, Index Anglo-latinus, Index locorum, Etymologia nominum, et Explicatio quorundam terminorum [John Ray]. Cantabrigiæ, 1660. 12mo. pp. 28, 182, 103.

2. (R. C. App. i.)—Appendix ad Catalogum Plantarum circa Cantabrigiam nascentium: continens addenda et emendata [John Ray]. Cantabrigiæ, 1663. 12mo. pp. 13.

3. (R. C. App. ii.)—Appendix ad Catalogum Plantarum circa Cantabrigiam nascentium: continens addenda et emendata. Editio secunda, aucta Plantis sexaginta [John Ray and Peter Dent]. Cantabrigiæ, 1685. 12mo. pp. 30.

4. (M. M.)—Methodus Plantarum circa Cantabrigiam nascentium [John Martyn]. Londini, 1727. 12mo. pp. 7, 132.

5. (M. Pl.)—Plantæ Cantabrigienses: or a Catalogue of the plants which grow wild in the county of Cambridge, dis-

posed according to the System of Linnæus.—Herbationes Cantabrigienses: or, Directions to the places where they may be found, comprehended in thirteen Botanical Excursions. To which are added lists of the more rare plants growing in many parts of England and Wales. By Thomas Martyn, M.A., Fellow of Sidney College, and Professor of Botany in Cambridge. London, 1763. 8vo. pp. 13, 14.

6. (Lyons)—Israelis Lyons, jun. Fasciculus Plantarum circa Cantabrigiam nascentium, quæ post Rajum observatæ fuere. Londini, 1763. 8vo. pp. 16, 56.

7. (Relh. ed. 1)—Ricardi Relhan, A.M., Collegii Regalis Capellani, Flora Cantabrigiensis exhibens plantas agro Cantabrigiensi indigenas, secundum systema sexuale digestas: cum characteribus genericis, diagnosi specierum, synonymis selectis, nominibus trivialibus, loco natali, tempore inflorescentiæ. Cantabrigiæ, 1785. 8vo. pp. 22, 490.

8. (Relh. ed. 1, Suppl. 1)—Floræ Cantabrigiensi Supplementum. Auctore Ricardo Relhan, A.M., Collegii Regalis Capellano. Cantabrigiæ, 1786. 8vo. pp. 39.

9. (Relh. ed. 1, Suppl. 2)—Floræ Cantabrigiensi Supplementum alterum. Cantabrigiæ, 1788. 8vo. pp. 36.

10. (Relh. ed. 1, Suppl. 3)—Floræ Cantabrigiensi Supplementum tertium. Cantabrigiæ, 1793. 8vo. pp. 44.

11. (Relh. ed. 2)—Flora Cantabrigiensis. Editio altera. Cantabrigiæ, 1802. 8vo. pp. 13, 568.

12. (Relh.)—Flora Cantabrigiensis. Editio tertia. Cantabrigiæ, 1820. 8vo. pp. 11, 597.

The following works only partially relate to Cambridgeshire, although they contain complete lists of the plants known to grow there.

13. (R. Cat. Angl. ed. 1)—Catalogus Plantarum Angliæ, et Insularum adjacentium : tum indigenas, tum in agris passim cultas complectens. In quo præter synonyma necessaria facultates quoque summatim traduntur, una cum Observationibus et Experimentis novis medicis et physicis. Opera Joannis Raii, A.M., et Societatis Regiæ Sodalis. Londini, 1670. 12mo. pp. 22, 359.

14. (R. Cat. Angl. ed. 2)—Catalogus Plantarum Angliæ, &c. Editio secunda, Plantis circiter quadraginta sex, et observationibus aliquam multis auctior. Londini, 1777. 12mo. pp. 28, 328.

15. (Hensl. Cat. ed. 1)—A Catalogue of British Plants, arranged according to the Natural System, with the Synonyms of De Candolle, Smith, and Lindley. By the Rev. J. S. Henslow, M.A., Professor of Botany in the University of Cambridge, 1829. 12mo. pp. 40.

16. (Hensl. Cat. ed. 2)—The same, with the addition of the Synonyms of Hooker. Cambridge, 1835. pp. 4, 61.

In quoting the above books I have used the contraction which is prefixed in brackets to each of their titles. In all cases the third edition of Relhan's *Flora* is intended, unless the number of the edition is stated.

PLAN.

A FEW explanations are desirable for the purpose of rendering the use of this book as easy as possible.

The names of the Orders are printed in large capitals, as *RANUNCULACEÆ;* those of the Genera in smaller capitals, as ANEMONE; and those of the Species in Egyptian type, and numbered consecutively in each genus, as

1. **A. Pulsatilla.**

The arrangement of the matter under each species is (1) the Specific, or as Linnæus more properly called it the Trivial, name, as *A. Pulsatilla;* then the English name, as *Pasque Flower,* when there is a real one, for the names of modern invention are intentionally omitted; (2) the old denominations of the plant as found in the works relating to Cambridgeshire of Ray, Lyons, the Martyns, and Relhan, with occasionally a reference to an earlier or a later author; (3) the general character of the places where the plant grows, as "Chalk-hills;" followed in the same line by its duration, as P., meaning perennial; and its usual time of flowering, as "April, May;" (4) the localities where the plant has been seen.

The localities (4) are arranged under the districts (of which an account will be found in the preceding TOPOGRAPHICAL

REMARKS) within which they lie, as, in the case of *A. Pulsatilla*, (1) Cambridge, (2) Royston, and (5) Burwell. In recording the localities, the author is responsible for all those which are followed by a full stop. Those names of places followed by a semicolon (;) were received from the correspondents whose initials succeed that stop; as, "Linton; W. H. C.," shews that the plant was found at Linton by the Rev. W. H. Coleman, but that the author has not himself observed it there nor seen a specimen from thence; had Mr Coleman shewn a specimen to him, a note of admiration (!) would have followed the word Linton. When several localities were communicated by the same correspondent they are separated from each other by semicolons, and the whole series followed by his initials. Those localities which rest solely upon the authority of the older botanists are printed in *Italics*, and followed by the abbreviated name of the author on whose authority they rest.

In order to shew the history of each plant as a native of this county, the earliest work in which it is known to have been recorded as such stands first among the synonyms; it is often that of Ray, but sometimes an earlier writer; occasionally also it is an author of later date than Relhan, such as Prof. Henslow, in one of the editions of his *Catalogue of British Plants*.

The following marks are used in this book, and require explanations. The first three of them always precede the name of the plant. They refer solely to the condition of the plant in Cambridgeshire.

* Certainly introduced, but *naturalized*.

† *Possibly introduced*, but now having the appearance of being a true native.

‡ *Probably introduced,* but admitting of some doubt upon the subject.

When the whole account of a plant is included within [ ], that species is considered as having the very slenderest or even no claim to a place in our Flora, although recorded by preceding writers.

The duration of plants is marked by the letters A., B., P., Sh., and T.; being the abbreviations of the terms Annual, Biennial, Perenuial, Shrub, and Tree, respectively.

CONTRIBUTORS.

THE following initial letters are used in place of the full names of those correspondents upon whose authority localities are recorded.

A. M. B.......Miss A. M. BARNARD, grand-niece of the late Sir J. E. Smith.
D. B..........Mr D. BRITTEN, of Royston.
J. B..........Mr JAMES BALDING, of Wisbech.
C. B. C. .....Mr C. B. CLARKE, M.A., Fellow of Queens' College.
J. C..........Mr JAMES CARTER, Surgeon, Cambridge.
W. H. C....Rev. W. H. COLEMAN, M.A., formerly of St John's College.
H. F.........Mr HENRY FORDHAM, of Royston.
G. S. G. ....Mr G. S. GIBSON, F.L.S., of Saffron Walden.
H. ....Prof. HENSLOW, M.A., F.L.S.
W. M.......Mr WILLIAM MARSHALL, Solicitor, of Ely.
N............Rev. W. W. NEWBOULD, M.A., late of Comberton.
A. P.........Mr ALGERNON PECKOVER, F.L.S. and family, of Wisbech.
E. S.........Mr EDWARD SKEPPER, of Bury St Edmunds.
R. B. S.....Mr R. B. SMART, Surgeon, late of Linton.

J. W.........Mr JOB WATSON, Surgeon, of Hemingford, Huntingdonshire.
N. W. ....Mrs NASH WOODHAM, of Shepreth.
S. W. W....Mr SAMUEL W. WANTON, M.A., formerly of St John's College.
T. Y..........Rev. T. YORK, M.A., Fellow of Queens' College, and Rector of Little Eversden.

ANCIENT AUTHORITIES.

J. F..........J. FISHER, formerly of Christ's College.
J. M. ....JOHN MARTYN, F.R.S., formerly Professor of Botany, Cambridge.
T. M. ....Rev. THOMAS MARTYN, M.A., formerly Fellow of Sidney Sussex College, and Professor of Botany, Cambridge.
Ray..........JOHN RAY, M.A., F.R.S., formerly Fellow of Trinity College, Cambridge.
Relh. ....Rev. R. RELHAN, M.A., F.R.S., F.L.S., formerly Chaplain of King's College.

GEOGRAPHICAL DISTRIBUTION OF PLANTS IN CAMBRIDGESHIRE.

	A. Chalk.		B. Clay.			C. Fen.			Page in Flora.
	Cambridge.	Royston.	Wimpole.	Cottenham.	Burwell.	Ely.	Chatteris.	Wisbech.	
	1	2	3	4	5	6	7	8	
Clematis Vitalba................................	1	2	3	4	5				1
Thalictrum saxatile.........................	1	2			5				1
T. flavum	1	2	3	4	5	6	7	8	2
Anemone Pulsatilla	1	2			5				2
A. nemorosa.....................................	1		3	4					2
Myosurus minimus	.		3						3
Ranunculus trichophyllus	1	2	3	4	5	6		8	3
R. Drouetii	1	2	3	4	5	6			3
R. heterophyllus	1		3	4	5	6	7	8	4
R. Baudotii				4					4
R. floribundus	1								4
R. circinatus	1		3	4	5	6	7		4
R. fluitans.......................................	1	2	3	4	5	6			5
R. hederaceus	1	2	3	.					5
R. sceleratus...................................	1	2	3	4	5	6	7	8	6
R. Flammula	1	2	3	4	5	6	7	8	6
R. Lingua..	1	2	.		5	6			6
R. Ficaria..	1	2	3	4	5	6	7	8	6
R. auricomus	1		3	4	5			8	7
R. acris...	1	2	3	4	5	6	7	8	7
R. repens	1	2	3	4	5	6	7	8	7
R. bulbosus	1	2	3	4	5	6		8	7
R. hirsutus			3	4					7
R. arvensis	1	2	3	4		6			8
R. parviflorus	.		3	4					8
Caltha palustris	1	2	3	4	5	6		8	8
Helleborus viridis	1		3		5				9
H. foetidus......................................	1	.							9
Aquilegia vulgaris	1	.	3		5				9
Delphinium Consolida	1	2	3	4	5	6			10
Berberis vulgaris...........................	.	2	3	4	5				10
Nymphæa alba...............................	1	2	.	4	5	6		8	11

d

	A.		B.			C.			PAGE.
	1	2	3	4	5	6	7	8	
Nuphar lutea	1	2	3	4	5	6	7	8	11
Papaver Argemone	1	2	3	4	5	6		8	11
P. hybridum	1	2	3	4	5				12
P. Rhœas	1	2	3	4	5	6	7	8	12
P. dubium	1	2	3	4	5	6	7		12
P. somniferum				4	5	.			12
Roemeria hybrida					5				13
Chelidonium majus	1	2	3	4	5		7	8	13
Fumaria capreolata	1		3		5			8	14
F. officinalis	1	2	3	4	5	6	7	8	14
F. micrantha	1	2							14
F. parviflora	1	2			5				15
F. Vaillantii	1	2			5				15
Cheiranthus Cheiri	1	2	3			6		8	15
Nasturtium officinale	1	2	3	4	5	6	7	8	16
N. sylvestre	1			4	5	6	7	8	16
N. palustre		.		4		6			16
Barbarea vulgaris	1	2	3	4	5	6		8	16
Arabis hirsuta	1				5			8	17
A. Turrita	1		3						17
Cardamine hirsuta			3			6		8	17
C. pratensis	1	2	3	4	5	6		8	18
Sisymbrium officinale	1	2	3	4	5	6	7	8	18
S. Irio	1							.	18
S. Sophia	1	2	3	4	5	6	7	8	19
S. thalianum	1		3	4					19
Alliaria officinalis	1	2	3	4	5	6		8	19
Erysimum cheiranthoides	1	2	3	4	5	6	7	8	19
Brassica campestris	1								20
B. Napus	1								20
Sinapis nigra			3	4	5	6	7		20
S. arvensis	1	2	3	4	5	6	7	8	21
S. alba	1	2	3	4	5	6	7	8	21
Diplotaxis muralis	1								21
Draba verna	1	2	3	4	5	6		8	22
Cochlearia anglica								8	22
Armoracia amphibia	1			4	5	6			23
Thlaspi arvense				.					23
Teesdalia nudicaulis			3						23
Iberis amara		2							23
Lepidium campestre	1						7		24
L. ruderale								8	24
L. latifolium								8	24
Capsella Bursa-pastoris	1	2	3	4	5	6	7	8	25
Senehiera Coronopus	1	2	3	4	5	6	7		25
Raphanus Raphanistrum	1	2	3	4	5	6		8	26
Reseda lutea	1	2	3	4	5				26
R. luteola	1	2	3	4	5			8	26
Helianthemum vulgare	1	2	3		5	6			27
Viola odorata	1	2	3	4	5			8	27
V. hirta	1	2	3	4	5				28
V. sylvatica	1	2	3	4	5				28

	A.		B.			C.			PAGE.
	1	2	3	4	5	6	7	8	
Viola canina			3		5	6	7		28
V. stagnina				4	5				28
V. tricolor	1	2	3	4	5	6	7	8	29
Drosera rotundifolia	1	2	3						29
D. intermedia	.								29
D. anglica									29
Parnassia palustris	1	2	3		5				30
Polygala vulgaris	1	2	3	4	5			8	30
Fraukenia lævis								.	30
Dianthus Caryophyllus					.			.	31
D. deltoides	1								31
Saponaria officinalis	1	2	3	4	5			8	31
Silene anglica					5				32
S. inflata	1	2	3	4	5		7	8	32
S. noctiflora	1	2	3	4	5	6	7	8	32
Lychnis Flos-cuculi	1	2	3	4	5	6			33
L. vespertina	1	2	3	4	5	6	7	8	33
L. Githago	1	2	3	4	5			8	33
Sagina procumbens	1		3	4	5	6	7	8	34
S. apetala	1		3	4	5	6	7	8	34
S. ciliata	1		3						34
S. nodosa	1	2	3		5		7	8	35
Alsine tenuifolia	1	2	3		5				35
Moehringia trinervis	1	2	3	4					36
Arenaria serpillifolia	1	2	3	4	5	6		8	36
A. leptoclados		2	3	4	5				36
Stellaria media	1	2	3	4	5	6	7	8	37
S. Holostea	1	2	3						37
S. glauca	1	2			5	6	7	.	37
S. graminea	1	2	3	4	5	6			38
S. uliginosa	1					6			38
Moenchia erecta			3	.					38
Malachium aquaticum	1	2	3	4	5	6	7	8	38
Cerastium glomeratum	1	2	3	4	5	6			39
C. triviale	1	2	3	4	5	6	7	8	39
C. semidecandrum	.	2	3	4	5	6			40
C. arvense	1	2	3	4	5			8	40
Malva moschata	1	2	3						40
M. sylvestris	1	2	3	4	5		7	8	41
M. rotundifolia	1	2	3	4	5	6		8	41
Althæa officinalis								8	41
Tilia parvifolia				4					42
Hypericum quadrangulum	1	2	3	4	5	6	7	8	42
H. perforatum	1	2	3	4	5				42
H. humifusum	1	2	3		5				43
H. hirsutum	1	2	3	4		6	7		43
H. pulchrum	1		3						43
H. elodes			3						43
Acer campestre	1	2	3	4	5	6	7	8	44
Geranium pratense	1	2	3	4	5				45
G. sanguineum	1								45
G. pyrenaicum	1		3						45

GEOGRAPHICAL DISTRIBUTION

	A.		B.			C.			PAGE.
	1	2	3	4	5	6	7	8	
Geranium pusillum	1	2	3	4	5	6			46
G. columbinum			3						46
G. dissectum	1	2	3		5	6	7	8	46
G. rotundifolium	.		.	.					46
G. molle	1	2	3	4	5	6	7	8	47
G. lucidum	1			4	5			8	47
G. Robertianum	1	2	3	4	5	6	7	8	47
Erodium cicutarium	1	2	3	4	5		7	8	48
E. moschatum			.		5			8	48
Linum perenne	1								48
L. catharticum	1	2	3	4	5	6	7	8	49
Radiola millegrana	1								49
Oxalis Acetosella	1								50
Euonymus europæus	1	2	3	4	5	6			50
Rhamnus catharticus	1	2	3	4	5				51
R. Frangula	1	2	3						51
Ulex europæus	1	2	3	4	5				51
U. nanus	.	2			5				51
Genista tinctoria	1		3	4					52
G. anglica	1		3	4					52
Sarothamnus scoparius	1		3		5				52
Ononis arvensis	1	2	3	4	5	6		8	53
O. campestris	1	2	3	4	5			8	53
Medicago sylvestris	1				5				53
M. falcata	1				5				53
M. lupulina	1	2	3	4	5	6	7	8	54
M. maculata	1		3	4	5				54
M. minima	.				5				54
Melilotus officinalis	1	2	3	4	5	6	7		55
M. arvensis	1	2	3		5				55
Trifolium pratense	1	2	3	4	5	6	7	8	55
T. medium	1	2	3						56
T. ochroleucum	1		3	4	5	6			56
T. arvense	1	2	3	4	5			8	56
T. striatum	1		3	4					57
T. scabrum	1		3	4	5				57
T. subterraneum			3						57
T. repens	1	2	3	4	5	6	7	8	57
T. fragiferum	1	2	3	4	5	6	7	8	58
T. procumbens	1	2	3	4	5			8	58
T. minus	1	2	3	4	5	6		8	58
T. filiforme			3	.		6			59
Lotus corniculatus	1	2	3	4	5	6	7	8	59
L. tenuis	1		3	4					59
L. major	1	2	3		5	6	7		60
Anthyllis Vulneraria	1	2	3	4	5				60
Astragalus hypoglottis	1	2			5				60
A. glycyphyllos	1	2	3	4	5				61
Vicia hirsuta	1		3	.			7		61
V. tetrasperma	1		3	4					62
V. gracilis			3	4					62
V. sylvatica	.								62

	A.		B.			C.			PAGE.
	1	2	3	4	5	6	7	8	
Vicia Cracca	1	2	3	4	5		7	8	63
V. sepium	1		3	4		6	7		63
V. sativa	1		3	4		6		8	63
V. lathyroides	1								64
Lathyrus Aphaca	1		3	4				8	64
L. Nissolia	.		.			.			64
L. pratensis	1	2	3	4	5	6	7	8	64
L. sylvestris	1		3		5				65
L. palustris			.		5				65
Ornithopus perpusillus	1		3		5				65
Hippocrepis comosa	1	2	3		5				66
Onobrychis sativa	1	2	3	4	5				66
Prunus communis	1	2	3	4	5	6	7	8	66
P. Padus			.					8	67
P. Avium	1			.				8	67
P. Cerasus	.								67
Spiræa Ulmaria	1	2	3	4	5		7	8	68
S. Filipendula	1	2	3	4	5				68
Sanguisorba officinalis	1		3	4				8	69
Poterium Sanguisorba	1	2	3	4	5				69
P. muricatum		2	3	4	5				69
Agrimonia Eupatoria	1	2	3	4	5	6		8	70
Alchemilla vulgaris	1		3						70
A. arvensis	1	2	3	4	5	6		8	70
Potentilla anserina	1	2	3	4	5	6	7	8	71
P. argentea	1		3						71
P. verna	1		3						71
P. reptans	1	2	3	4	5	6	7	8	71
P. Tormentilla	1	2	3	4	5				72
P. fragariastrum	1		3	4	5				72
Comarum palustre			3		5				72
Fragaria vesca	1	2	3	4			7		73
Rubus Idæus	1		3						73
R. thyrsoideus	1		3						73
R. discolor	1	2	3	4	5	6	7	8	73
R. Radula			3						74
R. Koehleri	1		3						74
R. diversifolius	1		3						74
R. Balfourianus	1		3	4	5				74
R. corylifolius	1	2	3	4	5	6	7	8	74
R. althæifolius	1	2	3	4	5				75
R. tuberculatus	1		3		5				75
R. cæsius	1	2	3	4	5	6	7	8	75
Geum urbanum	1	2	3	4	5	6	7	8	75
G. rivale	1		3						76
Rosa spinosissima					5				76
R. villosa	1		3						76
R. tomentosa	1		3	4					77
R. inodora	1			4	5				77
R. rubiginosa	1	2	3	4	.				77
R. canina	1	2	3	4	5	6	7	8	78
R. systyla	1		3						78

GEOGRAPHICAL DISTRIBUTION

	A.		B			C.			PAGE.
	1	2	3	4	5	6	7	8	
Rosa arvensis	1	2	3	4	5	6	7		78
Cratægus Oxyacantha	1	2	3	4	5	6	7	8	79
Pyrus communis	1		3						79
P. Malus	1	2	3	4	5	6		8	79
P. Aucuparia	1		3						79
P. torminalis			.						80
Lythrum Salicaria	1	2	3	4	5	6	7	8	80
L. Hyssopifolia	1			4	5	6			80
Peplis Portula			3						81
Epilobium hirsutum	1	2	3	4	5	6	7	8	81
E. parviflorum	1	2	3	4	5	6	7	8	81
E. montanum	1		3	4		6	7	8	81
E. tetragonum	1		3	4	5		7		82
E. obscurum			3						82
E. palustre	1	2	3					8	82
Circæa lutetiana	1	2	3		5				82
Myriophyllum verticillatum	1	2	3	4	5	6			83
M. spicatum	1		3	4	5	6	7	8	83
M. alterniflorum			3						83
Hippuris vulgaris	1	2	3	4	5	6	7	8	84
Bryonia dioica	1	2	3	4	5	6	7	8	84
Montia fontana			3	.		6			84
Herniaria glabra	1								85
Lepigonum rubrum	1	2	3			6			85
L. medium								8	85
L. marinum	1					6		8	86
Spergula arvensis	1		3	4	5				86
Scleranthus annuus	1		3	.	5				86
Sedum Telephium	.	.							87
S. album							.		87
S. dasyphyllum	1	.							87
S. acre	1	2	3	4	5	6		8	87
S. sexangulare				.	.	.			88
S. reflexum	.								88
Ribes Grossularia	1	2	3	4	5				89
R. nigrum	.								89
R. rubrum	.		3		5				89
Saxifraga granulata	1		3	4		6			89
S. tridactylites	1	2	3	4	5	6			90
Hydrocotyle vulgaris	1	2	3	4	5	6	7	8	90
Sanicula europæa	1	2	3	4					90
Cicuta virosa						.			91
Apium graveolens	1		3	4	5	6		8	91
Petroselinum segetum		2	3	4		6			92
Helosciadium nodiflorum	1	2	3	4	5	6	7	8	92
H. inundatum	1		3			6			92
Sison Amomum	1	2	3	4	5	6			93
Ægopodium Podagraria	1	2	3		5			8	93
Carum Carui	.			4	.			8	93
Bunium flexuosum	1		3	4					94
B. Bulbocastanum	1	2							94
Pimpinella magna	1		3	4			7		94

	A.		B.			C.			PAGE
	1	2	3	4	5	6	7	8	
Pimpinella Saxifraga	1	2	3	4	5				95
Sium latifolium			3	4	5	6	7	8	95
S. angustifolium	1	2	3	4	5	6	7		95
Bupleurum tenuissimum								8	96
B. rotundifolium	1	2	3	4	5				96
Œnanthe fistulosa	1		3	4	5	6	7	8	96
Œ. Lachenalii	1	2	3	4	5	6	7	8	97
Œ. silaifolia			3						97
Œ. Phellandrium			3	4	5	6	7	8	97
Œ. fluviatilis	1		3	4	5	6	7		98
Æthusa Cynapium	1	2	3	4	5	6	7	8	98
Fœniculum officinale					5				98
Seseli Libanotis	1								98
Silaus pratensis	1		3	4	5	6			99
Angelica sylvestris	1	2	3	4	5		7		99
Peucedanum palustre					5				99
Pastinaca sativa	1	2	3	4	5	6	7	8	100
Heracleum Sphondylium	1	2	3	4	5	6	7	8	100
Daucus Carota	1	2	3	4	5	6	7	8	100
Caucalis daucoides	1	2	3	4	5				101
C. latifolia					5				101
Torilis Anthriscus	1	2	3	4	5	6	7	8	101
T. infesta	1	2		4	5	6	7	8	102
T. nodosa	1	2	3	4	5	6	7	8	102
Scandix Pecten-veneris	1	2	3	4	5	6		8	102
Anthriscus sylvestris	1	2	3	4	5	6	7	8	103
A. vulgaris	1	2	3	4	5			8	103
Chærophyllum temulum	1	2	3	4	5	6	7	8	103
Conium maculatum	1	2	3	4	5	6	7	8	104
Smyrnium Olusatrum	1	2	3	4					104
Coriandrum sativum						6	7	8	105
Adoxa Moschatellina			3	4					105
Hedera Helix	1		3	4	5			8	105
Cornus sanguinea	1	2	3	4	5		7	8	105
Viscum album	1				5				106
Sambucus Ebulus			3	4	5				106
S. nigra	1	2	3	4	5	6	7	8	106
Viburnum Lantana	1	2	3	4	5	6			107
V. Opulus	1	2	3	4				8	107
Lonicera Caprifolium	1	2	3						107
L. Periclymenum	1	2	3	4	5	6	7	8	108
Sherardia arvensis	1	2	3	4	5	6	7	8	108
Asperula cynanchica	1	2	3		5				108
A. odorata			3						109
Galium cruciatum	1	2	3	4	5				109
G. tricorne	1	2	3	4	5				109
G. Aparine	1	2	3	4	5	6	7	8	110
G. anglicum					5				110
G. erectum	1	2	3	4	5				110
G. Mollugo	1	2	3	4	5			8	111
G. verum	1	2	3	4	5	6	7	8	111
G. saxatile	1		3		5				111

	A.		B.			C.			PAGE.
	1	2	3	4	5	6	7	8	
Galium uliginosum		2	3		5				111
G. palustre	1	2	3	4	5	6		8	112
G. elongatum		2	3	4	5	6	7	8	112
Centranthus ruber	1		.			.			113
Valeriana officinalis	1	2	3	4				8	113
V. sambucifolia	1	2	3		5				113
V. dioica	1	2	3		5				114
Valerianella olitoria	1		3						114
V. Auricula			3					8	114
V. dentata	1	2	3	4					115
Dipsacus sylvestris	1	2	3	4	5	6		8	115
D. pilosus	1		3		5				115
Knautia arvensis	1	2	3	4	5			8	116
Scabiosa succisa	1	2	3	4	5				116
S. Columbaria	1	2	3	4	5				116
Eupatorium cannabinum	1	2	3	4	5	6	7	8	117
Petasites vulgaris	1	2	3		5			8	117
Tussilago Farfara	1	2	3	4	5	6	7	8	118
Aster Tripolium								8	118
Erigeron acris	1	2	3		5			8	118
Bellis perennis	1	2	3	4	5	6	7	8	119
Solidago Virgaurea			3						119
Inula Helenium			3						119
I. Conyza	1	2			5				119
Pulicaria vulgaris	.	.		4		6			120
P. dysenterica	1	2	3	4	5	6	7	8	120
Bidens tripartita			3	4	5	6	7	8	120
B. cernua	1		3	4	5	6			121
Anthemis arvensis	1	2	3	4	5				121
A. Cotula	1	2	3	4	5	6	7		121
A. nobilis								8	122
Achillea Ptarmaca	1	.	3	4	5	6	7	8	122
A. Millefolium	1	2	3	4	5	6	7	8	122
Chrysanthemum Leucanthemum	1	2	3	4	5	6	7	8	122
C. segetum		2	3	4		6		8	123
Matricaria Parthenium		2	3	4	5		7		123
M. inodora	1	2	3	4		6	7		123
M. Chamomilla	1	2	3	4	5	6	7		124
Artemisia Absinthium	1	2	3	4	5	6		8	124
A. vulgaris	1	2	3	4	5	6	7	8	124
A. maritima								8	125
Tanacetum vulgare	.	2	3	4				8	125
Filago germanica	1	2	3	4	5	6	7	8	126
F. apiculata			3						126
F. spathulata	1	2	3		5		7		126
F. minima	1	2	3		5				126
Gnaphalium luteo-album		.							127
G. uliginosum		2	3			6	7	8	127
G. sylvaticum			3						127
Antennaria dioica	1		.	.	5				127
Senecio vulgaris	1	2	3	4	5	6	7	8	128
S. viscosus			.			.	.		128

	A.		B.			C.			PAGE.
	1	2	3	4	5	6	7	8	
Senecio sylvaticus	1		3						128
S. erucifolius	1	2	3	4	5	6	7	8	128
S. Jacobæa	1	2	3	4	5	6	7	8	129
S. aquaticus	1	2	3	4	5	6	7	8	129
S. paludosus					5	6			129
S. palustris					.	6	.	.	129
S. campestris	1				5				129
Carlina vulgaris	1	2	3	4	5			8	130
Arctium tomentosum			.	4			7	8	130
A majus	1	2	3	4	5	6	7		131
A. minus	1	2	3	4	5	6	7	8	131
A. pubens	1	2	3	4	5	6		8	131
Serratula tinctoria	1		3	4					132
Centaurea nigra	1	2	3	4	5	6	7	8	132
C. Cyanus	1	2	3	4	5			8	132
C. Scabiosa	1	2	3	4	5	6		8	133
C. solstitialis				4					133
C. Calcitrapa	1		3			6		8	133
Onopordum Acanthium	1	2	3	4		6			134
Carduus nutans	1	2	3	4	5	6	7	8	134
C. crispus	1	2	3	4	5	6	7		134
C. lanceolatus	1	2	3	4	5	6	7	8	135
C. eriophorus	1	2	3	4	5	6			135
C. arvensis	1	2	3	4	5	6	7	8	135
C. palustris	1	2	3	4	5	6	7		135
C. pratensis	1	2	.		5	6			135
C. acaulis	1	2	3	4	5	6		8	136
Silybum marianum	1			4				8	137
Lapsana communis	1	2	3	4	5	6	7	8	137
Arnoseris pusilla			3						137
Cichorium Intybus	1	2	3	4	5	6		8	137
Hypochœris glabra			.		5				138
H. radicata	1		3	4	5	6			138
H. maculata	1	.			5				138
Thrincia hirta	1	2	3	4	5	6	7		139
Apargia hispida	1	2	3	4	5	6		8	139
A. autumnalis	1	2	3	4	5	6	7	8	139
Tragopogon minor	1	2	3	4	5	6		8	140
Picris hieracioides	1	2	3	4				8	140
Helminthia echioides	1	2	3	4	5	6	7	8	140
Lactuca saligna	.		.	.					141
L. virosa	1					.			141
L. Scariola	1		.	.	.	6			141
L. muralis	1	2			.				141
Leontodon Taraxacum	1	2	3	4	5	6	7	8	142
L. palustre	1	2		4	5	6			142
Sonchus oleraceus	1	2	3	4	5	6	7	8	142
S. asper	1	2	3	4	5	6	7	8	143
S. arvensis	1	2	3	4	5	6	7	8	143
S. palustris						.			143
Crepis fœtida	1	2							143
C. virens	1	2	3	4	5	6	7	8	143

	A.		B.			C.			PAGE.
	1	2	3	4	5	6	7	8	
Crepis biennis	1				5				144
Hieracium Pilosella	1	2	3	4	5	6		8	144
H. murorum ?									144
H. umbellatum			3						144
H. boreale			3						145
Jasione montana	1		3						145
Campanula glomerata	1	2	3	4	5				145
C. latifolia	1		3			6			145
C. Trachelium	1		3	4					146
C. rotundifolia	1	2	3	4	5				146
Specularia hybrida	1	2	3	4	5	6		8	147
Calluna vulgaris	1	2	3		5				147
Erica Tetralix			3						147
E. cinerea			3						148
Vaccinium Oxycoccos			3						148
Monotropa Hypopitys			3	4					148
Ilex Aquifolium	1			4					148
Ligustrum vulgare	1	2	3	4	5			8	149
Fraxinus excelsior	1		3	4	5			8	149
Vinca minor	1	2	3	4	5			8	149
V. major	1	2	3	4					150
Chlora perfoliata	1	2	3	4	5				150
Erythræa pulchella	1		3	4	5			8	151
E. Centaurium	1	2	3	4	5			8	151
Gentiana Amarella	1	2	3	4	5				152
Villarsia nymphæoides				4		6	7	8	152
Menyanthes trifoliata	1	2	3		5				153
Convolvulus arvensis	1	2	3	4	5	6	7	8	153
C. sepium	1	2	3	4	5	6	7	8	153
Cuscuta europæa			3	4	5	6			154
C. Epithymum			3						154
C. Trifolii	1	2	3	4					154
Asperugo procumbens									155
Cynoglossum officinale	1		3	4	5	6			155
Lycopsis arvensis	1		3	4	5				156
Symphytum officinale	1	2	3	4	5	6	7	8	156
Echium vulgare	1	2	3	4	5	6			157
Lithospermum officinale	1	2	3	4	5				157
L. arvense	1	2	3	4	5			8	158
Myosotis palustris	1	2	3	4	5	6	7	8	158
M. cæspitosa			3	4	5	6		8	158
M. arvensis	1	2	3	4	5	6	7	8	158
M. collina	1		3	4					159
M. versicolor	1		3					8	159
Solanum nigrum	1		3	4	5	6		8	159
S. Dulcamara	1	2	3	4	5	6	7	8	160
Atropa Belladonna	1				5	6		8	160
Hyoscyamus niger	1	2	3	4	5	6		8	160
Datura Stramonium						6		8	161
Orobanche Rapum			3						161
O. elatior	1	2	3						161
O. Picridis			3						162

	A.		B.			C.			PAGE.
	1	2	3	4	5	6	7	8	
Orobanche minor	1	2			5				162
Verbascum Thapsus	1	2	3	4	5				162
V. nigrum	1	2	3		5				163
Antirrhinum majus	1			4	5	6			163
A. Orontium		2							164
Linaria Cymbalaria	1	2		.	5	6		8	164
L. Elatine	1	2	3	4	5				164
L. spuria	1	2	3	4	5				164
L. minor	1	2	3	4	5				165
L. vulgaris	1	2	3	4	5			8	165
Scrophularia nodosa	1		3	4	5		7		166
S. aquatica	1	2	3	4	5		7	8	166
Limosella aquatica			3	.					166
Melampyrum cristatum	1	2	3	4					167
M. pratense	.		3						167
Pedicularis palustris	1	2	3		5				167
P. sylvatica	1	2	3						168
Rhinanthus Crista-galli	1	2	3	4	5	6		8	168
Euphrasia officinalis	1	2	3	4	5	6		8	168
E. nemorosa	1		3	4	5				169
E. Odontites	1	2	3	4	5		7	8	169
Veronica scutellata	1	2	3	4	5	6	7		169
V. Anagallis	1	2	3	4	5	6	7	8	170
V. Beccabunga	1	2	3	4	5	6	7	8	170
V. Chamædrys	1	2	3	4	5	6		8	170
V. montana	1		3						170
V. officinalis	1	2	3	.	5				171
V. spicata	.				5				171
V. serpillifolia	1	2	3	4	5	6	7	8	171
V. arvensis	1	2	3	4	5	6	7	8	172
V. agrestis	1		3	4	5	6	7	8	172
V. polita	1	2	3	4	5	6		8	172
V. Buxbaumii	1	2	3					8	172
V. hederifolia	1	2	3	4	5	6		8	173
Mentha rotundifolia	.			.	.				173
M. sylvestris	1		3		.				173
M. viridis					5				174
M. piperita		.	3	4	5				174
M. aquatica	1	2	3	4	5	6	7	8	174
M. pratensis		.		.	.				174
M. sativa	1		3	4	5	.			175
M. arvensis	1	2	3	4	5	6	7	8	175
M. Pulegium			.	.				.	175
Lycopus europæus	1	2	3	4	5	6		8	176
Salvia verbenaca	1	2	3	4	5				176
Origanum vulgare	1	2	3	.					176
Thymus Serpillum	1	2	3	4	5				177
T. Chamædrys	1	2	3		5				177
Calamintha Nepeta	1	2	3		5				177
C. officinalis	1	2	3	4	5				178
C. Acinos	1	2			5				178
C. Clinopodium	1	2	3	4	5				178

	A.		B.			C.			PAGE
	1	2	3	4	5	6	7	8	
Scutellaria galericulata	.	2	3	4	5	6		8	179
Prunella vulgaris	1	2	3	4	5	6	7	8	179
Nepeta Cataria	1	2	3	4	5			8	180
N. Glecoma	1	2	3	4	5	6	7	8	180
Lamium amplexicaule	1	2	3	4	5				180
L. incisum			3			6		8	181
L. purpureum	1	2	3	4	5	6	7	8	181
L. album	1	2	3	4	5	6	7	8	181
L. Galeobdolon	1		3	4					181
Leonurus Cardiaca	1		3		.			.	182
Galeopsis Ladanum	1	2	3	4	5				182
G. Tetrahit	1		3	4	5		7	8	182
G. versicolor	1	2		4	5	6	7	8	183
Stachys Betonica	1		3	4					183
S. sylvatica	1	2	3	4	5	6	7	8	183
S. palustris	1	2	3	4	5		7	8	184
S. arvensis	1		3		5				184
Ballota fœtida	1	2	3	4	5	6	7	8	184
Marrubium vulgare	1	2	3	4		6	7	8	184
Teucrium Scorodonia			3			6			185
T. Scordium				4		6			185
Ajuga reptans	1	2	3	4	5			8	185
A. Chamæpitys		2							186
Verbena officinalis	1	2	3	4	5	6		8	186
Pinguicula vulgaris	1	2	3		5	.			186
Utricularia vulgaris	.	2		4	5	6		8	187
U. minor	.	.	3		5	.			187
Primula vulgaris	1	2	3	4					188
P. veris	1	2	3	4	5			8	188
P. elatior	1		3	4					188
Hottonia palustris	1	2	3	4	5	6	7	8	189
Lysimachia vulgaris	1	2	3	4	5	6	7	8	189
L. Nummularia	1		3	4	5	6	7	8	189
L. nemorum	.								190
Anagallis arvensis	1	2	3	4	5	6	7	8	190
A. tenella	1	2	3		5		7		190
Centunculus minimus			.						191
Glaux maritima								8	191
Samolus Valerandi	1	2	3	4	5	6	7	8	191
Statice Limonium								8	192
S. caspia								8	192
Armeria maritima								8	192
Plantago Coronopus			3	.				8	192
P. maritima								8	193
P. lanceolata	1	2	3	4	5	6	7	8	193
P. media	1	2	3	4	5	6			193
P. major	1	2	3	4	5	6	7	8	193
Littorella lacustris	.		.						194
Suæda maritima								8	194
Chenopodium olidum	1		3						195
C. polyspermum			3	.	5	6	7		195
C. urbicum	.		.	.					195

	A.		B.			C.			PAGE.
	1	2	3	4	5	6	7	8	
Chenopodium album	1	2	3	4	5	6	7	8	195
C. ficifolium	1	2	3	4	5	6	7	8	196
C. murale				4					196
C. hybridum						6			196
C. rubrum	1		3	4	5	6	7	8	197
C. Bonus-Henricus	1	2	3	4	5	6			197
Beta maritima									197
Salicornia herbacea								8	197
Atriplex littoralis								8	198
A. angustifolia	1	2	3	4	5	6	7	8	198
A. erecta	1	2	3	4		6	7	8	198
A. deltoidea	1		3	4	5	6	7	8	199
A. hastata	1		3	4	5	6	7		199
A. Babingtonii								8	199
Obione pedunculata									199
O. portulacoides								8	200
Rumex maritimus	1		3		5	6			200
R. palustris	1			4		6	7	8	200
R. conglomeratus	1	2	3	4	5	6	7	8	201
R. sanguineus	1	2	3	4	5	6			201
R. pulcher	1	2	3	4	5	6		8	201
R. obtusifolius	1	2	3	4	5	6	7	8	201
R. pratensis	1	2	3	4	5	6			202
R. crispus	1	2	3	4	5	6	7	8	202
R. Hydrolapathum	1	2	3	4	5	6	7	8	202
R. acetosa	1	2	3	4	5			8	203
R. Acetosella	1	2	3	4	5				203
Polygonum Bistorta	1	2	3		5				203
P. amphibium	1	2	3	4	5	6	7	8	204
P. lapathifolium	1	2	3	4	5	6	7	8	204
P. laxum	1			4	5	6		8	204
P. Persicaria	1	2	3	4	5	6	7	8	204
P. mite	1			4	5		7	8	205
P. Hydropiper	1		3	4	5				205
P. minus									205
P. aviculare	1	2	3	4	5	6	7	8	205
P. Convolvulus	1	2	3	4	5	6	7	8	205
Daphne Laureola	1		3	4		6		8	206
Thesium humifusum	1	2			5				206
Aristolochia Clematitis									207
Euphorbia Helioscopia	1	2	3	4	5	6	7	8	207
E. platyphylla			3	4					207
E. amygdaloides	1		3						208
E. Peplus	1	2	3	4	5	6		8	208
E. exigua	1	2	3	4	5				208
Mercurialis perennis	1		3	4					209
M. annua						6			209
Ceratophyllum demersum	1	2	3	4	5			8	209
Callitriche verna	1	2	3	4	5	6	7	8	210
C. platycarpa	1	2	3	4	5	6			210
Parietaria erecta	1			4		6			210
P. diffusa	1	2	3	4	5	6			211

GEOGRAPHICAL DISTRIBUTION

	A.		B.			C.			PAGE.
	1	2	3	4	5	6	7	8	
Urtica pilulifera								8	211
U. urens	1	2	3	4	5	6	7	8	211
U. dioica	1	2	3	4	5	6	7	8	211
Humulus Lupulus	1	2	3	4	5			8	212
Ulmus suberosa	1		3	4	5		7		212
U. glabra					5				212
U. montana?					5				212
Salix fragilis		2	3						213
S. Russelliana			3		5				213
S. alba		2	3	4	5				213
S. vitellina			.					8	213
S. undulata									213
S. triandra	1	2	.			.	7		214
S. Hoffmanniana	1	2			5				214
S. amygdalina				4		6			214
S. purpurea						.			214
S. Helix	1								214
S. rubra	1				.	6			214
S. Forbiana			3			.			215
S. viminalis		2	3	4					215
S. acuminata			.	.	5				215
S cinerea	1		3		5			8	215
S. aquatica	1		3				7		215
S. oleifolia	1							8	215
S. aurita	1	2	3	.	5				216
S. caprea	1	2	3	4	5		7	8	216
S. repens		2	3		5				216
Populus alba	1		3	.				8	216
P. canescens	1	.	3	4					217
P. tremula	.	2	3	4			7		217
P. nigra	1			4	5	.		8	217
Myrica Gale						6			217
Betula glutinosa	1		3		5				218
Alnus glutinosa	1	2	3		5				218
Fagus sylvatica	1		3	4	5				218
Quercus Robur	1	2	3	4					219
Corylus Avellana	1	2	3	4	5	6	7		219
Carpinus Betulus	1								219
Taxus baccata	1		3	4					220
Juniperus communis	1								220
Paris quadrifolia	1		3						221
Tamus communis	1	2	3	4	5	6		8	221
Hydrocharus Morsus-ranæ	1		.	4	5	6		8	222
Stratiotes aloides				4	5	6	7	8	222
Anacharis Alsinastrum	1		3	4	5	6	7		222
Orchis Morio	1	2	3	4	5	6		8	223
O. mascula	1		3	4				8	223
O. ustulata	1	2			5				224
O. maculata	1	2	3	4	5	6		8	224
O. latifolia		2	3						224
O. incarnata	1	2	3		5				225
O. pyramidalis	1	2	3	4	5			8	225

	A.		B.			C.			PAGE.
	1	2	3	4	5	6	7	8	
Gymnadenia conopsea	1	2	3	4	5				225
Aceras anthropophora	1		3						226
Habenaria viridis	1		3	4	5				226
H. chlorantha	1		3	4					226
Ophrys apifera	1	2	3	4	5				227
O. aranifera	1								227
O. muscifera	1	2	3		5				228
Herminium Monorchis	1								228
Spiranthes autumnalis	1	2						8	228
Listera ovata	1	2	3	4	5			8	229
Neottia Nidus-avis	1								229
Epipactis media	1				5				229
E. palustris	1	2			5				230
Cephalanthera grandiflora	1	2			5				230
Malaxis paludosa			3						231
Sturmia Loeselii					5				231
Iris Pseud-acorus	1	2	3	4	5	6		8	232
I. foetidissima	1		3	4	5				232
Narcissus Pseudo-narcissus		2	3					8	233
Convallaria majalis			3						234
Ruscus aculeatus				4	5				234
Fritillaria Meleagris									234
Ornithogalum umbellatum	1		3	4				8	234
O. pyrenaicum									235
Allium vineale	1	2	3	4	5				235
A. oleraceum			3						235
A. ursinum	1		3						236
Endymion nutans	1	2	3	4	5				236
Muscari racemosum	1								236
Colchicum autumnale									237
Narthecium ossifragum			3						237
Juncus effusus	1		3	4	5		7	8	238
J. conglomeratus			3	4	5	6	7	8	238
J. glaucus	1	2	3	4	5	6	7	8	238
J. obtusiflorus	1	2	3	4	5	6	7	8	238
J. acutiflorus	1		3	4	5				239
J. lamprocarpus	1		3		5		7		239
J. supinus			3	4	5	6	7	8	239
J. squarrosus			3						240
J. compressus			3		5			8	240
J. bufonius	1		3	4	5		7	8	240
Luzula sylvatica	1								240
L. pilosa	1		3						241
L. campestris	1	2	3					8	241
L. multiflora	1		3		5				241
Alisma Plantago	1	2	3	4	5	6	7	8	241
A. ranunculoides	1	2	3	4	5	6	7		242
Sagittaria sagittifolia	1	2	3	4	5	6	7	8	242
Butomus umbellatus			3	4	5	6		8	242
Triglochin maritimum								8	243
T. palustre	1	2	3	4	5		7	8	243
Typha latifolia	1	2	3	4	5		7	8	244

	A.		B.			C.			PAGE.
	1	2	3	4	5	6	7	8	
Typha angustifolia		2	3	4	5	6	.	8	244
Sparganium ramosum	1	2	3	4	5	6	7	8	245
S. simplex	1		3	4	5	6	7		245
S. minimum	1	2			5	6			245
Acorus Calamus			.	4	5	6			245
Arum maculatum	1	2	3	4	5	6	7	8	246
Lemna trisulca	1	2	3	4	5	6	7	8	246
L. minor	1	2	3	4	5	6	7	8	246
L. polyrrhiza	1		3	4	5	6	7		247
L. gibba	1	2	3	4	5	6	7	8	247
Potamogeton natans	1	2	3	4	5	6			247
P. plantagineus		2	3	4	5	6	7		248
P. rufescens					5				248
P. heterophyllus	1		3		5				248
P. lucens	1		3	4	5	6	7	8	248
P. prælongus	1		3	4	5	6			249
P. perfoliatus	1		3	4	5	6	7		249
P. crispus	1	2	3	4		6	7		249
P. zosterifolius			3		5	6			250
P. compressus	.			4	5				250
P. pusillus	.	2	3	4	5	6			250
P. pectinatus			3	4	5	6	7.	8	251
P. densus	1	2	3	4	5	6			251
Ruppia rostellata								.	251
Zannichellia palustris	1	2	3	4	5	6	7	8	252
Schœnus nigricans	1	2	3	4	5				252
Cladium Mariscus	1	2		4	5	6			252
Rhynchospora alba			3						253
Eleocharis palustris	1		3	4	5	6	7	8	253
E. multicaulis		2	3						253
E. acicularis					5	6			254
Scirpus maritimus						6		8	254
S. lacustris	1		3	4	5		7	8	254
S. Tabernæmontani			3					8	254
S. cæspitosus	.				5				255
S. pauciflorus	1	2	3						255
S. fluitans			3			6	7		255
S. setaceus	1		3		5				255
Blysmus compressus	1		.		.				255
Eriophorum angustifolium	1	2			5			8	256
E. latifolium	1								256
Carex dioica	.		3		.				256
C. pulicaris	.				5				257
C. disticha	1	2	3	4	5	6			257
C. vulpina	1		3	4	5	6	7	8	257
C. muricata	1	2	3	4	5	6		8	257
C. divulsa	.	2	3	4					258
C. teretiuscula	.	2	.						258
C. paniculata	1	2	3		5				258
C. axillaris	.								259
C. remota	1	2	3	.			7		259
C. stellulata		2	3	.	5				259

	A.		B.			C.			PAGE.
	1	2	3	4	5	6	7	8	
Carex curta			3						259
C. ovalis		2	3						260
C. stricta	1	2		4	5	6			260
C. acuta	1				5	6			260
C. vulgaris	1		3						260
C. pallescens	1		3						261
C. panicea	1	2	3		5				261
C. strigosa									261
C. pendula	1								261
C. præcox	1	2	3		5				261
C. pilulifera	1		3		5				262
C. glauca	1	2	3	4	5	6	7		262
C. flava	1	2	3		5				262
C. Œderi	1	2	3	4	5		7		262
C. fulva	1	2	3		5				263
C. distans	1		3	4					263
C. binervis	1	2	3						263
C. sylvatica	1		3	4					263
C. Pseudo-cyperus	1		3					8	264
C. filiformis	1			4	5				264
C. hirta	1		3	4	5	6			264
C. ampullacea	1		3		5	6			265
C. vesicaria					5	6			265
C. paludosa	1		3		5		7	8	265
C. riparia	1	2	3	4	5	6		8	265
Setaria viridis									266
Phalaris arundinacea	1		3	4	5	6		8	266
Anthoxanthum odoratum	1	2	3	4	5			8	267
Phleum asperum									267
P. Boehmeri	1								267
P. arenarium									267
P. pratense	1	2	3	4	5	6	7	8	268
Alopecurus pratensis	1	2	3	4	5	6			268
A. geniculatus	1	2	3	4	5	6	7	8	268
A. fulvus			3	4	5	6			269
A. agrestis	1	2	3	4	5	6		8	269
Nardus stricta			3						269
Milium effusum	1		3				7		269
Phragmites communis	1	2	3	4	5	6	7	8	270
Calamagrostis lanceolata		2			5				270
C. Epigejos	1	2	3						270
Apera Spica-venti	1		3						270
A. interrupta	1				5				271
Agrostis canina			3	4	5				271
A. vulgaris	1	2	3						271
A. alba	1	2	3	4	5	6	7	8	271
Holcus lanatus	1	2	3	4	5	6	7	8	272
H. mollis			3						272
Aira cæspitosa	1	2	3	4	5	6	7		272
A. flexuosa			3						272
A. caryophyllea	1		3						273
A. præcox	1	2	3		5				273

	A.		B.			C.			PAGE.
	1	2	3	4	5	6	7	8	
Trisetum flavescens	1	2	3	4	5	6		8	273
Avena fatua		2	3	4	5	6		8	273
A. pratensis	1	2	3	4	5				274
A. pubescens	1	2	3	4	5				274
Arrhenatherum avenaceum	1	2	3	4	5	6	7	8	274
Triodia decumbens	1	2	3	.	5				275
Koehleria cristata	1	2	3	4				8	275
Melica uniflora	1								275
Molinia cærulea	1	2	3		5				276
Poa annua	1	2	3	4	5	6	7	8	276
P. nemoralis	1		3	4					276
P. trivialis	1	2	3	4	5	6		8	276
P. pratensis	1	2	3	4	5	6			277
P. compressa	1	2	3	4	5	6			277
Glyceria aquatica	1	2	3	4	5	6	7	8	277
G. fluitans	1	2	3	4	5	6	7	8	278
G. plicata	1	2	3	4	5	6	7		278
G. pedicellata	1					6			278
Sclerochloa maritima								8	278
S. distans								8	278
S. rigida	1	2	3	4	5	6			279
S. loliacea								.	279
Briza media	1	2	3	4	5	6		8	279
Catabrosa aquatica	1		3	4	5				280
Cynosurus cristatus	1	2	3	4	5			8	280
Dactylis glomerata	1	2	3	4	5	6	7	8	280
Festuca sciuroides	1	2	3		5	6			280
F. Myurus	1		3						281
F. ovina	1	2	3	4	5	6		8	281
F. duriuscula			3		5				281
F. rubra	1		3	4				8	281
F. gigantea	1	2	3	4	5		7		282
F. arundinacea	1	2	3	4	5	6		8	282
F. pratensis	1	2	3	4	5	6		8	282
F. loliacea	1		3		5	6			283
Bromus erectus	1	2	3		5				283
B. asper	1	2	3	4	5				283
B. sterilis	1	2	3	4	5	6	7	8	283
Serrafalcus secalinus	1			.	5	.			284
S. commutatus		2	3	4	5				284
S. racemosus	1	2	3		5				284
S. mollis	1	2	3	4	5	6	7	8	285
S. arvensis		2	3	4					285
Brachypodium sylvaticum	1	2	3	4	5				285
B. pinnatum	1	2	3	4	5				285
Triticum caninum	1	2	3	4	5		7		286
T. repens	1	2	3	4	5	6	7		286
T. pungens								8	286
Hordeum pratense	1	2	3	4	5	6	7	8	287
H. murinum	1	2	3	4	5	6	7	8	287
H. maritimum								8	287
Lepturus incurvatus				.				8	287

IN CAMBRIDGESHIRE.

	A.		B.			C.			PAGE.
	1	2	3	4	5	6	7	8	
Lolium perenne	1	2	3	4	5	6	7	8	288
L. temulentum	1		3						288
L. arvense	.		3						288
Equisetum arvense	1	2	3	4	5	6	7	8	289
E. Telmateja			3			6			289
E. sylvaticum			.	.					289
E. limosum	1	2	3	4	5				289
E. fluviatile			3			6	7		290
E. palustre	1	2	3	4	5	6			290
E. hyemale			.			.			290
Polypodium vulgare	1	2	3	4		.		.	290
Lastrea Thelypteris					5	6			291
L. Oreopteris			.						291
L. Filix-mas	1		3	4	5			8	292
L. spinulosa	1								292
L. dilatata	1		3						292
Polystichum aculeatum			.						293
Athyrium Filix-fœmina			3	4					293
Asplenium Adiantum-nigrum	1		.	4	5			.	293
A. Trichomanes	.	2		4		6		.	293
A. Ruta-muraria	1	2	.	4		.		8	294
Scolopendrium vulgare	.	.	3	4		6			294
Blechnum boreale			3						294
Pteris aquilina	1		3	4		6			295
Botrychium Lunaria	1								295
Ophioglossum vulgatum	1		3	4	5				296
Lycopodium clavatum			3						296
L. inundatum			3						297
Chara flexilis			3		5				297
C. syncarpa			3						297
C. tenuissima					5				297
C. polysperma	1		3	4					297
C. vulgaris	1	2	3		5	6			298
C. hispida	1	.	3	4	5	6	7		298

N.B. This Table is intended to shew the distribution of the plants in Cambridgeshire. It also forms a list of the desiderata of each Botanical District. The numbers opposite to the name of each plant are those of the Districts in which the plant has been found. The full stops (.) mark the Districts in which there are ancient, but unconfirmed localities for the plant. The blank spaces shew that the plant has not been met with in the Districts thus left unmarked.

The number at the end of each line is that of the page where the full account of the plant is given.

A few plants are noticed in the body of the work, but omitted in this list on account of their having no claims to be included in our Flora.

This list also forms a systematic Index to the contents of the book.

It is hoped that this Table will promote the completion of our Flora by shewing, at a glance, what are its chief defects.

FLORA

OF

CAMBRIDGESHIRE.

DICOTYLÉDONES OR EXÓGENÆ.

RANUNCULACEÆ.

CLÉMATIS Linn.

C. Vitálba Linn. *Traveller's Joy.*

Viorna, R. C. 177. *Clematis sylvestris latifolia*, M. M. 60. *C. Vitalba*, M. Pl. 13. Relh. 220.

Hedges on a gravelly or chalky soil. Sh. June.

Abundant in the (1) Cambridge, (2) Royston, and (3) Wimpole Districts, except on the clay.—4. Near the Observatory.—5. Landwade. Upware.

THALICTRUM Linn.

1. **T. saxátile** Schleich. (See Appendix I.)

T. flexuosum, Ann. Nat. Hist. ser. 2. xi. 268.

Hedgebanks on a chalky soil. P. June, July.

1. Allington Hill. Little Trees Hill, Gogmagogs. Fulbourn; N.—2. Mr. Fordham states that *T. minus* is found in the field by the sand-pit plantation at Odsey and by the railway-arch there; it is probably *T. saxatile*.—5. Roadside between Newmarket and Snailwell.

2. **T. flávum** Linn. *Meadow Rue.*

T. majus, R. C. 161. M. M. 20. *T. flavum*, M. Pl. 13. Relh. 220.

Very wet places. P. June, July.

1. Cow-Fen; W. H. C. Wilbraham Fen. Near Quy Bridge. Hinton; H.—2. Sawston Fen!; N.—3. Grantchester Meadows; Trumpington Spinney; W. H. C. Harlton!; N.—4. Brick-pits by the Observatory. Waterbeach Fen. Mare Way. Batesbite; W. H. C. *On the Chesterton side of the river opposite Barnwell; King's Hedges;* Relh.—5. Quy, Bottisham, Burwell and Wicken Fens.—6. Thetford. West Fen, and Roswell Pits, Ely.—7. Doddington. Chatteris.—8. World's End Lane, Wisbech; A.P.

ANEMÓNE Linn.

1. **A. Pulsatílla** Linn. *Pasque Flower.*

Pulsatilla vulgaris, Ger. Herb. 309 (A.D. 1597). *Pulsatilla*, R. C. 128. M. M. 60. *A. Pulsatilla*, M. Pl. 12. Relh. 219.

Chalk Hills. P. April, May.

1. Gogmagog Hills, especially in the Park. Furze Hills, Hildersham. Westhoe Park, near Linton; W. H. C. *Barrington Hill, near Linton* (now ploughed up); Relh.—2. On balks near Ickleton; G. S. G. On a broad balk halfway from Royston to Foulmire; T. F.—5. Devil's Ditch.

2. **A. nemorósa** Linn. *Wind Flower.*

Anemone nemorum, R. C. 12. *Anemonoides*, M. M. 60. *A. nemorosa*, M. Pl. 12. Relh. 219.

Woods and Thickets. P. March to May.

1. Abington Park; S. W. W. South end of Devil's Ditch. Yen Hall Wood, West Wratting. Westley Wood.—3. In most of the woods, as Whitwell, Gamlingay, Hardwick, &c.—4. Madingley Wood.

MYOSÚRUS Linn.

1. M. mínimus Linn. *Mouse-tail.*

Myosurus, R. C. 102. *Myosurus minimus*, M. M. 58. M. Pl. 8. Relh. 134.

Fields on a moist gravelly soil, rare. A. June, July.

1. *Stourbridge Fair Green;* Relh.— 3. Gamlingay in 1837!; H. *Eversden, just out of the village towards Cambridge;* J. M. *Haslingfield, at the outskirts of the village on the road to Cambridge, plentifully;* Relh.—4. *By highway from Oakington to the Huntingdon road;* Ray.

RANÚNCULUS Linn.

1. R. trichophyllus Chaix. *Water-Fennel. Millefoil.*

Millefolium aquaticum Ranunculi flore et capitulo, R. C. 99. *Ranunculus fœniculi folio breviore,* M. M. 58. *R. aquatilis* β. M. Pl. 13. Relh. 225.

Ponds and ditches, not common. P. May, June.

1. Wilbraham and Cow Fens. Wood Ditton. Shudy Camps. Balsham. Hinton. West Wratting. Brinkley. Coldham's Lane, Cambridge.—2. Ickleton. Bassingbourn. Sawston!; N.—3. Hardwick. Little Eversden. Bourn-Brook. Sheep's-Green, Cambridge.—4. Dry Drayton. By bridge in Cuckoo Lane, Oakington.—5. Bottisham and Wicken Fens. — 6. West Fen and Roswell Pits, Ely. Witcham.—8. Wisbech.—Some of these stations may belong to *R. Drouétii.*

2. R. Drouétii F. Schultz.

Ann. Nat. Hist. ser. 2. xvi. 393.

First found and distinguished by the Rev. W. W. Newbould in 1846.

Ditches, not very common. P. May, June.

1. Burrough-Green. Coldham's Lane, Cambridge.— 2. Ditch by the lane leading from Hauxton to the turnpike

road to Newton.—3. Near the brook at Comberton. Hardwick. Little Eversden.—4. Brick-pit by Ely road, to the north of Chesterton.—5. Pit by the road from Horningsey to Clayhythe. Waterbeach Fen.—6. Haddenham.

3. R. heterophyllus Fr. *Water Crowfoot.*

R. aquatilis, R. C. 130. *Ranunculoides foliis variis*, M. M. 57. *R. aquatilis a.* M. Pl. 15. Relh. 225.

Ponds and streams. P. May, June.

1. Wood Ditton. Spring Head, Temple, Wilbraham. Brinkley. Burrough-Green. Hinton.—3. Sheep's-Green, Cambridge. Eltisley. Long Stow!; Bourn; Kingston; Toft; Croxton; N.—4. Ditch by Madingley road near Cumming's garden. Histon. Pit by Cuckoo Lane, Oakington. By Mare Way. Fen Drayton.—5. Pit by road beyond Horningsey.—6. Haddenham!; N. By Aldreth Bridge. Witchford. Witcham.—7. Chatteris. Doddington.—8. Wisbech!; A. P.

4. R. Baudótii Godr.

I suspect that a plant growing in a brick-pit by the Ely road, to the north of Chesterton, belongs to this species, although no floating leaves were found when I first noticed it on June 10, 1859.

5. R. floribúndus Bab.

Ponds, rare. P. June to September.
First noticed by C. C. B. in 1856.
1. Abundant in pits near West Wratting.

6. R. circinátus Sibth.

R. aquatilis γ. M. Pl. 13. Relh. 255.

Streams and ponds. P. June to August.

1. Hinton Brook. Hobson-conduit Stream, and Cow Fen, Cambridge. Wilbraham. Little Linton.—3. Sheep's-

Green, Cambridge. Coton. Harlton. Barrington. Hauxton. Wimpole.—4. In the river near Chesterton. Mare Way. Waterbeach Fen. Dry Drayton. Swavesey.—5. Fen Ditton. Reche Lode. Wicken and Snailwell Fens.—6. Roswell Pits, Ely. Aldreth. Witchford. Witcham. Near Sandy's Cut; N.—7. Vermuden's Drain, near Chatteris.

7. R. fluitans Lam.

Ranunculo sive Polyanthemo aquatili albo affine Millefolium maratriphyllum fluitans, R. C. App. ii. 15. *Ranunculoides fœniculi folio longiore*, M. M. 58. *Ranunculus aquatilis* δ. M. Pl. 13. Relh. 225.

Rivers, streams, and rarely in ponds connected with them. P. June, July.

1. Hinton-brook. Wilbraham; N.—2. Meldreth; N. Bassingbourn; D. B.—3. Sheep's-Green, Cambridge. Harston!; Barrington; Haslingfield; N.—4. River at Chesterton.—5. Horningsey. Upware. Snailwell Fen.—6. Roswell Pits, Ely.

8. R. hederáceus Linn. *Ivy-leaved Water Crowfoot.*

R. hederaceus rivulorum se extendens atra macula notatus, R. C. 131. *Ranunculoides*, M. M. 57. *R. hederaceus*, M. Pl. 13. Relh. 225.

Shallow ponds and mud, rare. P. June to August.

1. Little Wilbraham. In the pond which gives rise to the brook, by the road from Fulbourn to Teversham; W. H. C. *Jesus Green and Hinton;* Relh.—2. Triplow.—3. Gamlingay. Coton. Barton!; H. Near Little Eversden Church; W. H. C. Hardwick; Comberton; Eltisley; N.—4. *By Madingley Road, just out of Cambridge;* Ray. *Near the Castle Hill;* T. M.

9. R. scelerátus Linn. *Celery-leaved Crowfoot.*

R. aquatilis rotundifolius, R. C. 130. M. M. 56. *R. sceleratus*, M. Pl. 12. Relh. 222.

By ditches and ponds. A. June to September.

1. Empty Common, Cambridge. Wood Ditton. Hinton. Wilbraham. West Wratting.—2. Pond at Kneesworth House; D. B.—3 and 4. Abundant.—5. Wicken, Snailwell, and Bottisham Fens. Horningsey.—6. Ely. Aldreth.—7. Doddington Turf-fen.—8. Elm. Wisbech.

10. R. Flámmula Linn. *Lesser Spearwort.*

R. flammeus minor and *R. f. serratus*, R. C. 131. M. M. 57. *R. Flammula*, M. Pl. 12. Relh. 221.

Wet places. P. June to August.

Probably common in such spots throughout the county. It has been found in all the districts, and is abundant in the Fens.

11. R. Língua Linn. *Greater Spearwort.*

R. flammeus major, R. C. 131. M. M. 56. *R. Lingua*, M. Pl. 21. Relh. 221.

Marshes and Fens, rare. P. June, July.

1. By Quy water. Shelford; H. *Teversham Moor;* Ray.—2. Triplow peat-holes. Sawston Fen, but perhaps now lost; G. S. G.—3. *Gamlingay;* Relh.—5. Bottisham and Wicken Fens. Swaffham Prior !; H. Swaffham Bulbeck. Quy Fen; S. W. W. Anglesey Abbey.—6. Ely. Near Sandy's Cut; N. *Stretham and Aldreth;* Relh.

12. R. Ficária Linn. *Pilewort.*

Chelidonium minus, R. C. 33. *Ficaria*, M. M. 60. *Ficaria verna*, M. Pl. 13. *R. Ficaria*, Relh. 222.

Common in damp shady places. P. April, May

Found in all the districts.

13. R. auricomus Linn. *Goldilocks.*

R. C. 131. M. M. 57. M. Pl. 12. Relh. 222.

Open woods. P. April, May.

1. Hinton. Brinkley. Wood Ditton!; H.—3. Common.—4. Moorbarns Thicket. Madingley. Chesterton; W. H. C.—5. Fen Ditton.—8. Wisbech; J. B.

14. R. ácris Linn.

R. rectus, non repens flore simplici luteo, R. C. App. i. 8. *R. pratensis erectus acris,* M. M. 57. *R. acris,* M. Pl. 12. Relh. 224.

Meadows and pastures. P. June, July.

Common in all the districts.

15. R. répens Linn.

R. pratensis repens, R. C. 132. M. M. 57. *R. repens,* M. Pl. 12. Relh. 223.

Damp waste places and pastures. P. May to August.

Common throughout the county.

16. R. bulbósus Linn.

R. C. 130. M. Pl. 12. Relh. 223. *R. tuberosus major,* M. M. 56.

Meadows and pastures. P. May.

Common in the (1) Cambridge, (2) Royston, (3) Wimpole, (4) Cottenham, and (5) Burwell Districts.—6. Ely.—8. Wisbech; A. P.

17. R. hirsútus Curt.

R. rectus foliis pallidioribus, R. C. App. i. 8. M. M. 57. *R. bulbosus β.* M. Pl. 12. *R. hirsutus,* Relh. 223.

Damp waste ground. A. June to October.

3. To the south of the turnpike road at the end of the village of Barton. Comberton; Eversden; Harlton; Cal-

decot; between Haslingfield and Harston; N.—4. Cottenham Fen. On way to Balsar's Hill from Willingham. *Chesterton;* J. M.

†18. R. arvénsis Linn.

R. arvorum, R. C. 130. M. M. 56. *R. arvensis*, M. Pl. 13. Relh. 224.

Corn-fields, possibly introduced. A. June.

Common in the (1) Cambridge, (2) Royston, (3) Wimpole, and (4) Cottenham Districts.—6. Ely. Sutton. Witchford. Stuntney; N.—8. Wisbech.

19. R. parviflórus Linn.

R. hirsutus annuus flore minimo, R. C. App. i. 8. M. M. 57. *R. parviflorus*, M. Pl. 12. Relh. 224.

Hedge-banks in dry places. A. May, June.

1. *Trumpington; Shelford;* Relh.—3. Comberton. Haslingfield. Hardwick. In the lane near the church at Caldecot. Lane below Grantchester Church; W. H. C. Harlton; Orwell; Bourn; Little Gransden; N. Coton, sparingly; C. B. C. Under a hedge by the road from Toft to Kingston. *Gamlingay;* Relh.—4. Dry Drayton; N. *In the hedge by the church at Madingley, and by the road to Drayton;* J. M.

CALTHA Linn.

C. palústris Linn. *Marsh Marigold.*

R. C. 25. M. M. 65. M. Pl. 13. Relh. 227.

Marshy places. P. March, April.

1. Fulbourn. Hinton. Wood Ditton. Wilbraham. Cow Fen, Cambridge.—2. Triplow peat-holes. Octagon pond in Wimpole Avenue; N. Bassingbourn; D. B. Shepreth; N. W.—3. Sheep's-Green, Cambridge. Bourn Brook. Comberton. Barrington. Eltisley. Gamlingay. Harlton;

Great Eversden; N.—4. Waterbeach.—5. Horningsey. Ditton. Wicken. Snailwell.—6. By Sandy's Cut, Ely; N.—8. Wisbech; A. P.

HELLÉBORUS Linn.

‡1. **H. víridis** Linn. *Green Hellebore. Bear's-foot.*

H. niger hortensis flore viridi, R. C. 73. M. M. 64. *H. viridis*, M. Pl. 13. Relh. 226.

Hedges and thickets, probably introduced. P. March, April. Believed to be "really wild" in Hertfordshire.

1. Banks of the Camp at Granham's Farm, Shelford.—3. In the Brook Grove at Haslingfield; King's Grove, Barton; S. W. W. In an old orchard and in a field near the stump of cross at Coton, extirpated since 1830. *In the copses at Whitwell where the Narcissus grows;* Ray.—5. In a hedge at Biggin Closes near Ditton; S. W. W. *In the close adjoining the parsonage garden at Ditton;* J. M.

‡2. **H. fœtidus** Linn. *Stinking Hellebore.*

Helleboraster maximus, R. C. App. i. 5. *H. niger fœtidus*, M. M. 64. *H. fœtidus*, M. Pl. 13. Relh. 226.

Near houses, probably introduced.

1. Chalk-pit close at Hinton, very recently extirpated. In some closes near the old water-mill at Fulbourn. *In the second close on the right hand as you go from Hinton church to Fulbourn;* Ray.—2. *Triplow;* Relh.

AQUILÉGIA Linn.

‡1. **A. vulgáris** Linn. *Columbine.*

A. sylvestris, R. C. 14. *A. officinarum*, M. M. 66. *A. vulgaris*, M. Pl. 12. Relh. 217.

Fields and hedges, a doubtful native. P. May, June.

1. In the park, Gogmagog Hills; S. W. W. Stetchworth. Wood Ditton. West of Burrough-Green church.

Behind Mr Townley's grounds, Fulbourn; W. H. C. *In a little thicket at the hither end of Teversham Moor;* Ray. *Hinton;* Relh.—2. *Triplow;* Relh.—3. White Wood, Gamlingay; S. W. W. *Hatley St George;* Relh.—5. Devil's Ditch. Swaffham Fen; H. *Anglesey Abbey;* Relh.

DELPHÍNIUM Linn.

*1. **D. Consólida** Linn. *Larkspur.*

D. segetum flore cœruleo, Dill. in Ray's Syn. ed. 3. 273. *D. arvense,* M. M. 66. *D. Consolida,* M. Pl. 12. Relh. 217.

Differs from the true *D. Consolida* by having a downy capsule.

First noticed by Mr J. Sherard shortly before 1724 in great plenty in the corn at Swaffham. Afterwards became very abundant in corn-fields, but is now disappearing through improved farming. A. June, July.

Chalky fields in the (1) Cambridge, (2) Royston, (3) Wimpole, (4) Cottenham, and (5) Burwell districts.—6. Ely.

BERBERIDACEÆ.

BÉRBERIS Linn.

1. **B. vulgáris** Linn. *Barberry. Pipperidge Bush.*

Oxyacantha, R. C. 110. *B. officinarum,* M. M. 122. *B. officinalis.* M. Pl. 8. Relh. 145.

Hedges. Sh. May, June.

1. *Hinton;* Relh. *Hildersham;* J. M. *Balsham;* J. F.—2. Near Crishall Grange. *Triplow;* Relh.—3. Malton; D. B. *Grantchester;* Relh.—4. By path to Chesterton church, extirpated since 1845.—5. In a close by the pond, Fen Ditton; S. W. W. Swaffham Prior; Bottisham; H.

NYMPHÆACEÆ.

NYMPHÆA Linn.

N. álba Linn. *White Water Lily.*

R. C. 104. M. Pl. 12. Relh. 215. *Nenuphar album officinarum,* M. M. 99.

Rivers, ditches, and ponds. P. July.

1. Near the old mill at Fulbourn; W. H. C. *Between Barnwell and Hinton; Hinton Moor* (now drained); Relh. *Teversham Moor;* Ray.—2. Melbourn Common. Triplow peat-holes.—3. *Between Toft and Kingston and beyond Toft Bridge:* Relh.—4. In the old Ouse near Aldreth Bridge.—5. Bottisham and Wicken Fens. Ponds near the mill at Anglesey Abbey; S. W. W.—6. Ely.—8. Wisbech; J. B.

NÚPHAR Sm.

N. lútea Sm. *Yellow Water Lily. Brandy Bottle.*

Nymphæa lutea, R. C. 104. M. Pl. 12. Relh. 214. *Nymphæa lutea officinarum,* M. M. 12.

Rivers, ditches, and ponds. P. July.

1. River Cam.—2. Melbourn Common. Whittlesford; N. Shepreth; H. F.—3. Sheep's-Green, Cambridge. Bourn Brook. Wimpole. Barrington. Malton. Comberton; Toft; Tadlow; N. Little Eversden; T. Y.—4. King's Hedges. Waterbeach. Old Ouse near Aldreth Bridge. Ouse near Swavesey.—5, 6. Common in the Fens.—7. In the old Bedford river near Earith !; N.—8. Wisbech; J. B.

PAPAVERACEÆ.

PAPÁVER Linn.

† 1. P. Argémone Linn.

Argemone capitulo longiore, R. C. 15. *P. Argemone,* M. Pl. 12. Relh. 213.

Corn-fields. A. June, July.

Common, except in the Fens, in the (1) Cambridge, (2) Royston, (3) Wimpole, (4) Cottenham, and (5) Burwell districts.—6. Sutton. Witchford.—8. Wisbech.

† 2. P. hybridum Linn.

Argemone capitulo rotundiore, R. C. 15. *P. hybridum*, M. Pl. 12. Relh. 212.

Corn-fields, rarely on hedge-banks. A. June, July.

Not uncommon in the (1) Cambridge, (2) Royston, (3) Wimpole, (4) Cottenham and (5) Burwell districts.

† 3. P. Rhœas Linn. *Red Poppy. Corn Rose.*

R. C. 111. M. Pl. 12. Relh. 214. *P. erraticum officinarum*, M. M. 82.

Corn-fields. A. June, July.

Common in all the districts.

† 4. P. dubium Linn. (See Appendix II.)

P. laciniato folio, capitulo longiore glabro, R. Fasc. Stirp. Brit. 18. *Argemone capitulo longiore glabro*, R. Hist. ii. 856. *P. dubium*, Lyons, 37. M. Pl. 12. Relh. 213.

First found by Mr. P. Dent, and sent by him to Ray.

Corn-fields and hedge-banks. A. June, July.

Not uncommon in the (1) Cambridge, (2) Royston, and (3) Wimpole districts.—4. Girton. Fen Drayton. Cambridge.—5. Newmarket. Chippenham (see Appendix).—6. Witcham.—7. Chatteris.

‡ 5. P. somníferum Linn.

P. spontaneum sylvestre, R. C. 112. M. M. 82. *P. somniferum*, M. Pl. 12. Relh. 214.

Said to have been formerly largely cultivated in the Fens, where it sometimes comes up when the banks are deeply turned over.—4. On the banks of the closes which separate

Denny Abbey from the Ely road!; H. *Waterbeach Fen;* *Rampton;* Relh.—5. Bottisham Fen.—6. *In many places on the banks in the Isle of Ely;* Ray.

ROEMÉRIA Cand.

‡1. **R. hybrida** Cand.

Papaver corniculatum violaceum, R. C. 111. *Glaucium flore violaceo*, M. M. 82. *Chelidonium hybridum*, M. Pl. 12. Relh. ed. 1, 201. *Glaucium violaceum*, Relh. ed. 3, 211.

Corn-fields. A. June.

5. Between Swaffham Prior and Burwell.—Confined to this county.

GLAÚCIUM Tourn.

[1. **G. lúteum** Scop. *Horned Poppy.*

Hensl. Cat. ed. 1. 2.

Henslow marks this plant as a native of this county. He gives no authority for it, and is now unable to supply me with any information relative to it. Probably there was some mistake, as the plant is not known to have been found in the only likely place, namely, by the river below Wisbech.]

CHELIDÓNIUM Linn.

‡1. **C. majus** Linn. *Celandine.*

R. C. 33. M. M. 82. M. Pl. 12. Relh. 211.

Hedge-banks. Never seen except near houses. Probably not a native, although tolerably common. P. May to August.

1. Great Wilbraham; S. W. W. Fulbourn. Fen Ditton; W. H. C. Linton; R. B. S. Dullingham. Brinkley. Burrough-Green. Hinton; N. *Cow Fen;* J. F. Teversham; J. M.—2. Paddock near the Heath at Royston, rare; D. B.—3. White Wood, Gamlingay. Comberton. Arring-

ton. Harston; Toft; Bourn; N. Barton; W. H. C.—
4. Dry Drayton. Chesterton.—5. Biggin Abbey; S. W. W. Newmarket; N.—7. Doddington.—8. By the Canal between Elm and Wisbech. North Brink, Wisbech; A. P.

FUMARIACEÆ.

Fumária Linn.

? 1. F. capreoláta Linn.

F. major scandens flore pallidiore, R. C. App. ii. 7. R. Cat. Angl. ed. 1. 122. M. M. 37. *F. capreolata*, M. Pl. 16. Relh. 286.

Cultivated ground, rare. A. June to September.

There is much doubt about this plant as a native of Cambridgeshire. All that has been so named may probably be the rampant form of *F. officinalis;* or those found at Elm and near Wisbech and near Reche, may have been one or more of the species usually confounded under the name of *F. capreolata.*

1. Corn-field by the Gogmagog Hills; W. H. C. He now thinks that it was only *F. officinalis. On the borders of the Gogmagog Hills towards Hinton;* Ray.—3. Wimpole; S. W. W. *Gamlingay;* J. M.—5. One of the forms was seen near Reche Lode in 1841; N.—8. Near Oxborough Hall, Elm. Banks of the canal near Elm; J. B.

†2. F. officinális Linn. *Common Fumitory.*

Fumaria, R. C. 57. *F. officinarum*, M. M. 37. *F. officinalis*, M. Pl. 16. Relh. 286.

Fields and waste places, common. A. May to September.

Probably common in all the districts, although I have few stations recorded in the Fens.

‡ 3. F. micrántha Lag.

Fields, rare. A. June to September.

1. In a field at the south-west entrance of Fulbourn, where it was shown to me by some person whose name is unfortunately not recorded, in 1848. Near Allington Hill. —2. Morden Heath Farm; N.

4. F. parviflóra Lam.

Hensl. Cat. ed. 1, 2.
Chalky fields, rare. A. June to September.
First found by Prof. Henslow in May 1826.

1. Fields between Hinton and Gogmagog Hills, especially in the field beyond the reservoir of the Waterworks. Streetway Hill. Hildersham. Newmarket. Linton; G. S. G. —2. Morden Heath Farm!; N. North-east of Royston; H. F.—5. Kennet!; N. Swaffham; Rev. J. Downes. Bottisham; H. Exning; E. S. Newmarket; N.

5. F. Vailllántii Lois.

Chalky Fields. A. June to September.
First found by Prof. Henslow in May 1831, but not then distinguished from *F. parviflora.*

1. With *F. parviflora* in the field near the Reservoir at Hinton, and on the Gogmagog Hills. Near Allington Hill. Hildersham.—2. Morden Heath Farm!; N.—5. North-west of Newmarket. Bottisham!; H.

CRUCIFERÆ.

CHEIRÁNTHUS Linn.

*1. C. Cheiri Linn. *Wallflower.*

Leucoium luteum vulgare, R. C. 86. *L. luteum officinarum,* M. M. 78. *Ch. Cheiri,* M. Pl. 15. Relh. ed. 1, 252. *Ch. fruticulosus;* Relh. ed. 3. 269.

On old walls, introduced. P. April, May.

1. Barnwell Abbey and near the river there. Old walls in Cambridge.—2. Churchyard wall at Royston; D. B.—

3. On old wall at Castle End, Cambridge; W. H. C.—6. Ely.
—8. Wisbech; J. B.

Nastúrtium R. Br.

1. N. officinále R. Br. *Water Cress.*

N. aquaticum officinarum, R. C. 103. M. M. 80. *Sisymbrium Nasturtium*, M. Pl. 15. Relh. 265.

In running water, common. P. June, July.

Found in all the districts, but seems to be least plentiful in the Fens.

2. N. sylvéstre R. Br.

Eruca aquatica, R. C. 50. *Radicula aquatica foliis in profundas lacinias divisis*, M. M. 81. *Sisymbrium sylvestre*, M. Pl. 15. Relh. 266.

River-banks and wet places. P. June to August.

1. By the river below Barnwell and opposite to Chesterton. Dullingham. Stetchworth.—4. By the river below Waterbeach. Cottenham. By the river Ouse near Fen Drayton and by ditches there.—5. By the river below Clayhythe, especially at Bottisham Lock. Reche Lode. Horningsey.—6. Ely.—7. Doddington. Chatteris.—8. Wisbech; A. P.

3. N. palústre Cand.

N. terrestre. Relh. ed. 1. App. i. 14; ed. iii. 266.

Wet places, rare. P. June to September.

2. *In the village of Hauxton;* Relh.—4. By the river at Clayhythe and Ditton. Quarry at Upware.—6. Ely. Haddenham. Witchford.

Barbaréa R. Br.

1. B. vulgáris R. Br. *Yellow Rocket.*

Barbarea, R. C. 19. M. M. 79. *Erysimum Barbarea*, M. Pl. 15. Relh. 268.

By ditches and streams. B. ? May to August.

Common, although apparently less so in the Fens, where I have but few stations recorded, and none in the (7) Chatteris District.

A′RABIS Linn.

1. A. hirsúta R. Br.

Turritis hirsuta, M. Pl. 15. Lyons, 42. Relh. 271.

First found by Prof. J. Martyn.

Walls and dry banks; also on the black soil in the Fens. B. June to August.

1. Dale Moor near Sawston; H. *Roadside just beyond the turnpike at Bourn Bridge;* J. M.—5. In Quy and Bottisham Fens. On the Park-wall at Chippenham. *Walls at the Red Lion Inn, Newmarket;* Relh.—8. River-bank at Wisbech; A. P.—The plant of the Fens is the *A. sagittata* of De Candolle and the French authors.

*2. A. Turríta Linn.

M. Pl. 15. Lyons, 42. Relh. 270.

Introduced since the time of Ray. First noticed by Prof. T. Martyn in 1763.

Walls. B. May.

1. Old walls about Trinity and St John's Colleges, less abundant now than formerly, owing to recent repairs.—3. Lately established near the brook in the walks of St John's College.

[Ray records *Turritis* (Cat. 172) as "found where flax did grow near Cambridge;" but adds that he had never found it. It is the *Turritis glabra* of Linnæus, and is probably not a native of this county.]

CARDAMÍNE Linn.

1. C. hirsúta Linn.

Relh. ed. ii. 255; ed. iii. 264.

First found by Mr Skrimshire.

In rather damp places, rare. A. May to August.

3. St John's College walks. Queens' College garden; C. B. C. Toft; N.—6. West Fen, Ely.—8. Wisbech!; A. P. *Between Wisbech and Newton;* Mr W. Skrimshire.

2. C. praténsis Linn. *Cuckoo-flower. Lady's Smock.*

Cardamine, R. C. 27. *C. officinarum,* M. M. 80. *C. pratensis,* M. Pl. 15. Relh. 265.

Damp meadows. P. May.

Common, probably, throughout the county. I have but few stations recorded in the Fens; none in the (7) Chatteris District.

β. *C. dentata* Schult.

Bab. Man. ed. 1. 21.

This form used to grow at Hinton, but alterations have destroyed it.

A double-flowered state of *C. pratensis* is not uncommon.

SISYMBRIUM Linn.

1. S. officinále Scop. *Hedge Mustard.*

Erysimum Dioscoridis, R. C. 50. *E. officinarum,* M. M. 79. *E. officinale,* M. Pl. 15. Relh. 268.

Banks and waste ground. A. June, July.

Common, probably, in all the districts. Few localities are recorded in the Fens.

2. S. I'rio Linn.

Relh. ed. ii. 258; ed. iii. 267.

About towns, on walls and dry rubbish. A. July, August.

1. At Barnwell in 1818; J. W.—8. *Wisbech before* 1802. Mr W. Skrimshire.

3. S. Sophía Linn. *Flixweed.*

S. chirurgorum, R. C. 158. *S. chirurgorum officinarum,* M. M. 80. *S. Sophia,* M. Pl. 15. Relh. 267.

Waste ground. A. June to August.

1. Barnwell. Wilbraham. Shuckburgh Castle and Newmarket. Dullingham.—2. Royston; Foxton; N.—3. Malton; Comberton; N.—4. Near the Observatory. Chesterton. Histon. Mare Way.—5. Newmarket. Snailwell. Quy Road. Upware.—6. Ely. Sutton.—7. Doddington. Chatteris. Maney Station; N.—8. Wisbech. Eastrey Station; N.

4. S. thaliánum Gaud.

Pilosella siliquata major et minor, R. C. App. i. 7. *Sophia minor præcox foliis integris,* M. M. 80. *Arabis thaliana,* M. Pl. 15. Relh. 270.

Gravelly places, walls and banks, rare. A. April and May; September and October.

1. Barrington Hill; H.—3. Gamlingay.—4. Gravel-pits near the Observatory. Oakington; N.

ALLIARIA Adans.

1. A. officinalis Andrzj. *Sauce-alone. Jack-by-the-hedge.*

Alliaria, R. C. 6. M. M. 78. *Erysimum Alliaria,* M. Pl. 15. Relh. 268.

Hedge-banks. B. May and June.

Common, probably, in the whole county, although I have no stations recorded in (7) Chatteris District and only Wisbech in District 8.

ERÝSIMUM Linn.

1. E. cheiránthoïdes Linn. *Treacle Wormseed.*

Camelina, R. C. 26. *Alliaria angustifolia siliquis brevioribus,* M. M. 78. *E. cheiranthoides,* M. Pl. 15. Relh. 269.

Cultivated ground. A doubtful native, except in the Fens, where it seems to be spontaneous. A. June to August. Common throughout the county.

Brássica Linn.

‡1. B. campéstris Linn.

There are two forms of this species, viz.:

α. *B. campestris* Linn. Hensl. Cat. ed. ii. 8. *Napus sylvestris.* R. C. 102. *Wild Navew.*

β. *B. Rapa* Linn. *Rapum vel Rapa*, R. C. 134. *Rapum officinarum*, M. M. 79. *B. Rapa*, M. Pl. 16. Relh. 272. Also *Rapum radice oblonga*, R. C. 134. M. M. 79. *B. Rapa* β. M. Pl. 15. *Turnip.*

Fields, occasionally found. A very doubtful nature; probably always, certainly the var. β, accidently introduced. A. or B. July, August.

‡2. B. Nápus Linn. *Rape. Coleseed.*

Rapum sylvestre, R. C. 134. M. M. 79. *B. Napus*, M. Pl. 15. Relh. 271.

Fields, probably not a native. A. or B. May, June.
1. Allington Hill.

Sinápis Linn.

1. S. nígra Linn. *Black Mustard.*

Sinapi II. *sive vulgare*, R. C. 155. *S. officinarum*, M. M. 79. *S. nigra*, M. Pl. 15. Relh. 273.

Willowy river-banks; rarely in fields, where it is accidental. A. June to August.

3. *On banks newly cast up behind the colleges;* Ray. In small quantity at Harlton; N.—4. Cottenham and Waterbeach Fens. By the Mare Way. Westwick; N.—5. Clayhythe. Upware.—6. Roswell Pits, Ely. In the village of

Witcham. Stuntney; N.—7. Wisbech. By the railway between Whittlesey and Eastrey; N.

2. S. arvénsis Linn. *Charlock.*

Rapistrum arvorum, R. C. 132. *Rapum sylvestre seminibus lucidis nigris*, M. M. 79. *S. arvensis*, M. Pl. 15. Relh. 272.

Corn-fields. A. June to October.

Common throughout the county.

The farmers tell Mr Newbould that this plant and *S. alba* do not like precisely the same soil. At Orwell the fields on one side of a road are full of this plant, on the other of *S. alba*, although not quite exclusively in either case. Apparently the latter prefers chalky land, the former clay.

3. S. alba Linn. *White Mustard.*

Rapistrum luteum siliqua hirsuta articulata, R. C. 133. *S. nigra*, M. Pl. 15. Lyons, 43. Relh. 272.

Chalky corn-fields. A. July.

Common in the (1) Cambridge, (2) Royston, and (3) Wimpole Districts.—4. Dry Drayton. Madingley,—5. Swaffham Prior!; H. Chippenham.—6. Haddenham.—7. Doddington.—8. Wisbech, on the silt; A. P.

Diplotáxis Cand.

1. D. murális Cand.

Waste gravelly ground. A. August, September.

1. Gravel-pits near the junction of Wort's Causeway with the Wool-street, where it was first noticed in this county by C. C. B. in 1846. A little below Shardelow's Well near Fulbourn. A weed on waste ground at the Cambridge Botanical Garden, but how introduced there is not known.

Alýssum Linn.

[1. A. calycínum Linn.

Cultivated land. Not a native. A. May, June.

1. In a field opposite to the Railway Station at Sixmile-bottom, were it was first noticed in the county in 1855 by C. C. B. Near Bartlow, and between Linton and Hildersham; G. S. G.—4. Gravel-pits near the Observatory.]

Drába Linn.

1. D. vérna Linn. *Whitlow-grass.*

Paronychia vulgaris, R. C. 113. *Draba vulgaris caule nudo Polygoni folio hirsuto*, M. M. 78. *D. verna*, M. Pl. 14. Relh. 260.

Walls and dry banks, common. A. March to May.

Probably common throughout the county, although I have no station recorded in the (7) Chatteris District.

At least two of the plants distinguished as species on the continent are to be found in this county, but their names and localities have not been satisfactorily determined.

Cochleária Linn.

1. C. ánglica Linn. *Scurvy-grass.*

Relh. ed. ii. 253; ed. 3. 262.

First found by Mr Skrimshire.

Salt marshes. A. May.

8. By the river below Wisbech.

Armorácia Rupp.

[1. A. rusticána Rupp. *Horse Radish.*

Raphanus rusticanus, R. C. 132. *Cochlearia*, M. M. 15. *Cochlearia Armoracia*, M. Pl. 15. Relh. 262.

Waste ground near houses. Not a native. P. May.

1. Cambridge!; H.—4. Histon. Cottenham. *Magdalene College Close;* Ray.—6. Ely.]

2. A. amphíbia Koch.

Raphanus aquaticus, R. C. 132, *Sisymbrium amphibium*, M. Pl. 15. Relh. 267.

In very wet places. P. June to August.

1. Roadside beyond Barnwell, and on the bank of the river.—4. Waterbeach Fen. Near the west ferry at Over; N.—5. Horningsey. By Bottisham Lode. Upware. Wicken Fen.—6. Ely. Witchford. Stretham!; Rev. H. Baber.

[*Camelina sativa* has been occasionally found in crops, but is not naturalized here. It is the *Alyssum sativum* of Relh. ed. ii. 260; ed. iii. 260.]

THLÁSPI Linn.

[1. T. arvénse Linn. *Penny Cress*.

Relh. ed. i. 247; ed. iii. 261.

Fields and roadsides. A. May to August.

4. *A little way below the great Gull in Waterbeach Fen;* Relh.]

TEESDÁLIA R. Br.

1. T. nudicaúlis R. Br.

Bursa pastoris minor, R. C. 24. *Iberis petræa foliis Bursæ pastoris*, M. M. 81. *Iberis nudicaulis*, M. Pl. 12. Relh. 263.

Sandy and gravelly places. A. May, June.

2. *Litlington Field, on the balks between the chalk-pit and Limbury Hill, and on the way to Morden;* Relh.— 3. Gamlingay Heath, chiefly near the borders of the county towards Potton.

IBÉRIS Linn.

1. T. amára Linn. *Candytuft*.

Watson's New Botanist's Guide, ii. 599.

First noticed in the county by the Rev. W. H. Coleman, in 1837.

Chalky Fields. A. July.

2. Near Royston to the right of the road to Melbourn!; W. H. C. By the Icknield Way, both to the east and west of Royston; near Morden Heath Farm!; Between Known's Folly and Melbourn; N. By Penny-loaf Hill, Odsey; H. F.

Lepídium Linn.

[**L. Drába** (Linn.) was noticed in 1857 and 1858 by the Railway Station at Oakington! by Mr Newbould. It will probably establish itself there.]

‡ 1. **L. campéstre** R. Br.

Hensl. Cat. ed. i. 3.
Dry gravelly soil. B. June to August.

1. Bank on the left hand of the road to Hinton (Mill Road), just before the entrance to the fields, sparingly in 1825; H. Not seen recently.—7. Near Meadlands Farm by the upper end of the Old Bedford River in 1844 or 1845, sparingly; N.

2. **L. ruderále** Linn.

M. Pl. 14. Lyons, 42. Relh. 261.
First found by Prof. J. Martyn.
Waste ground near the sea. A. May, June.

8. River-bank both above and below Wisbech!; H. *Near the turnpike at Tidd Gout!;* Relh.

3. **L. latifólium** Linn. *Dittander.*

R. C. App. ii. 12. M. Pl. 14. Relh. ed. ii. 252; ed. iii. 261. *L. officinarum*, M. M. 80.
Near the sea. P. July, August.

Mr Peter Dent introduced this plant into his edition of Ray's Appendix, as found "in a little close on the right hand of Maids' Causeway on the way to Barnwell," a place now built over. Prof. T. Martyn doubts the correctness of the name, and Relhan omits it from his Flora until again discovered by Mr Skrimshire.

8. *Leverington near Wisbech;* Mr. Skrimshire.

Capsélla Vent.

1. C. Bursa-pastoris Cand. *Shepherd's Purse. Pick-purse. Caseweed.*

Bursa-pastoris, R. C. 24. M. M. 81. *Thlaspi Bursa-pastoris,* M. Pl. 14. Relh. 262.

Waste and cultivated ground. A. March to October.
Common throughout the county.

Senebiéra Pers.

1. S. Corónopus Poiret. *Swine's-cress.*

Coronopus Ruellii, R. C. 39. Relh. ed. 2, 254; ed. 3, 263. *Carara,* M. M. 81. *Cochlearia Coronopus,* M. Pl. 14. Relh. ed. i. 248.

Waste ground, especially on roadsides. A. June to September.

Common throughout the county, although few stations have been recorded in the Fens.

Isátis Linn.

[1. I. tinctória Linn. *Dyer's Woad.*

Glastum sativum, R. C. 62. *Glastum,* M. M. 82. *Isatis tinctoria,* M. Pl. 15. Relh. 264.

Cultivated land, not a native, nor naturalized. B. July.

6. *New barns near Ely;* Relh. Cultivated there for one year, but then discontinued at the request of the parish, for fear of increasing the number of paupers; W. M. *Planted about Littleport;* Ray.—8. *Wisbech;* Mr Woodward in Bot. Guide, i. 59.]

Ráphanus Linn.

1. R. Raphanístrum Linn. *Jointed Charlock.*

Rapistrum luteum siliqua glabra articulata, R. C. 132.
Raphanus Raphanistrum, M. Pl. 15. Rclh. 273.

Cultivated land. A. June, July.

Common in the (1) Cambridge, (2) Royston, (3) Wimpole, and (4) Cottenham Districts.—5. Chippenham. Quy Road.—6. Sutton.—8. Between Peterborough and Whittlesey; N.

RESEDACEÆ.

Reséda Linn.

1. R. lútea Linn. *Wild Mignonette.*

R. vulgaris, R. C. 138. M. M. 20. *R. lutea,* M. Pl. 11. Relh. 190.

Waste chalky land. B. June to August.

Abundant on the chalk in the (1) Cambridge and (2) Royston Districts.—3. Haslingfield. Orwell. Little Eversden. Harston; N.—4. Chesterton.—5. Horningsey. Quy Road. Newmarket. Chippenham.

2. R. Lutéola Linn. *Weld.*

Luteola, R. C. 92. M. M. 20. *R. luteola,* M. Pl. 11. Relh. 189.

Waste chalky ground. B. July, August.

Rather common in the (1) Cambridge, (2) Royston, and (3) Wimpole Districts.—4. Chesterton Road.—5. Exning. Horningsey. Quy Road.—8. Wisbech; A. P.

CISTACEÆ.

HELIÁNTHEMUM Gaert.

1. H. vulgáre Gaert. *Rock-rose.*

Chamæcistus, R. C. 31. *H. vulgare flore luteo*, M. M. 93. *Cistus Helianthemum*, M. Pl. 12. Relh. 216.

Banks on a chalky soil. P. July to September.

1. Gogmagog Hills. Fulbourn. Linton. Bartlow. Hildersham. Fleam Dyke. Balsham. Hinton. Shuckburgh Castle, Newmarket Heath. Burrough-Green. Six-mile-Bottom.—2. Bran Ditch near Heydon Grange, but in this county. Whittlesford. Kneesworth. By Newmarket Road, Royston; D. B. Odsey; H. F. Shepreth; N. W. *Triplow;* J. M.—3. Kingston; N. *Gamlingay Heath;* J. M.—5. Devil's Ditch. Chippenham. Newmarket.—6. *Haddenham;* J. M.

VIOLACEÆ.

VÍOLA Linn.

1. V. odoráta Linn. *Sweet Violet.*

V. nigra seu purpurea and *V. martia alba odora*, R. C. 176. *V. martia officinarum* and *V. martia alba*, M. M. 97. *V. odorata*, M. Pl. 20. Relh. 92.

Groves and hedge-banks. P. March, April.

Common in the (1) Cambridge, (2) Royston, (3) Wimpole, and (4) Cottenham Districts.—5. Chippenham.—8. Wisbech; A. P.

2. V. hírta Linn.

V. martia hirsuta inodora, R. Fasc. Stirp. Brit. 21. *V. hirta*, Lyons, 49. Relh. 92.

First noticed by Mr Dale.

Thickets and hedge-banks in chalky places. P. April, May.

1. Common.—2. Hauxton. Morden Heath Farm!; N. Odsey; H. F.—3. Common.—4. Madingley. Near Observatory; W. H. C. Honey Hill near Childerley; Papworth St Everard; N. *Moor Barns;* Relh.

β. *calcarea* Bab.

1. Gogmagog Hills, especially beyond the Park. Devil's Ditch.—5. Devil's Ditch.

3. V. sylvática Fries. *Wood Violet.*

(See Appendix III.)

V. canina sylvestris, R. C. 176. M. M. 97. *V. canina,* M. Pl. 20. Relh. 93.

Hedge-banks and in thickets. P. April, May.

1. Common.—2. Shepreth. Royston; D. B.—3. Common.—4. Madingley.—5. Fen Ditton; N.

A larger-flowered form, supposed by some to be a distinct species and called *V. Riviniana,* is found in the Rivey near Linton; and near Toft Bridge, and in Kingston and Hayley Woods.

4. V. canína Linn. *Dog Violet.*

V. flavicornis, Hensl. Cat. ed. 1, 3.

Heaths and peaty places. P. April, May.

3. Gamlingay Heath.—5. Bottisham Fen.—6. West Fen, Ely.—7. Doddington Turf-Fen.

5. V. stagnína Kit.

V. lactea, Hensl. Cat. ed. 1, 3.

Peat-bogs. P. May, June.

4. Once grew near Lockspit Hall, Cottenham.—5. Bottisham, Wicken, and White Fens.

6. V. tricolor Linn. *Heart's-ease. Pansies.*

R. C. 177. M. Pl. 20. Relh. 93. *V. bicolor arvensis,* M. M. 98.

Our plant is the *V. arvensis* Murr.

Cultivated and waste ground. A. May to October.

Appears to be common throughout the county, although I have very few stations recorded in the Fens.

DROSERACEÆ.

Drósera Linn.

1. D. rotúndifolia Linn. *Sundew.*

Ros solis folio rotundo, R. C. 139. *Ros solis officinarum*, M. M. 97. *D. rotundifolia*, M. Pl. 8. Relh. 133.

Boggy places. P. July, August.

1. *Hinton, Teversham, and Fulbourn Moors;* Relh.— 2. Melbourn and Foulmire Commons.—3. Gamlingay.

2. D. intermédia Hayn.

Rorella sive Ros solis foliis oblongis, R. C. 139. *Ros solis folio oblongo*, M. M. 97. *D. longifolia*, M. Pl. 8. Relh. 134.

Boggy places. A. July, August.

1. *Hinton Moor;* Ray. *Teversham and Sawston Moors;* Relh. Not found since those places have been drained.

3. D. ánglica Huds.

Relh. ed. 2, 129; ed. 3, 134.

Boggy places. A. July, August.

1. *Sawston and Hinton Moors;* Relh. Not found since those places were drained.

PARNÁSSIA Linn.

1. P. palústris Linn. *Grass of Parnassus.*

Gramen Parnassi, R. C. 70. *Parnassia vulgaris et palustris,* M. M. 96. *P. palustris,* M. Pl. 7. Relh. 130.

Boggy places. P. August to October.

1. Shelford Common, until very recently. Teversham. *Hinton and Trumpington Moors;* Ray. *Linton;* Relh.— 2. Peat-holes near Triplow. Dernford Fen; S. W. W. Sawston Fen; N. Foulmire Moor; N. W.—3. Gamlingay. —5. Quy. Bottisham!; Anglesey Abbey!; H.

POLYGALACEÆ.

POLÝGALA Linn.

1. P. vulgáris Linn. *Milkwort.*

Polygala, R. C. 121. M. M. 76. *P. vulgaris,* M. Pl. 16. Relh. 287.

Dry pastures and peaty meadows. P. June to September.

1. Fulbourn. Teversham. Gogmagog Hills. Hinton. Wilbraham. Balsham. Six-mile-Bottom. Fleam Dyke. Devil's Ditch.—2. Melbourn Common. Heydon Ditch. Triplow. Sawston Fen; N. Litlington. Kneesworth. Royston. Shepreth; N. W.—3. Gamlingay. Eltisley, and throughout the western part of the District.—4. Chesterton.—5. Bottisham and Wicken Fens. Chippenham. Snailwell Heath. Devil's Ditch. High Ditch Lane, Fen Ditton. —8. Wisbech; J. B.

FRANKENIACEÆ.

FRANKÉNIA Linn.

1. F. lævis Linn.

M. Pl. 8. Lyons, 32. Relh. 146.

Salt marshes. P. August.

8. *Tidd Gout near Wisbech;* J. M.

CARYOPHYLLACEÆ.

Diánthus Linn.

* D. Caryophyllus Linn. *Clove Pink.*

Relh. ed. 2, 166; ed. 3, 173.
Old walls. P. June.

5. *Chippenham Park wall;* Relh.—8. *Leverington!;* Relh.

2. **D. deltoïdes** Linn. *Maiden Pink.*

Caryophyllus virgineus, R. C. App. ii. 4. *Caryophyllus minor repens nostras*, M. M. 91. *Dianthus deltoides*, M. Pl. 10. Relh. 174.

Dry hills. P. June to September.

1. Hildersham, on the Furze-hill next to Linton!; Rev. Dr Cookson.—3. *Near White Wood, Gamlingay;* J. M.

Saponária Linn.

‡ 1. **S. officinális** Linn. *Soapwort.*

Saponaria, R. C. 150. *S. officinarum*, M. M. 93. *S. officinalis*, M. Pl. 10. Relh. 173.

Near houses. Probably planted formerly. P. August.

1. Hinton, near the entrance by the footway from Cambridge to the spring. Barnwell Abbey; S. W. W. *Shelford;* J. M.—2. By the river at Ickleton; G. S. G. *Whittlesford;* J. M.—3. Grantchester; S. W. W. *In lane between Trumpington church and London road; Comberton;* Relh.—4. On the way to Cottenham from Histon; Rev. S. Hiley. Near Cambridge; W. H. C. *Madingley;* J. M.—5. Paper Mills, near Cambridge; S. W. W.—8. Horse-shoe Corner, Wisbech.

Siléne Linn.

1. S. ánglica Linn. *English Catchfly.*

Lychnis sylvestris annua angustifolia flore rubente, R. C. App. ii. 12. R. Cat. Angl. ed. 1, 202. *L. flore albo minimo*, M. M. 91. *S. anglica*, M. Pl. 10. Relh. 174.

Sandy and gravelly fields. A. June to October.

5. Chippenham Gravel-pit. Between Fordham and Freckenham. *Near the gravel-pits as you go to the nearest windmill on the north side of Newmarket;* Ray.

2. S. infláta Sm. *Bladder Campion. Spatling Poppy. White Bottle.*

Behen album, R. C. 20. *B. album officinarum*, M. M. 92. *Cucubalus Behen*, M. Pl. 10. Relh. ed. 1, 168. *S. inflata*, Relh. ed. 3, 175.

Borders of fields and gravel-pits. P. June to August.

Tolerably common except in the Fens. I have no station for it in the (6) Ely district.—7. Chatteris.—8. Near Whittlesey.

3. S. noctiflóra Linn.

Relh. ed. 1, Suppl. i. 13; ed. 3, 175.

Gravelly and chalky fields. A. July, August.

First found by Relhan in 1786.

1. Gogmagog Hills. Balsham. Hildersham; G. S. G. Sawston; Hinxton; N. *Catlidge Hall;* Relh.—2. Hauxton; Harston; N.—3. Common in the western part; N.—4. Opposite to Upware. Childerley; N.—5. Bottisham!; Anglesey Abbey; H. Chippenham. Horningsey. Upware. Exning; E. S. *Near the Devil's Ditch;* Relh.—6. Near Sandy's Cut; Stuntney; N.—7. Chatteris. Doddington.—8. Wisbech. Foul Anchour.

Lychnis Linn.

1. L. Flos-cúculi Linn. *Cuckoo-flower. Wild Williams. Meadow Pink. Ragged Robin.*

Armerius sylvestris, R. C. 15. *Flos-cuculi pratensis*, M. M. 92. *L. Flos-cuculi*, M. Pl. 10. Relh. 183.

Wet meadows and bogs. P. May, June.

Probably thinly scattered throughout the county, although I have but few stations recorded in the Fens, and none in the (8) Wisbech District.

2. L. vespertína Sibth. *White Campion.*

L. sylvestris flore albo, R. C. 92. M. M. 91. *L. dioica*, M. Pl. 10. Relh. ed. 1, 177. *L. dioica* β. Relh. ed. 2, 117; ed. 3, 184.

Hedges and arable fields. B? June to September.
Common throughout the county.

3. L. Githa´go Lam. *Corn Cockle.*

Pseudomelanthium, R. C. 127. M. M. 92. *Agrostemma Githago*, M. Pl. 10. Relh. 183.

Corn-fields. A. June to August.

Common in the (1) Cambridge, (2) Royston, and (3) Wimpole districts.—4. By Huntingdon road, near Cambridge.—5. Chippenham. Newmarket Heath. Horningsey. Upware.—8. Wisbech; J. B.

Sagína Linn.

1. S. procúmbens Linn. *Pearlwort.*

Saxifraga Anglica facie Seseli pratensis R. C. 151. *Alsinella muscoso flore repens*, M. M. 83. *S. procumbens*, M. Pl. 4. Relh. 70.

Waste spots which are rather damp. P. May to September.

1. Stourbridge Fair Green; Mr A. G. More. *Gogmagog Hills;* Ray.—3. Gamlingay. Yard of Almshouses at the gate of Wimpole Park. Between the stones by the school at Toft!; N.—4. Chesterton; N. *Hill of Health;* J. M.—5. Kennet Heath. On a wall at Wood Ditton. Newmarket. —6. West Fen, Ely. Haddenham Sand-pit.—7. Doddington Turf-fen.—8. Foul Anchour.—Perhaps introduced at the Wimpole Gate and Toft stations.

2. S. apétala Linn.

Saxifraga graminea pusilla flore parvo herbido et muscoso, R. C. 151. *S. procumbens* β. M. Pl. 4. *S. apetala,* Relh. ed. 1, App. i. 10; ed. 3, 71.

Dry, gravelly, and sandy places, and on walls. A. May to September.

1. Walls and walks at Cambridge. Shuckburgh Castle, Newmarket Heath. Dullingham.—3. On the walks and in the New Court of St John's College. Comberton. Toft. Caldecot. Kingston. Gamlingay. Arrington; Caxton; Tadlow; Eversden; N.—4. Chesterton. Willingham. Cottenham. Swavesey. Fen Drayton. Oakington; Elsworth; N. *Church-yard wall at Milton,* Relh.—5. Kennet Heath. Chippenham Park wall. Newmarket; N. Bottisham!; H.—6. Witcham.—7. Doddington.—8. Wisbech.

3. S. ciliáta Fries.

Botan. Gazette, i. 176.

First noticed in this county by C. C. B. in 1849, at Gamlingay.

Dry, sandy, and gravelly places. A. May, June.

1. Furze-hills, Hildersham; G. S. G. and N. The latter botanist rather doubts the correctness of the name.—3. On the hedge-bank by the road leading from the site of the old mansion to the brick-pits at Gamlingay.

A plant which seems to be the *S. densa* (Jord.), was found at Wisbech ! by Prof. Henslow. It was first published in Bab. Man. ed. 4, 50.

4. S. nodósa E. Meyer. *Knotted Spurrey.*

Alsine palustris foliis tenuissimis, sive Saxifraga palustris alsinefolia, R. C. 9. *Saxifraga minor foliis Knawel flore majusculo albo*, M. M. 95. *Spergula nodosa*, M. Pl. 10. Relh. 186.

Wet, sandy, and peaty places. P. July, August.

1. On the common one mile from Cambridge towards London. *Teversham and Hinton Moors;* Ray.—2. Ickleton; G. S. G.—3. Gamlingay.—5. Bottisham and Wicken Fens. Kennet Heath. Quy Fen; S. W. W. Newmarket. Quarry near Upware.—7. Doddington Turf-fen.—8. Opposite Guiherne on the south bank of the river Nene. Wisbech.

ALSÍNE Wahl.

1. A. tenuifólia Wahl.

R. C. 9. *Spergula tenuifolia elatior*, M. M. 95. *Arenaria tenuifolia*, M. Pl. 10. Relh. 179.

Dry, sandy, and chalky places, and on walls. A. May, June.

1. Wall of Fulbourn churchyard. Near Allington Hill. Dullingham, on and about the railway. Burrough-Green, on an old wall. Six-mile-Bottom. Gogmagog Hills; Great Wilbraham; S. W. W. Hildersham; G. S. G.—2. On a wall at Little Shelford. Triplow; W. H. C. Odsey; H. F.

Ickleton; C. B. C.—3. Orwell; N.—5. Chippenham Park wall. Snailwell. Devil's Ditch; W. H. C.

Mœhringia Linn.

1. M. trinérvis Clair.

Arenaria fontana credita flosculorum foliolis non divisis, R. C. App. i. 3. *A. plantaginis folio,* R. C. App. ii. 2. *Spergula plantaginis folio,* M. M. 95. *Arenaria trinervis,* M. Pl. 10. Relh. 178.

Damp shady places and ditch sides. A. May, June.

1. Wood Ditton. South end of Devil's Ditch. West Wratting. *Balsham;* Ray. *Linton;* Relh.—2. Elmdon! H. —3. Gamlingay. Long Stow; N.—4. Chesterton. King's Hedges.

Arenaria Linn.

1. A. serpillifólia Linn.

Alsine minor multicaulis, R. C. 9. *Spergula multicaulis,* M. M. 95. *A. serpillifolia,* M. Pl. 10. Relh. 178.

Dry places and walls. A. June to August.

Probably a common plant, although some of the localities may belong to *A. leptoclados.*—1. Cambridge. Hinton. Wilbraham. Shudy Camps. Six-mile-Bottom. Hildersham. Newmarket. Weston Colville. Brinkley.—2. Shepreth. Little Shelford. Foxton.—3. Cambridge. Barton; Harlton; N.—4. Swavesey. Chesterton. Impington. Landbeach.—5. Chippenham. Horningsey.—6. Haddenham. Ely. Sutton. Witchford. Witcham.—8. Wisbech.

2. A. leptocládos Guss. (See Appendix IV.)

First detected in this county by Mr Newbould.

2. Royston. Steeple Morden. Foxton. Near Ashwell.— 3. Gamlingay. Orwell. Cambridge. Harston. Shepreth.

Comberton. Toft.—4. Oakington.—5. Devil's Ditch.—I am indebted to the Rev. W. W. Newbould for all these localities.

STELLÁRIA Linn.

1. S. média Wither. *Chickweed.*

Alsine media, R. C. 9. M. Pl. 7. Relh. ed. 1, 128. *A. officinarum,* M. M. 94. *Stellaria media,* Relh. ed. 3, 176.

Rich land both waste and cultivated. A. March to September.

Common throughout the county.

2. S. Holóstea Linn. *Stitchwort.*

Holosteum vernum flore majore, R. C. 76. *Alsine pratensis gramineo folio ampliore,* M. M. 94. *S. Holostea,* M. Pl. 10. Relh. 176.

Woods and hedges. P. April to June.

1. Abington Park; S. W. W. Devil's Ditch. Shudy Camps. West Wratting. Weston Colville. Brinkley. Deersley's Wood, Newmarket. — 2. Foulmire. — 3. Gamlingay. Eversden and Kingston Woods. Near the brook, Kingston!; Toft; Caldecot; Long Stow; N. Whitwell; S. W. W. Bourn; W. H. C. Arrington; D. B.

3. S. glaúca Wither.

S. graminea β. M. Pl. 10. Relh. ed. 1, 170. *S. glauca,* Relh. ed. 3, 178.

First found by Prof. J. Martyn at Gamlingay.

Marshes. P. May to July.

1. Meldreth; D. B.—2. Gamlingay!; H., but probably now extinct.—5. Bottisham and Wicken Fens. Upware quarry. Burwell!; H. Baitesbite; S. W. W.—6. Ely. *Plentiful in the Isle;* Ray.—7. Doddington Turf-fen. Carter's Bridge. *Chatteris;* J.M.—8. *Bardolph Fen, Wisbech;* Relh., but that is in Norfolk, I believe.

4. S. gramínea Linn.

Holostei Ruellii diversitas, R. C. 76. *Alsine pratensis gramineo folio angustiore*, M. M. 94. *S. graminea*, M. Pl. 10. Relh. 177.

Heathy and bushy places. P. May to August.

1. *Cow Fen;* Relh.—2. Royston; D. B.—3. Gamlingay.—4. Girton. Between Cottenham and Histon. Fen Drayton.—5. Kennet Heath.—6. Ely. Sutton. Witcham. *Mepal;* J. M.

5. S. uliginósa Murr. *Water Chickweed.*

Alsine longifolia uliginosis proveniens locis, R. C. 8. M. M. 94. *S. graminea* γ. M. Pl. 10. Relh. ed. 1, 170. *S. uliginosa*, Relh. ed. 3, 178.

Wet places. A. May, June.

1. Cow Fen. The common at one mile from Cambridge towards London; H.—3. Gamlingay. Sheep's-Green, Cambridge; W. H. C.—6. Ely.

MOENCHIA Ehrh.

1. M. erécta Sm.

Holosteum tetrapetalon sive Alsine tetrapetalos Caryophylloïdes, R. C. App. ii. 9. R. Cat. Angl. ed. 2, 163. *Alsinella foliis caryophylleis*, M. M. 84. *Sagina erecta*, M. Pl. 4. Relh. 71.

Dry gravelly and sandy places. A. May, June.

3. By the road from White Wood to Gamlingay.—4. *Plentifully on the Hill of Health in* 1805; Relh.

MALÁCHIUM Fries.

1. M. aquáticum Fries. *Great Chickweed.*

Alsine major, R. C. 9. *A. aquatica major*, R. C. 7. M. M. 94. *Cerastium aquaticum*, M. Pl. 10. Relh. 185.

Ditches, river-banks and wet places. P. July, August.

1. Long Drove, Wilbraham Fen.—2. Sawston; G. S. G. Abundant at Meldreth; D. B. Cow Fen.—3. Grantchester Lane. Sheep's-Green, Cambridge. Barrington. Toft; Tadlow; N.—4. Near Upware. Waterbeach Fen. Mare Way. By the Ouse near Swavesey.—5. Near Upware. Wicken. Horningsey. Bottisham Lock. Fen Ditton!; H.—6. By lower way to Roswell Pits, Ely. Aldreth. Witchford. Barraway. Near Sandy's Cut; N.—7. Chatteris. Doddington. —8. By Wisbech Canal.

CERÁSTIUM Linn.

1. C. glomerátum Thuil.

Alsine hirsuta altera viscosa, R. C. App. i. 3. M. M. 94[1]. *C. viscosum*, M. Pl. 10. Relh. ed. 1, 178. *C. vulgatum*, Relh. ed. 2, 177; ed. 3, 184.

Fields and banks. A. April to September.

1. About Cambridge. Hinton.—2. Foulmire. Ickleton. —3. About Cambridge. Gamlingay. Barton. Hardwick. Comberton; Toft; Bourn; N.—4. Chesterton. Oakington; N.—5. Horningsey.—6. Ely. Sutton.

2. C. triviále Link.

Alsine hirsuta Myosotis, R. C. 8. *Myosotis arvensis hirsuta flore parvo*, M. M. 95. *C. vulgatum*, M. Pl. 10. Relh. ed. 1, 178. *C. viscosum*, Relh. ed. 2, 177; ed. 3, 184.

Waste places, old walls, banks. A. or B. April to September.

Probably common throughout the county.

---

[1] I think with Fries, that these plants of Ray and Martyn are our *C. glomeratum*, and that the same authors' names quoted under *C. triviale* are there correctly placed. Smith transfers them with the Linnæan names.

Common in the (1) Cambridge, (2) Royston, (3) Wimpole, (4) Cottenham, and (5) Burwell Districts.—6. Haddenham. Witchford. Sutton. Ely.—7. Chatteris. Doddington.—8. Wisbech.

3. C. semidecándrum Linn.

Lyons, 34. M. Pl. 10. Relh. 185.

First noticed by Mr Lyons.

Tops of walls and dry banks. A. April, May.

1. Gogmagog Hills. Wool-street. *By 1st milestone on the Hills road* (opposite the Railway Station); Lyons.—2. Hauxton.—3. Gamlingay. Harston; N.—4. Castle End, Cambridge. On walls at Chesterton. *Hill of Health;* J. M.—5. Chippenham Park wall.—6. Haddenham.

4. C. arvénse Linn.

Auricula muris pulchro flore albo, R. C. 19. *Myosotis arvensis subhirsuta flore majore*, M. M. 94. *C. arvense*, M. Pl. 10. Relh. 185.

Sandy, gravelly, and especially chalky places. P. April to August.

1. Gogmagog Hills. Above the Chalk-pit at Hinton. Linton. Hildersham. Weston Colville. Brinkley. Six-mile-Bottom. Newmarket Heath. Hinxton; N.—2. Heydon Ditch. Meggot's Mound. Ickleton. Royston. Whittlesford. Steeple Morden; N.—3. Gamlingay.—4. Castle Hill, Cambridge. *Hill of Health;* J. M.—5. Chippenham. Snailwell Heath. Swaffham Bulbeck!; H. Near the quarry, Upware. Newmarket; Bottisham; N.—8. Wisbech!; J. B.

MALVACEÆ.

MÁLVA Linn.

1. **M. moschátá** Linn. *Musk Mallow. Vervain Mallow.*

Alcea vulgaris, R. C. App. i. 3. *M. verbenaca*, M. M. 55. *M. Alcea*, M. Pl. 16. *M. moschata*, Relh. 282.

Gravelly hedge-banks and borders of fields. P. July, August.

1. Roadside between Brinkley and Six-mile-Bottom. Balsham Wood; R. B. S. *Linton;* Relh.—2. Hedge-bank sparingly between Royston and Bassingbourn; D. B. Odsey; H. F.—3. Above Eversden quarry; N. *Kingston Wood;* Ray.

2. **M. sylvéstris** Linn. *Common Mallow.*

M. vulgaris, R. C. 94. *M. vulgaris officinarum,* M. M. 55. *M. sylvestris,* M. Pl. 16. Relh. 282.

Roadsides and waste places. P. June to September. Common throughout the county.

3. **M. rotúndifolia** Linn. *Dwarf Mallow.*

M. sylvestris minor, R. C. 95. M. M. 55. *M. rotundifolia,* M. Pl. 16. Relh. 282.

Waste ground. P. or B. June to September.

Common throughout the county except on the Peat Soil. Therefore rare in the Fens.—6. In the Isle of Ely (proper) in many places.—7. No locality known to me.—8. Wisbech. Bridge at Four Gouts.

ALTHÆA Linn.

1. **A. officinális** Linn. *Marsh Mallow.*

Althæa, R. C. 9. *A. officinarum* and *A. vulgaris folio retuso brevi,* M. M. 55. *A. officinalis,* M. Pl. 16. Relh. 281.

Marshes, especially near the sea. P. August, September.

3. *Banks of a ditch between the osier-holt by Cow Fen and Trumpington Meadow;* Relh.—8. By the Roman bank at Newton. Elm; Upwell; W. M. By the road from Wisbech to Peterborough, and between it and the river Nene. Abundant about Wisbech; A. P. *Twenty-foot Drain near March by a bridge on the road to Wisbech;* Relh.

TILIACEÆ.

Tília Linn.

[1. **T. europæa** Linn. *Lime Tree.*

M. Pl. 12. Lyons, 38. Relh. 215.

Planted about Cambridge, Wimpole, &c. Tree. July.]

‡2. **T. parvifolia** Ehrh.

T. europæa β. M. Pl. 12. Lyons, 38. *T. parvifolia,* Relh. 215.

Woods. Tree. August.

3. *White Wood, Gamlingay;* T. M.—4. King's Hedges.

HYPERICACEÆ.

Hypéricum Linn.

1. **H. quadrángulum** Linn. *St Peter's Wort.*

Ascyron, R. C. 16. *H. Ascyron dictum caule quadrangulo,* M. M. 93. *H. quadrangulum,* M. Pl. 17. Relh. 307.

Wet places by ditches and streams. P. July.

Probably common throughout the county, although I have very few stations recorded in the Fens.

2. **H. perforátum** Linn. *St John's Wort.*

Hypericum, R. C. 79. *H. officinarum,* M. M. 93. *H. perforatum,* M. Pl. 17. Relh. 307.

Groves and hedges. P. July, August.

Common on the chalk and gravel in the (1) Cambridge, (2) Royston, and (3) Wimpole Districts.—4. Fen Drayton. Bird's Pastures, Childerley; N.—5. Horningsey. Swaffham Bulbeck!; H. Chippenham. Landwade.

3. H. humifúsum Linn.

H. minus supinum, R. C. App. ii. 10. R. Cat. Angl. ed. 2, 168. *H. humifusum*, M. Pl. 17. Relh. 308.

Gravelly and sandy places. P. July.

1. *Chalk-pit-close, Hinton;* Relh. On the first of the furze hills next Hildersham, towards Juniper Hill; Ray.— 2. Triplow; G. S. G.—3. Gamlingay.—5. Newmarket; N.

4. H. hirsútum Linn.

H. majus seu Androsœmum Matthioli, R. C. App. i. 6. M. M. 93. *H. hirsutum*, M. Pl. 17. Relh. 308.

Groves and hedges. P. July, August.

1. Devil's Ditch. Shudy Camps. Balsham. Borley Wood. Great Wilbraham. West Wratting. Weston Colville. Brinkley. Fulbourn. Teversham.—2. Hinxton; N. Odsey; H. F.—3. Eversden, Hardwick, Kingston, Eltisley and other woods; also thickets and hedge-banks in the western part of the district.—4. Madingley. Moorbarns Thicket. Childerley; Papworth; N.—6. Ely. Sutton. Witcham.— 7. Chatteris. Doddington.

5. H. púlchrum Linn.

H. minus erectum, R. C. 78. M. M. 93. *H. pulchrum*, M. Pl. 17. Relh. 308.

Dry heaths and banks. P. June, July.

1. Hinton.—3. Gamlingay.

6. H. elódes Linn. *Round-leaved St Peter's-wort.*

Ascyrum supinum villosum palustre, R. C. 17. *H. palustre supinum tomentosum*, M. M. 93. *H. elodes*, M. Pl. 17. Relh. 309.

Bogs. P. July, August.

3. Grew in the bogs at Gamlingay, which were drained in 1843.

[*H. hircinum.* There is a note upon an unpublished drawing prepared for *Eng. Bot.* which states that Relhan found this plant growing at Impington "by the side of a pond near the great house in immense quantity" in 1799. I do not know if the plant still continues there, as is probable, but it certainly has no claim to be considered as a native.]

ACERACEÆ.

A'cer Linn.

1. A. campéstre Linn. *Maple.*

Hedges and thickets. Tree. May, June.

A. minus, R. C. 3. M. M. 123. *A. campestre*, M. Pl. 23. Relh. 161.

Probably a native. Common in the (1) Cambridge, (2) Royston, (3) Wimpole, and (4) Cottenham Districts.— 5. Quy Road. Chippenham. Wicken.—6. Witchford. Stuntney; N.—7. Doddington Wood.—8. Wisbech; J. B.

*2. A. Pseudo-platanus Linn. *Sycamore.*

A. majus, R. C. 2. M. M. 123. *A. Pseudo-platanus*, M. Pl. 23. Relh. 160.

Planted about Cambridge, Wimpole, Wisbech, &c.

GERANIACEÆ.

Gerá́nium Linn.

[1. G. phæum Linn.

Relh. ed. 1, Suppl. ii. 13; ed. 3, 277.

Woods and thickets. P. May, June.

1. *Found by the Rev. R. Forby near the fence of the first close on the east side of the churchyard at Teversham, about* 1788.]

2. G. praténse Linn. *Crowfoot Crane's-bill.*

G. batrachioïdes, R. C. 61. M. M. 96. *G. pratense*, M. Pl. 15. Relh. 277.

Moist pastures. P. June to August.

1. In the chalk-pit-close, Hinton. Wood Ditton. West Wratting. Stapleford; S. W. W.—2. By the railway near Sawston!; Near the Railway Station, Meldreth; N.—3. Near the wood, Eltisley. Warren between Trumpington and the river; Left-hand side of the last field by the footpath to Coton; W. H. C. Caxton; Bourn; N. Papworth St Everard; T. Y. Comberton.—4. Howe's Closes. Oakington; S. W. W. Girton; N. Between the river and the footpath to Chesterton Church; Closes by Madingley Wood; W. H. C. *Histon; Hill of Health; Moorbarns Thicket;* J. M.—5. Horningsey. Bottisham; H. *Fen Ditton;* J. M.

3. G. sanguíneum Linn.

G. hematodes, R. C. 61. M. M. 96. *G. sanguineum*, M. Pl. 16. Relh. 280.

Dry places. P. July.

1. Devil's Ditch, near Stetchworth. Wood Ditton. Church Meadow, Balsham; R. B. S. *In a wood between Stetchworth and Chitley* (Cheveley?); Relh.

†4. G. pyrenaicum Linn.

Relh. ed. 3, 279.

First noticed by the Rev. W. Pulling.

Roadsides. P. June, July.

1. Hedge on right-hand side of Linton road, a little beyond Red Cross Turnpike, and in the adjoining hedge at right angles to it. Wool-street between Balsham and Linton; R. B. S. *Back of Barnwell* (in a place that is now built over), Rev. W. Pulling; Relh.—3. To the right of footpath through Newnham Closes; S. W. W.

5. **G. pusíllum** Linn. *Least Dove's-foot.*

G. malachoïdes sive columbinum minimum, R. C. 61.
G. malachoïdes seu minimum, M. M. 96. *G. pusillum,* M.
Pl. 15. Relh. 279.

Cultivated and waste gravelly ground. A. June to September.

1. New Botanic garden ground, Cambridge. Wilbraham. Fulbourn. Shudy Camps. Hinton.—2. Royston.
—3. Gamlingay. Eltisley. Great Eversden. Barton.—
4. Fen Drayton. Swavesey. Chesterton. Impington. *Hill of Health;* Ray.—5. Horningsey. Swaffham Bulbeck!; H. Chippenham.—6. Witchford. Witcham. Sutton.

‡ 6. **G. columbínum** Linn.

Cultivated ground. A. June, July.

3. Once found, in 1857, in a cultivated field at Harlton; N.

7. **G. disséctum** Linn. *Dove's-foot.*

G. columbinum majus dissectis foliis, R. C. 61. M. M.
96. *G. dissectum,* M. Pl. 16. Relh. 280.

Dry banks and waste places. A. Juue to August.

Tolerably abundant on gravel and chalk throughout the (1) Cambridge, (2) Wimpole, aud (3) Cottenham Districts, but has not been noticed close to Cambridge.—5. Horningsey. Swaffham Prior. Upware. Newmarket. Chippenham.—6. Ely. Witchford. Witcham. Sutton.—7. Chatteris. Doddington.—8. Wisbech. Foul Anchour.

8. **G. rotundifólium** Linn.

Lyons, 44. M. Pl. 15. Relh. ed. 1, 261; ed. 3, 279.
Old walls and waste places. A. June, July.

FLORA OF CAMBRIDGESHIRE. 47

1. *Paper Mills*, J. M.; Lyons.—3. *White Wood, Gamlingay*, J. M.; Lyons.—4. *Trinity Conduit Head;* Lyons.

In his first edition Relhan gives all these stations for the plant; omits it altogether in his second edition; and restores only that of White Wood in the third. It seems probable therefore that he found it between 1802 and 1820 in that place, and that it was lost before 1802 from the others.

3. G. mólle Linn.

G. columbinum, R. C. 61. *G. columbinum officinarum*, M. M. 96. *G. molle*, M. Pl. 15. Relh. 278.

Cultivated and waste ground. A. April to August.

Common in the (1) Cambridge, (2) Royston, (3) Wimpole, and (4) Cottenham Districts.—5. Horningsey. Ditton!; H. Chippenham. — 6. Ely. Witchford. Wentworth. Stuntney; N.—7. Chatteris. Doddington.—8. Wisbech.

10. G. lúcidum Linn.

G. saxatile, R. C. 62. M. M. 96. *G. lucidum*, M. Pl. 15. Relh. 278.

Rather damp but exposed banks. A. May to August.

1. On the bank of Midsummer Common next to Barnwell.—4. Near the footpath to Chesterton church.—5. Quy!; H.—8. Wisbech; A. P.

11. G. Robértianum Linn. *Herb Robert.*

R. C. 62. M. Pl. 15. Relh. 278. *G. Robertianum officinarum*, M. M. 96.

Damp shady banks. A. May to September.

1. Common.—2. Triplow. Royston. Steeple Morden; N. Odsey; A. M. B.—3, 4. Common.—5. Chippenham. Upware.—6. Ely. Witcham. Sutton.—7. Doddington.—8. Wisbech; J. B.

Eródium L'Herit.

1. E. cicutárium Sm. *Crane's-bill.*

Geranium arvense vel minus, R. C. 61. *G. cicutæ folio inodorum,* M. M. 96. *G. cicutarium,* M. Pl. 15. Relh. ed. 1, 260. *E. cicutarium,* Relh. ed. 2, 266; ed. 3, 276.

Dry gravelly or chalky fields. A. June to September.

1. Hildersham. Shudy Camps. Six-mile-Bottom. Shuckburgh Castle, Newmarket Heath. Gogmagog Hills.—2. Royston. Odsey; H. F.—3. Gamlingay. Eltisley. Harston!; N.—4. Castle Hill, Cambridge; S. W. W. Girton. Cottenham. *Hill of Health;* J. M.—5. Chippenham. Quy!; H. Newmarket; N.—7. Chatteris.—8. Wisbech; J. B.

Relhan records the *E. pimpinellæfolium* (Sibth.) as a native of the county, and I fancy that I have myself seen it at some of the above-mentioned stations.

‡ 2. E. moschátum Sm.

Geranium moschatum, Lyons, 44. M. Pl. 15.

Waste ground, scarcely a native. A. June, July.

3. *Near Gamlingay Park, Mr C. Miller and Prof. T. Martyn;* Lyons. 5. Between the Inn and the steam-engine at Upware.—8. Leverington and Parson's Drove!; Rev. M. J. Berkeley.

LINACEÆ.

Línum Linn.

[1. L. usitatíssimum Linn. *Flax.*

L. sativum, R. C. 89. *L. officinarum,* M. M. 97. *L. usitatissimum,* M. Pl. 7. Relh. 132.

Fields, occasionally, but not a native. A. July.]

2. L. perénne Linn. *Wild Blue Flax.*

L. sylvestre radice perenni flore cœruleo, R.C. 89. M. M. 97. *L. perenne,* M. Pl. 7. Relh. 132.

β. *L. sylvestre cœruleum perenne procumbens, flore et capitulo minore*, R. Syn. 362. *L. sylvestre cœruleum perenne nostras*, M. M. 97.

Chalky places. P. June, July.

1. Hinton, by the road above the chalk-pit; and between that place and Fulbourn. Gogmagog Hills. Shelford Common; H. Near milestone, marked 14, on the Newmarket Road near Hinxton, abundantly; N. Bourn Bridge; G.S.G.

3. L. cathárticum Linn. *Mil-mountain.*

L. sylvestre catharticum, R. C. 90. *L. catharticum officinarum*, M. M. 97. *L. catharticum*, M. Pl. 8. Relh. 133.

In both dry and fenny places. A. June to August. Common throughout the county.

Radíola Gmel.

1. R. millegrána Sm.

Damp, sandy places. A. July to September.
1. Near Newmarket, Oct. 1, 1821; J. W.

BALSAMINACEÆ.

Impátiens Linn.

[* 1. I. parviflóra Cand.

Road-sides, an escape from gardens. A. July, August.

2. Plentiful in a lane leading to the fen at Sawston! in 1856; In the village of Duxford; N.—4. By the road to Chesterton; Mr W. Walton, M.A.]

OXALIDACEÆ.

Oxális Linn.

1. **O. Acetosélla** Linn. *Wood Sorrel.*

Trifolium acetosum vulgare, R. C. 165. *Oxys alba,* M. M. 74. *O. Acetosella,* M. Pl. 10. Relh. 182.

Damp woods and shady places. P. May.

1. Devil's Ditch; H. Wood Ditton Park Wood; S. W. W. *Grove near Burrough-Green Church; About Cheveley Park;* Ray. *Balsham;* J. M.

CELASTRACEÆ.

Euónymus Linn.

1. **E. europæus** Linn. *Spindle Tree. Prickwood.*

Euonymus, R. C. 50. M. M. 122. *E. europæus,* M. Pl. 5. Relh. 100.

Woods and hedges. Sh. May.

1. Between Linton and Bartlow. Wood Ditton Park Wood. Balsham. Long Pasture, near Hildersham. Westley Wood. Bartlow Wood; R. B. S. By the common at one mile from Cambridge towards London. Near old water-mill at Fulbourn; S. W. W.—2. *Whittlesford;* J. M. 3. Barton Road. Grantchester. Hardwick Wood. Eltisley. Gamlingay. Comberton; Caldecot; Long Stow; Hayley Wood; N. Kingston and Eversden Woods.—4. Madingley Wood. 5. Clayhythe. Upware.—6. Ely; H.

RHAMNACEÆ.

Rhámnus Linn.

1. **R. cathárticus** Linn. *Buckthorn.*

R. C. 138. M. Pl. 5. Relh. 99. *Rhamnus,* M. M. 122. Hedges on a chalky soil. Sh. May, June.

1. Chalk-pit-close, Hinton. Near old mill, Fulbourn. Devil's Ditch. Allington Hill. West Wratting. Linton. Shudy Camps. *Teversham;* J. M.—2. Not far from Ashwell; N. Bury Lane, Melbourne, D. B.—3. Whitwell. Barton Road. And throughout the western part of the district.—4. Moor Barns; W. H. C. Madingley Wood. Childerley; Elsworth; N.—5. Quy Road. Bottisham!; Swaffham!; H. Wicken Fen. Near Paper Mills; N.

2. **R. Frángula** Linn. *Black Alder.*

Alnus nigra baccifera, R. C. App. ii. 1. *Frangula officinarum*, M. M. 121. *R. Frangula*, M. Pl. 5. Relh. 99.

Wet woods and thickets on gravel. Sh. May, June.

1. Closes near the old water-mill, at Fulbourn, and by footpath to Wilbraham. Long Pasture, Hildersham.—2. Odsey; A. M. B.—3. White Wood, Gamlingay.

LEGUMINOSÆ.

Ulex Linn.

1. **U. europæus** Linn. *Furze. Whin. Gorse.*

Genista spinosa vulgaris, R. C. 59. *Scorpius*, M. M. 124. *U. europæus*, M. Pl. 16. Relh. 289.

Heaths and banks. Sh. February to June.

1. Shuckburgh Castle and other neglected spots on Newmarket Heath. Furze-hills, Hildersham. Dullingham. Brinkley. *Gogmagog Hills;* J. M.—2. By road-sides on what was once Triplow Heath.—3. Gamlingay. To the west of Eversden Manor House.—4. Near the spring in Moor Barns Thicket. By Madingley Wood.—5. Chippenham. Kennet Heath.

2. **U. nánus** Forst. *Dwarf Furze.*

U. europæus β. Relh. ed. 1, 270. *U. nanus*, Relh. ed. 2, 277; ed. 3, 289.

Heaths. Sh. August to November.

1. *Barrington Hill, near Linton;* Relh., a place now ploughed up.—2. Rarely on Triplow Heath; G. S. G.— 5. *Newmarket Heath;* Rclh.

GENÍSTA Linn.

1. **G. tinctória** Linn. *Green-weed. Dyer's-weed. Wood-waxen.*

Genistella infectoria, R. C. 60. *Coroneola,* M. M. 124. *G. tinctoria,* M. Pl. 16. Relh. 288.

Pastures on a clay soil. Sh. July to September.

1. Furze-hills, Hildersham. Stetchworth!; Wood Ditton; H. Cheveley; Ray.—3. By Hayley Wood, near Long Stow!; Bourn; N.—4. Near Two-pot-house, Childerley; H. and N. *On a bushy common on the north side of Madingley;* Ray.

2. **G. ánglica** Linn. *Needle Furze. Petty Whin.*

Genistella aculeata, R. C. 60. M. M. 124. *G. anglica,* M. Pl. 16. Relh. 288.

Heaths. Sh. May, June.

1. *Furze-hills, Hildersham;* J. M. — 3. Gamlingay. Bourn; Long Stow; Near Hayley Wood; N.—4. Honey Hill, Childerley; N.

SAROTHÁMNUS Wimm.

1. **S. scopárius** Koch. *Broom.*

Genista, R. C. 60. *G. officinarum,* M. M. 125. *Spartium scoparium,* M. Pl. 16. Relh. 287.

Gravelly, heathy places. Sh. May, June.

1. Wood Ditton. Shuckburgh Castle, Newmarket Heath. Allington Hill. Dullingham Gravel-pits; S. W. W. Devil's Ditch at Stetchworth; H.—3. Abundant at Gam-

lingay.—5. Chippenham. Near Four-mile Stables, Newmarket Heath.

ONÓNIS Linn.

1. O. arvénsis Linn. *Rest-harrow.*

Anonis non spinosa purpurea, R. C. 13. M. M. 90. *O. arvensis*, M. Pl. 16. *O. arvensis* a. Relh. 289.

Sandy and gravelly places. Sh. June to September.

1. Not unfrequent in the (1) Cambridge, (2) Royston, (3) Wimpole, and (4) Cottenham Districts.—5. Upware. Quy. Fen Ditton. Chippenham.—6. Stretham; Rev. H. Baber.—8. Wisbech; J. B.

2. O. campéstris Koch.

Anonis seu Ononis, R. C. 12. *Anonis officinarum*, M. M. 90. *O. spinosa*, M. Pl. 16. *O. arvensis* β. Relh. 289.

Barren, wettish ground. Sh. June to September.

Common in the (1) Cambridge, (2) Royston, (3) Wimpole, (4) Cottenham, and (5) Burwell Districts.—8. Riverbank at Wisbech; A. P.

MEDICÁGO Linn.

1. M. sylvéstris Fries.

Sandy and gravelly places. P. June, July.

1. Furze-hills, Hildersham.—5. Gravel-pit at Chippenham; where it was first noticed in the county in 1852, by C. C. B.

2. M. falcáta Linn. *Yellow Medick.*

Trifolium sylvestre luteum siliqua cornuta, vel Medica frutescens, R. C. 167. *Medica sylvestris*, M. M. 90. *M. falcata*, M. Pl. 17. Relh. 304.

1. Wilbraham!; H. *Linton;* J. M. *Bourn Bridge; Near the river-side between Cambridge and Trumpington;*

Ray.—5. Gravel-pit at Chippenham, and other spots there. Snailwell. *Near Quy Church;* Ray.

[Henslow gives *M. sativa* as a naturalized plant in this county in both editions of his Catalogue, but I doubt its permanence in any place.]

3. M. lupulína Linn. *Black Medick. Little Yellow Trefoil.*

T. luteum minimum, R. C. 167. *Melilotus minor*, M. M. 89. *M. lupulina*, M. Pl. 17. Relh. 305.

Waste ground and fields. A. May to August.

Common, except in the Fens.— 6. Ely. Stuntney.— 7. Doddington. Chatteris.—8. Wisbech. Whittlesey; N.

4. M. maculáta Sibth. *Heart Trefoil. Clover.*

Trifolium cochleatum folio cordato maculato, R. C. 166. *Medica echinata glabra cum maculis nigricantibus*, M. M. 90. *M. arabica*, M. Pl. 17. *M. polymorpha arabica*, Relh. ed. 1, 286. *M. polymorpha a.* Relh. ed. 2, 292; ed. 3, 305.

On a gravelly soil. A. May to August.

1. Cambridge. Hinton.—3. College-walks, and fields near Cambridge. Comberton.—4. Near the Observatory. Madingley.—5. Horningsey.

5. M. mínima Lam. *Smallest Hedge-hog Trefoil.*

Trifolium echinatum arvense, R. C. 166. *M. echinata minor*, M. M. 90. *M. arabica β.* M. Pl. 17. *M. polymorpha β. minima*, Relh. ed. 2, 292; ed. 3, 305.

Sandy and gravelly places. A. May.

1. *In an old gravel-pit near Wilbraham church*, Ray.— 5. Gravel-pit at Chippenham. Between Fordham and Freckenham. *Near Newmarket;* Mr D. Turner in Bot. Guide. *Gravel-pit near windmill, north of Newmarket;* Ray.

LEGUMINOSÆ. 55

MELILÓTUS Lam.

1. M. officinális Willd. *Melilot.*

M. vulgaris, R. C. 96. *M. officinarum,* M. M. 89. *Trifolium Melilotus-officinalis,* M. Pl. 17. Relh. ed. 1, 278. *Trifolium officinale,* Relh. ed. 2, 286; ed. 3, 298.

Road-sides, borders of fields, and other waste places. B. ? June, July.

1. Dullingham. Hinton. Balsham.—2. Whaddon. Morden Heath Farm!; Royston; N. Odsey; A. M. B. —3. Common.—4. Common.—5. Upware. Horningsey.— 6. Stretham. Roswell Pits, Ely.—7. Chatteris.

2. M. arvénsis Willd.

Ann. Nat. Hist. Ser. 2, ii. 294.

First noticed in this county by C. C. B. in 1848.

Road-sides, borders of fields, and other waste places. B. June, July.

1. Hills Road, Cambridge. Hildersham. Shudy Camps. Between Newmarket and Saxon Street. Balsham; R. B. S. Bartlow; G. S. G. Deserted railway near Hinxton; N.— 2. Gravel-pit near Whittlesford Railway Station. Sawston!; Shepreth; Melbourn; Royston; Near Ashwell; N.— 3. Comberton; Orwell; Kingston; Harlton; N.—5. Chippenham. Bottisham Lode; Rev. F. J. A. Hort (Phytol. iii. 804).

TRIFÓLIUM Linn.

‡ 1. T. praténse Linn. *Honeysuckle Trefoil. Purple Clover.*

R. C. 168. M. Pl. 17. Relh. 300. *T. vulgare officinarum,* M. M. 88.

Meadows and pastures.

Found in all the districts, but perhaps always an escape from cultivation.

2. T. médium Linn.

T. majus flore purpureo, R. C. 168. *T. pratense purpureum majus*, M. M. 88. *T. alpestre*, Relh. ed. 1, 281. *T. medium*, M. Pl. 17. Relh. ed. 3, 300.

Pastures on a clay soil. P. June to September.

1. *Hinton; Wood Ditton;* Relh. — 2. Sawston!; N.— 3. By Hayley Wood, near Long Stow! N. *In an inclosed ground near the Cam, not far from Newnham, by the footway to Grantchester;* Ray.

3. T. ochroleúcum Linn.

T. pratense hirsutum majus flore albo-sulphureo, R. C. 167. M. M. 88. *T. ochroleucum*, M. Pl. 17. Relh. 299.

Rather dry places, by road-sides and on banks. P. June, July.

1. Teversham; N. Shudy Camps. Balsham. Borley Wood. Hildersham. *Hinton;* Ray.—3. Coton. Harlton. Comberton. Toft. By Eversden Wood. Hardwick. Croydon. Wimpole. In the old chalk-pit at Haslingfield. Gamlingay. Caldecot. Barrington. Near Kingston Wood. Bourn; Long Stow; By Hayley Wood; Caxton; Eversden; N. *Gravel-pits near Paradise, Cambridge;* Relh.—4. Madingley road; also by the wood, and outside the Park. King's Hedges. Dry Drayton. Childerley; N. Papworth St Agnes; T. Y.—5. Quy; H.—6. Ely.

4. T. arvénse Linn. *Hare's-foot Trefoil.*

Lagopus vulgaris tenuifolius, R. C. 83. *Lagopus officinarum*, M. M. 89. *T. arvense*, M. Pl. 17. Relh. 301.

Sandy and gravelly places. A. July to September.

1. Furze-hills, Hildersham; Shuckburgh Castle, Newmarket Heath. Stetchworth; N. *Coldham's Common;* Relh.—2. Odsey; A. M. B.—3. Gamlingay. Near Cambridge.—4. Near the Observatory. *Hill of Health;* J. M. —5. Chippenham.—8. By Shire Drain, near Foul Anchour.

5. T. striatum Linn.

Trifolium dilute purpureum glomerulis florum oblongis sine pediculis caulibus adnatis, R. C. 168. M. M. 89. *T. striatum*, M. Pl. 17. Relh. 301.

Sandy and gravelly places. A. June, July.

1. Hildersham; G. S. G.—3. Gamlingay. College-walks, Cambridge.—4. Near the Observatory. Madingley. *In all the closes between Cambridge and Chesterton church;* Ray. (N.B. This is probably the plant intended by *T. glomeratum* in Martyn's *Plantæ*, p. 29.)

6. T. scabrum Linn.

Lyons, 46. M. Pl. 17. Relh. 301.

First found by Mr Lyons.

Sandy and gravelly places. A. May to July.

1. Shuckburgh Castle, Newmarket Heath. Hildersham Furze-hills. Wilbraham Gravel-pits; H.—3. Near Cambridge. Gamlingay.—4. Near the Observatory. Swavesey.—5. Chippenham Gravel-pit. *Newmarket;* Relh.

7. T. subterraneum Linn.

T. pumilum supinum flosculis longis albis, R. C. App. i. 9; Cat. Angl. ed. 1, 306. M. M. 88. *T. subterraneum*, M. Pl. 17. Relh. 299.

Sandy places. P. May, June.

3. By the road from White Wood to the village of Gamlingay, and elsewhere on the former heath.

8. T. répens Linn. *Dutch or White Clover.*

T. pratense album, R. C. 167. M. M. 88. *T. repens*, M. Pl. 17. Relh. 299.

Waste ground and pastures. P. May to September.

1. Cambridge. Shudy Camps. Balsham. Newmarket. Hildersham.—2. Ickleton. Foxton; N. Odsey; A. M. B. —3. Cambridge. Eversden. Wimpole. Gamlingay. Toft.

Caldecot. Harlton; Bourn; N.—4. Girton. Histon. Long Stanton.—5. Chippenham. Newmarket.—6. Stuntney; N. —7. Doddington. Chatteris.—8. Wisbech.

This plant is often cultivated, and therefore may be found occasionally in most parishes. A form in which the pod is changed into a leaf is not unfrequent.

9. T. fragiferum Linn. *Strawberry Trefoil.*

R. C. 166. M. Pl. 17. Relh. 302. *Cystitriphyllum palustre vulgare*, M. M. 89.

Damp places by roads or hedges. P. July, August.

1. The common at one mile from Cambridge, towards London. *Parker's Piece;* T. M.—2. Steeple Morden; N. —3. By Barton road. By St Neots road. St John's College-walks. And many other places.—4. By the Huntingdon road. Between Oakington and Long Stanton. Elsworth; N. Papworth; T. Y.—5. Snailwell Fen.—6. Stuntney; N.—7. Doddington Turf-fen.—8. East side of the river near Foul Anchour. River-side below Wisbech.

10. T. procúmbens Linn. *Hop Trefoil.*

T. luteum lupinum, R. C. 166. M. M. 89. *T. agrarium*, Relh. ed. 1, 283. *T. procumbens*, Relh. ed. 3, 302.

Dry pastures, banks, gravel-pits. A. June to August.

1. Cambridge. Hildersham. Near Hinxton; N. Shudy Camps. Six-mile-Bottom. Dullingham. Shuckburgh Castle, Newmarket Heath.—2. Whittlesford. To the east of Royston. Odsey; A. M. B.—3. Wimpole, and—4. Cottenham Districts abundantly.—5. Chippenham. To the east of Newmarket; N.—8. Wisbech. Foul Anchour.

11. T. minus Sm. *Lesser Hop Trefoil.*

T. lupulinum alterum minus, R. C. 166. M. M. 89. *T. procumbens*, M. Pl. 17. Relh. ed. 1, 283. *T. minus*, Relh. ed. 2, 290; ed. 3, 303.

Dry, gravelly places. A. June to August.

1. Stetchworth. Shudy Camps. Balsham. Near Hinxton; N. Hildersham. Newmarket Heath.—2. Shepreth. Royston.—3. Cambridge. Gamlingay. And many other places.— 4. Near the Observatory. Cuckoo Lane, Histon. Impington. Chesterton. Childerley; Dry Drayton; N. —5. Chippenham.—6. Ely. Haddenham. Wentworth.— 8. Wisbech.

12. **T. filifórme** Linn.

Lyons, 46. M. Pl. 17. Relh. 303.
First found by Mr Lyons or Prof. J. Martyn.
Gravelly and sandy places. A. June, July.

3. Gamlingay.—4. *Gravel Hill, near the Observatory; Hill of Health;* Relh.—6. Ely; H.

Lótus Linn.

1. **L. corniculátus** Linn. *Bird's-foot Trefoil.*

Trifolium corniculatum primum, R. C. 165. *L. corniculata glabra minor*, M. M. 90. *L. corniculatus* a. M. Pl. 17. Relh. 303.

Banks, pastures, road-sides. P. July, August.

Common, except in the Fens.—5. Quy road. Horningsey. Chippenham. 6. Ely Witcham. Witchford. Sutton.—7. Doddington.—8. Wisbech.

2. **L. ténuis** Sm.

Trifolium corniculatum minus angustioribus foliis fruticosius, R. C. App. ii. 18. *L. pentaphyllos minor angustioribus foliis fruticosior*, M. M. 90. *L. corniculatus* β. M. Pl. 17. Relh. 304.

Banks and fields. P. July, August, but coming into flower rather later than *L. corniculatus.*

1. Hinton; N.—3. Coton fields. Abundant at Kingston. Caldecot; Toft; Bourn; Caxton; Barrington; Great

and Little Eversden; By Hayley Wood; Gamlingay; N.—
4. Cuckoo Lane, near Rampton. By the Huntingdon road;
W. H. C. Dry Drayton; Childerley; Papworth St Agnes;
N. Elsworth; T. Y.

3. L. májor Scop.

Trifolium corniculatum tertium, R. C. 165. *L. pentaphyllos flore majore*, M. M. 90. *L. corniculatus* γ. M. Pl. 17. Relh. 304.

Wet, bushy places. P. July, August.

1. *Trumpington Moor;* Relh.—2. Gatwell End; Steeple Morden; N.—3. Gamlingay. *Hatley St George;* Ray.—5. Snailwell Fen. By the brook below Chippenham.—6. Ely. —7. Doddington Wood.

ANTHÝLLIS Linn.

1. A. Vulnerária Linn. *Kidney Vetch. Lady's Finger.*

A. leguminosa, R. C. 13. *Vulneraria rustica*, M. M. 87. *A. Vulneraria*, M. Pl. 16. Relh. 290.

Dry, chalky ground. P. June to August.

Common in the (1) Cambridge, (2) Royston, and (3) Wimpole Districts.—4. Near the Observatory.—5. Devil's Ditch. Snailwell Heath. Chippenham. Anglesey Abbey!; H.

ASTRÁGALUS Linn.

1. A. hypoglóttis Linn.

Glaux Dioscoridis, R. C. 63. *A. incanus parvus purpureus nostras*, M. M. 88. *A. arenarius*, M. Pl. 17. Relh. ed. 1, 278. *A. hypoglottis*, Relh. ed. 3, 297.

Chalky and gravelly places. P. June, July.

1. Gogmagog Hills. Fleam Dyke. Devil's Ditch. Near Allington Hill. Furze Hills, Hildersham. Brinkley. Bal-

sham; R. B. S. Babraham chalk-pit; S. W. W.—2. Meggot's Mount. Heydon Ditch. Triplow Heath; G. S. G. Plantations at Odsey and Steeple Morden; H. F.—5. Devil's Ditch. Between Newmarket and Chippenham. Between Hare Park and Swaffham Prior; W. H. C.

2. **A. glycyphýllos** Linn. *Wild Liquorice.*

Glaux vulgaris, R. C. 63. *A. luteus, perennis, procumbens, vulgaris sive sylvestris,* M. M. 88. *A. glycyphyllos,* M. Pl. 17. Relh. 297.

Hedge-banks and thickets on a chalky soil. P. June.

1. Chalk-pit-close, Hinton. Teversham; H. Hildersham; On a bushy hill at the entrance of Linton from Cambridge; W. H. C. Babraham; G. S. G.—2. By the road-side opposite Guilden Morden Hall; H. F. *Shelford;* Relh.—3. Lane between Trumpington Church and the river. Hardwick; N. Whitwell; W. H. C. Left-hand side of Barton road at one mile beyond the turnpike; W. H. C. and N.—4. *Castle Hill, Cambridge;* Ray. *Madingley; Impington;* Relh.—5. By road to Clayhythe from Horningsey, and at Biggin. *Newmarket Heath, by the road on the Cambridge side of the Ditch, for two miles, plentifully;* Relh.

Vícia Linn.

1. **V. hirsúta** Koch. *Hairy Tare. Tinetare.*

V. parva sive Cracca minor cum multis siliquis hirsutis, R. C. 175. *Aracus sive Cracca minor,* R. C. 15. *Cracca minor,* M. M. 85. *Ervum hirsutum,* M. Pl. 16. Relh. 295.

Bushy places and in corn-fields. A. June to August.

1. Shudy Camps. Balsham Wood. Linton Wood; S. W. W. *Barnwell Gravel-pits;* Relh.—3. Gamlingay. *Kingston Wood; Hatley;* J. M.—4. *Madingley Wood;* J. M. —7. Doddington.

2. V. tetraspérma Moench. *Smooth Tare.*

Viciæ sive Craccæ minimæ species cum siliquis glabris, R. C. 175. *Cracca minor siliquis singularibus flosculis cærulescentibus,* M. M. 85. *Ervum tetraspermum,* M. Pl. 16. Relh. 295.

Bushy places. A. June to August.

1. Balsham Wood. Shuckburgh Castle, Newmarket Heath.—3. Longstow; N. Eversden and Kingston (perhaps *V. gracilis*); W. H. C.—4. Westwick; N. Madingley (perhaps *V. gracilis*); W. H. C. *By the road to Oakington;* Ray. *Histon;* Relh.

3. V. grácilis Loisel.

Borders of fields and amongst crops. A. June to August.

3. Kingston !; In a field to the north of Eversden Wood; About Comberton !; Caldecot; Little Eversden !; Caxton; N.—4. By the road to Histon. Behind Madingley Park. Mare Way.

Probably the *E. tetraspermum, var.* 2, of Withering (Bot. Arr. ed. 2, 781), found by Mr Woodward "on a remarkably dry gravel" near Cambridge, was *V. gracilis.* If not so, the plant was first found in the county by Mr W. O. Newnham in 1845.

4. V. sylvática Linn.

Eng. Bot. t. 79. Relh. ed. 1, Suppl. iii. 5; ed. 3, 293.
Woods. P. July, August.

1. *Hall Wood near Wood Ditton !;* Relh. The wood does not now exist. It was first found by the Rev. John Hemsted, and the specimen represented in Eng. Bot. sent by him in 1792.

5. V. Crácca Linn. *Tufted Vetch.*

Aracus, R. C. 14. *Cracca*, M. M. 85. *V. Cracca*, M. Pl. 16. Relh. 293.

Damp bushy places and hedges. P. June to August.

Common in the (1) Cambridge, (2) Royston, (3) Wimpole and (4) Cottenham Districts.—5. Chippenham. Bottisham Fen. Horningsey. Wicken Fen. —7. Doddington Wood.—8. Wisbech.

6. V. sépium Linn. *Bush Vetch.*

V. maxima dumetorum, R. C. 176. M. M. 85. *V. sepium*, M. Pl. 16. Relh. 294.

Shady, bushy places. P. June to August.

1. Wood Ditton. West Wratting. Weston Colville. Burrough-Green. Rivey near Linton; W. H. C.—3. Near Cambridge. Eversden and Kingston Woods. Bourn; Hayley Wood; N.—4. Graveley; N.—6. Witcham.—7. Doddington Wood.

‡7. V. satíva Linn. *Common Vetch or Tare.*

Aracus sive Cracca major, R. C. 14. *Vicia*, R. C. 175. *V. officinarum*, M. M. 85. *V. sativa*, M. Pl. 16. Relh. 294.

Borders of fields, but probably introduced. P. May, June.

1. Balsham. Shuckburgh Castle, Newmarket Heath.—3. Gamlingay. Cambridge.—4. Near the Observatory.—6. Ely.

β. *V. angustifólia* Roth.

V. sylvestris sive Cracca major, R. Syn. 321. *V. sativa* β. Relh. 294.

Dry and sandy places.

8. World's End, near Wisbech!; J. B.

γ. *V. Bobártii* Forst.

V. sylvestris, flore ruberrimo, siliqua longa nigra, R. Syn. 321.

Sandy and gravelly places.
3. Gamlingay.

γ. Was first noticed by C. C. B. in 1835; β. by Mr Balding in 1859; they are both of spontaneous growth.

?8. V. lathyroïdes Linn.

Gravelly and sandy places. A. May, June.

1. A doubtful native. Mr R. B. Smart has a record of having found it near Balsham, but does not remember doing so.

Láthyrus Linn.

1. L. Áphaca Linn. *Yellow Vetchling.*

Aphaca, R. C. 13. M. M. 85. *L. Aphaca,* M. Pl. 16. Relh. 291.

Hedge-banks. A. May to August.

1. Hills Road, Cambridge. Hinton. Teversham; H. *Newmarket;* Rev. J. Hemsted (Bot. Guide). Balsham; R. B. S.—3. By the footpath to Coton. Orwell; Comberton; Footpath to Grantchester; N.—4. By the road-side between the Observatory and Moor Barns.—8. Wisbech; J. B.

2. L. Nissólia Linn. *Crimson Grass Vetch.*

Catananche leguminosa quorundam, R. C. App. i. 4. *Nissolia vulgaris,* M. M. 87. *L. Nissolia,* M. Pl. 16. Relh. 291.

Bushy places. A. June.

3. *On the left-hand side of the road to Barton, near the House-in-the-fields;* Relh.—6. *Found by Mr P. Dent near Haddenham;* Ray.

3. L. praténsis Linn. *Tare Everlasting.*

L. luteus sylvestris dumetorum, R. C. 85. M. M. 84. *L. pratensis,* M. Pl. 16. Relh. 291.

Moist meadows. P. July, August.

Common in the (1) Cambridge, (3) Wimpole, and **(4)** Cottenham Districts.—2. Meldreth; D. B.—5. Chippenham. Horningsey.—6. Haddenham. Witchford.—7. Doddington Wood.—8. Wisbech.

4. **L. sylvéstris** Linn. *Everlasting Pea.*

L. major latifolius, R. C. 85. M. M. 84. *L. sylvestris*, M. Pl. 16. Lyons, 45. Relh. 292. *L. latifolius*, M. Pl. 16. Relh. 292.

Woods and thickets. P. July to September.

1. Borley Wood. Wooded part of Devil's Ditch; H. Linton Wood; S. W. W. *Castle Camps, on the way to Bartlow,* Mr Dale (Ray Syn. 319).—3. Gamlingay Wood; H. Hardwick and Kingston Woods; N. Eversden Wood; S. W. W. *Hatley St George;* Relh.—5. Quy Road.

The *L. sylvestris* and *L. latifolius* of Relhan are the same plant.

5. **L. palústris** Linn.

M. Pl. 16, and 42. Relh. 292.

Boggy places. P. June, July.

3. *In a bushy close near the church, Little Eversden;* J. M.—5. Burwell and Wicken Fens. Anglesey Abbey; H.

ORNÍTHOPUS Linn.

1. **O. perpusíllus** Linn. *Small Bird's-foot.*

Ornithopodium minus, R. C. 110. M. M. 88. *O. perpusillus*, M. Pl. 16. Relh. 295.

Sandy and gravelly places. A. May to July.

1. Chippenham; W. H. C.—3. Gamlingay Heath.—5. To the east of Newmarket; N.

LEGUMINOSÆ.

Hippocrépis Linn.

1. H. comósa Linn. *Horseshoe Vetch.*

Ferrum equinum Germanicum siliquis in summitate, R. C. App. ii. 6; Cat. Angl. 100. M. M. 87. *Polygala Cortusi,* R. C. 121. *H. comosa,* M. Pl. 16. Relh. 296.

Chalky banks. P. May to August.

1. Gogmagog Hills. Hinton. Fleam Dyke. Woolstreet. Hildersham. Devil's Ditch. Allington Hill. Six-mile-Bottom. Balsham; R. B. S. *Linton;* Relh.—2. Royston. Litlington. Kneesworth. Odsey; H. F.—3. In an old chalk-pit at Haslingfield.' Harlton; Wimpole; N.—5. *Chippenham;* Relh. Devil's Ditch.

Onobrýchis Gaert.

1. O. satíva Lam. *Medick-fitchling. Cock's-head. Saintfoin.*

Onobrychis sive Caput gallinarum, Ger. Herb. 1064. *Onobrychis,* R. C. 105. M. M. 86. *Hedysarum Onobrychis,* M. Pl. 17. Relh. 296.

Chalky places. P. June, July.

1. Common throughout the chalk-district; as on the Gogmagog Hills, Devil's Ditch, &c.; formerly much more abundant.—2. Throughout the chalk-district.—3. Old chalk-pit at Haslingfield. Harlton; Eversden; N.—4. Chalk-pit near Madingley; W. H. C.—5. Chippenham.

ROSACEÆ.

Prúnus Linn.

1. P. commúnis Huds.

a. P. spinosa Linn. *Sloe. Blackthorn.*

P. sylvestris, R. C. 127. *P. sylvestris officinarum,* M. M. 120. *P. spinosa,* M. Pl. 11. Relh. 196.

Hedges and thickets. Sh. April, May.

Abundant, except in the Fens, where it is found on the islands occasionally.

β. *P. insititia* Linn. *Bullace.*

P. sylvestris fructu majore albo and *P. s. f. m. nigro*, R. C. App. ii. 15. M. M. 120. *P. insititia*, M. Pl. 11. Relh. 196.

Hedges. Sh. April, May.

Tolerably abundant, except in the Fens.

*γ. *P. domestica* Linn. *Wild Plum Tree.*

Woods and hedges. T. April, May.

2. Shelford!; H.— 3. Harlton; Hayley Wood; N.— 4. Chesterton; Impington; H. Elsworth; N.

It is also likely to be found in other places, but is probably an escape from cultivation.

‡2. **P. Pádus** Linn. *Bird Cherry.*

Relh. ed. 1, Suppl. i. 13; ed. 3, 195.

Woods and hedges. T. May.

3. *Hedges near Wimpole;* Ray.—8. Hedgerows near Wisbech; A. P.—It may be a native at Wimpole, but scarcely can be one at Wisbech.

*3. **P. Avium** Linn. *Wild Cherry.*

Cerasus sylvestris fructu rubro, R. C. App. ii. 5; Cat. Angl. 64. M. M. 120. *P. avium*, M. Pl. 11. *P. Cerasus*, Relh. 195.

Woods and thickets. T. May.

1. Chalk-pit-close, Hinton (only one tree remains. Ray gives this locality for the Wild Cherry).—4. *In a thicket nigh Elsworth Wood;* Ray.—8. In hedgerows near Wisbech!; A. P.

4. **P. Cérasus** Linn. *Dwarf Cherry Tree.*

Chamæcerasus, R. C. App. ii. 5. *Cerasus pumila*, M. M. 120. *P. avium pumila*, M. Pl. 11. *P. Cerasus β.* Relh. 195.

Hedges. Sh. May.

1. *In some closes of Teversham going from the church towards Gains*, P. Dent in R. C. App. (A.D. 1685). Dr J. Fisher records it as there after A.D. 1770.

Spiræa Linn.

1. **S. Ulmária** Linn. *Meadow-sweet.*

Ulmaria, R. C. 178. M. M. 65. *S. Ulmaria*, M. Pl. 11; Relh. 199.

Damp meadows and by water. P. June to August.

1. Wood Ditton. Brinkley. Hildersham. Shudy Camps. Balsham and Borley Woods.—2. Triplow. Guilden Morden; Gatwell-end, near Steeple Morden; N.—Shepreth; N. W. Whittlesford. Malton; D. B.—3. Caldecot. Toft. Kingston. Eversden. Eltisley. Croydon. Barrington. Gamlingay. Harlton; Caxton; Longstow; Croxton; N.—4. Long Stanton. Histon. Cottenham Fen. Madingley. Oakington. Elsworth; Papworth St Everard; T. Y.—5. Snailwell. Horningsey. Wicken Fen. Chippenham.—7. Doddington Wood. Between March and the 16-foot river; N.—8. Wisbech; J. B.

2. **S. Filipéndula** Linn. *Common Dropwort.*

Filipendula, R. C. 53. M. M. 60. *S. Filipendula*, M. Pl. 11. Relh. 199.

Chalky pastures. P. June, July.

1. Gogmagog Hills. Hinton. Devil's Ditch. Allington Hill. Six-mile-Bottom. Brinkley. Fleam Dyke. Balsham; R. B. S. Sawston; N.—2. Litlington. Between Whittlesford and Shelford. Hauxton. Triplow; Hinxton; N. Newmarket Road, Royston; D. B. Odsey; H. F.—3. Eltisley. Common in the western part of this district; N. —4. Madingley Park. Childerley; Elsworth; N. Cottenham

Fen. *Moor Barns; Girton;* J. M.—5. To the east of Newmarket; N.

Sanguisórba Linn.

1. **S. officinális** Linn. *Great Burnet.*

Pimpinella sylvestris, R. C. 118. *S. major flore spadiceo,* M. M. 20. *S. officinalis,* M. Pl. 3. Relh. 65.

Damp meadows. P. June to August.

1. In a plantation near the old water-mill, Fulbourn; W. H. C. and S. W. W. *Shelford;* J. M.—3. Amongst willows by the bridge at Toft. Hayley Wood near Longstow; In the Moat at Caxton; N. *In the same grove at Whitwell, where the Narcissus grows;* Ray.—4. Fen Drayton. Willingham. Plentiful in Cottenham Fen by the way towards Aldreth. *King's Hedges;* Relh. *Long Stanton;* Ray.—8. Between Peterborough and Whittlesey near the railway; N.

Potérium Linn.

1. **P. Sanguisórba** Linn. *Small Burnet.*

Pimpinella vulgaris sive minor, R. C. 118. *Sanguisorba officinarum,* M. M. 80. *P. Sanguisorba,* M. Pl. 22. Relh. 394.

Dry chalky places. P. June to August.

Common in the (1) Cambridge, (2) Royston, (3) Wimpole, and (4) Cottenham Districts, where the chalk is near the surface.—5. Newmarket. Horningsey.

†2. **P. muricátum** Spach.

Botan. Gaz. i. 224.

Found for the first time in this county, on almost the same day, by Mr W. Mathews and the Rev. W. W. Newbould, in the year 1849.

Dry chalky places at the sides of fields. P. June, July.

Perhaps not a true native, but introduced with seeds.—
2. Limbury Hill, near Royston. Between Known's Folly and Melbourn; N.—3. Fox-hole-down Farm, Barrington!; Longstow!; By Hardwick Wood; Caxton; Between the quarry at Eversden and the Mare Way; N.—4. Gravel Hill near the Observatory.—5. Chippenham Avenue. Between Chippenham and Badlingham.

Agrimónia Linn.

1. A. Eupatória Linn. *Agrimony.*

Agrimonia, R. C. 5. *A. officinarum*, M. M. 36. *A. Eupatoria*, M. Pl. 11. Relh. 189.

Fields and road-sides. P. June, July.

Common in (1) Cambridge, (2) Royston, (3) Wimpole, and (4) Cottenham Districts.—5. Quy Road. Horningsey. Chippenham.—6. Haddenham. Stuntney; N.—8. Wisbech; J. B.

Alchemílla Linn.

1. A. vulgáris Linn. *Lady's Mantle.*

Alchemilla, R. C. App. ii. 1; Cat. Angl. ed. 2, 11. *A. officinarum*, M. M. 18. *A. vulgaris*, M. Pl. 4. Relh. 66.

Rather damp, sandy, and gravelly places. P. June to August.

1. Linton. In a field near the wood, Balsham; R. B. S.; it was also found there by Ray or Dent formerly.—3. Gamlingay Wood.

2. A. arvénsis Linn. *Parsley Piert.*

Perchpier Anglorum, R. C. 116. *Percepier Anglorum*, M. M. 18. *Aphanes arvensis*, M. M. 4. Relh. ed. 1, 69. *Alchemilla arvensis*, Relh. ed. 3, 66.

Sandy and gravelly fields. A. May to August.

1. Cambridge. Hinton. Six-mile-Bottom. Brinkley. Dullingham. Newmarket. Near Hinxton; N.—2. Ickleton. Shelford. Royston. Melbourn; H. F. Odsey; A. M. B. —3. Cambridge. Gamlingay. Haslingfield.—4. Fen Drayton. Near the Observatory.—5. Newmarket. Chippenham. Exning.—6. Ely. Sutton. Witcham.—8. Wisbech; J. B.

POTENTÍLLA Linn.

1. P. anserína Linn. *Silver-weed. Wild Tansy.*

Argentina, R. C. 15. *Pentaphylloïdes argentea dicta*, M. M. 59. *P. anserina*, M. Pl. 11. Relh. 204.

Road-sides and damp ground. P. June, July.

Common in all the Districts.

2. P. argéntea Linn.

Pentaphyllum erectum foliis profunde sectis subtus argenteis flore luteo, R.C. App. ii. 14. *Quinquefolium folio argenteo*, M. M. 59. *P. argentea*, M. Pl. 12. Relh. 205.

Dry gravelly places. P. June, July.

1. Furze Hills, Hildersham. St Peter's College grove; S. W. W. *Many places about Linton;* Ray or Dent.— 3. Gamlingay.

3. P. vérna Linn.

Relh. ed. 1, 197; ed. 3, 205.

Dry places. P. April, May.

1. In the plantation on the north side of, and on the bank of the Wool-street, near Vandlebury. *In the chalk-pit near the hill with trees on it* [Little Trees Hill?] *at the Gogmagogs;* Relh.—3. *By White Wood, Gamlingay;* Relh. Newmarket Heath; W. H. C.

4. P. réptans Linn. *Five-leaved Grass.*

Pentaphyllum vulgare, R. C. 116. *Quinquefolium officinarum*, M. M. 59. *P. reptans*, M. Pl. 12. Relh. 205.

72 ROSACEÆ.

Road-sides and banks. P. June to September.

Probably common throughout the county, although less abundant in the Fens.

5. P. Tormentilla Nesl. *Tormentil*.

Tormentilla, R. C. 163. *Tormentilla officinarum*, M. M. 56. *T. erecta*, M. Pl. 12. Relh. ed. 1, 198. *T. officinalis*, Relh. ed. 3, 206.

Barren pastures and heathy places. P. June to August.

1. Sawston Moor; S. W. W. Teversham and Fulbourn Moors; W. H. C. Gogmagog Hills. Hildersham.—2. Shelford. Triplow; Morden Heath Farm!; Sawston; N. Royston; D. B.—3. Gamlingay. Near Cambridge. Croydon. East Hatley; N.—4. Madingley. King's Hedges; W. H. C. —5. Horningsey. Snailwell Heath. Chippenham. To the east of Newmarket; N.

β. *P. procumbens* Sibth.

3. Hayley Wood near Longstow!; N.

6. P. fragariástrum Ehrh. *Barren Strawberry*.

Fragaria minimè vesca, R. C. 55. *Comaroïdes*, M. M. 59. *F. sterilis*, M. Pl. 11. Relh. 204.

Woods and banks. P. April, May.

1. Devil's Ditch. Burrough-Green. Brinkley. Dullingham. Bartlow; H. Fulbourn; Babraham; W. H. C. Linton; Relh.—3. Gamlingay. Eversden; Hayley Wood; Bourn!; Kingston; Hardwick Wood; N.—4. Madingley Wood; W. H. C.—5. Bottisham; H.

CÓMARUM Linn.

1. C. palústre Linn. *Marsh Cinquefoil. Purple Marshlocks.*

Pentaphylloïdes rubrum palustre, R. C. 115. *P. palustre rubrum*, M. M. 59. *C. palustre*, M. Pl. 12. Relh. 207.

Very marshy or boggy places. P. July.

3. In the bogs near the old pond, Gamlingay.—5. Wicken Fen, at the side nearest to Upware.

FRAGÁRIA Linn.

1. F. vésca Linn. *Strawberry.*

Fragaria, R. C. 54. *F. officinarum,* M. M. 59. *F. vesca,* M. Pl. 11. Relh. 232.

Woods and thickets. P. May, June.

1. Fulbourn. Linton. Wood Ditton. Newmarket Heath. West Wratting. Weston Colville. Balsham and Borley Woods.—2. Odsey; A. M. B. *Triplow;* Relh.—3. Near to and in White Wood, Gamlingay. Eversden and Kingston Woods. Wimpole. Longstow; N. *In a lane* (now obliterated) *near House-in-the-fields;* Relh.—4. Madingley Wood. —7. Doddington Wood.

RÚBUS Linn.

1. R. Idæus Linn. *Raspberry.*

Relh. ed. 1, Suppl. iii. 5; ed. 3, 202.

Thickets. Sh. June.

1. Plantation at the north-east corner of the pasture behind the House on the Gogmagogs; W. H. C.—3. White Wood near Gamlingay.

2. R. thyrsoïdeus Wimm.

First detected in the county in 1856 by C. C. B.

Hedges. Sh. July, August.

1. Coldham's Lane.—3. Barrington. Burrell's Walk, Cambridge. Toft. Kingston.

3. R. discolor W. and N.

Rubus, R. C. 141. *R. vulgaris officinarum,* M. M. 122. *R. fruticosus,* M. Pl. 11. Relh. 235.

Hedges. Sh. July, August.
Abundant throughout the county.

4. R. Rádula Weihe.

First found by Mr W. Mathews, jun. in 1851.
Woods. Sh. July, August.
3. Eversden Wood!; Mr W. Mathews.

5. R. Koéhleri Weihe.

γ. *R. pallidus* Weihe.
Hedges and thickets. Sh. July, August.
1. By the Wool-street near Balsham, first found in 1854.
—3. Caldecot, and westward of that place, tolerably abundant.

6. R. diversifólius Lindl.

R. fusco-ater, a. Bab. Man. ed. 4. 103.
Hedges. Sh. July, August.
1. Behind Hildersham, first found in 1855.
3. Caldecot. Kingston.

7. R. Balfouriánus Blox.

Woods and hedges. Sh. July, August.
1. Burrough-Green. Teversham. Near the church at Hinton.—3. Toft, first found in 1848.—4. By the footway to Chesterton.—5. Wicken Fen.

8. R. corylifólius Sm.

Relh. ed. 2, 195; ed. 3, 203.
Hedges and thickets. Sh. June to August.
α. *R. sublustris* Lees.
β. *conjungens* Bab.
The two varieties are thinly scattered throughout the (1) Cambridge, (2) Royston, (3) Wimpole, and (4) Cottenham

Districts.—5. Chippenham. Horningsey. Upware. Swaffham Bulbeck. Fen Ditton.—6. Ely.—7. Doddington. Chatteris.

γ. *purpureus* Bab.
1. Stetchworth.—3. Kingston.—8. Wisbech.

9. R. althæifólius Host. (see Appendix V.)

Hedges. Sh. July, August.

Common about Cambridge and elsewhere in the (1) Cambridge, (3) Wimpole, and (4) Cottenham Districts.—2. Hauxton. Whittlesford.—5. Upware. Swaffham Bulbeck.

First brought to me from Eversden, by Mr Newbould, in 1847.

10. R. tuberculátus Bab. MS. (see Appendix V.)

Hedges. Sh. July, August.

1. Wood Ditton, where it was first noticed in 1851. Coldham's Lane. Near the chalk-pit-close, Hinton.—3. Barton Road. Grantchester. Caldecot.—5. By road from Upware to Wicken.

11. R. cæsius Linn. *Dewberry.*

R. minor fructu cœruleo, R. C. 140. M. M. 122. *R. cæsius*, M. Pl. 11. Relh. 202.

Hedges and heaths. Sh. June to September.

Generally, although thinly, distributed throughout the county.

Géum Linn.

1. G. urbánum Linn. *Wood Avens.*

Caryophyllata, R. C. 29. *C. officinarum*, M. M. 58. *G. urbanum*, M. Pl. 12. Relh. 206.

Hedge-banks and thickets. P. June to August.

Common in the (1) Cambridge, (2) Royston, (3) Wimpole,

and (4) Cottenham Districts.—5. Horningsey. Chippenham.—6. Ely. Stuntney; N.—7. Doddington. Chatteris.—8. Wisbech.

?2. G. intermédium Ehrh.

Caryophyllata vulgaris flore majore, R. C. App. i. 4. M. M. 58. *G. urbanum* β, M. Pl. 12. *G. urbanum* β, *sylvestre.* Hensl. Cat. ed. 1, 7; ed. 2, 17.

Damp woods. P. June, July.

1. Wooded part of the Devil's Ditch near Stetchworth in 1830.

There is much doubt concerning the rank of this plant, which may perhaps be a hybrid.

3. G. rivále Linn. *Water Avens.*

Relh. ed. 1, 199; ed. 3, 207.

Damp woods. P. June, July.

First found in this county by Dr Chevallier, formerly Master of St John's College.

1. Wood Ditton Park Wood. Wooded part of the Devil's Ditch. *Hall and Catledge Woods; Shudy Camps;* Relh.—3. Gamlingay Wood.

Rósa Linn.

‡1. R. spinosíssima Linn.

Hensl. Cat. ed. 1. 8.

Sandy and chalky heaths. Sh. May.

5. In a hedge by White Drove-way, and other hedges in the fens, near Swaffham Prior; H. in 1826.

2. R. villósa Linn.

M. Pl. 11. Lyons, 37. Relh. 200.

First found by Prof. J. Martyn.

Hedges. Sh. June, July.

1. Between the Trumpington road and Cow Fen.—3. Grantchester Lane; H. Barton road. *White Wood, Gamlingay;* J. M.

3. R. tomentósa Sm.

Relh. ed. 2, 193; ed. 3, 202.
Hedges. Sh. June, July.
1. Hinton. Teversham. Hildersham; W. H. C. Newmarket Heath.—3. Grantchester Lane. Coton. Kingston Wood; W. H. C. Wimpole.—4. Madingley Wood.

Smith states that *R. subglobosa* (*R. Sherardi*) was found in this county by the Rev. J. Holme (Eng. Fl. ii. 385). It is a variety of this species.

4. R. inodóra Fries.

Hedges and thickets. Sh. June, July.
First found by C. C. B. at Snailwell in 1852.
1. Hinton.—4. Swavesey. Behind Madingley Park.—5. Snailwell.

5. R. rubiginósa Linn. *Sweet-Briar.*

R. sylvestris odora, R. Cat. Angl. ed. 1, 266. M. Pl. 11. Lyons, 36. Relh. 201.
Hedges and bushy places. Sh. June, July.
1. Wood Ditton. Gogmagog Hills. Newmarket Heath. Near old water-mill, Fulbourn, but planted; S. W. W. Hinton. Behind the Temple, Great Wilbraham. Hildersham and the Furze Hills.—2. Whittlesford. Triplow.—3. Whitwell; S. W. W. To the west of Eversden Wood; W. H. C. Toft, an escape from gardens; N.—4. Behind Madingley Park.—5. *Fen Ditton;* J. M.

This plant is probably not a true native in many of these places.

6. **R. canína** Linn. *Briar. Dog-Rose. Hep-tree.*

R. sylvestris, R. C. 139. *R. canina officinarum,* M. M. 117. *R. canina,* M. Pl. 11. Relh. 201.

Hedges and thickets. Sh. June, July.

Common throughout the county, although less abundant in the Fens.

β. *R. sarmentácea* Woods.

1. Fulbourn.—4. Dry Drayton.

γ. *R. circulósa* Woods.

1. Hinton. Fulbourn. Balsham.—3. Wimpole. Haslingfield.—4. Near the Observatory.—5. Quy Road.—8. Wisbech.

ε. *R. Försteri* Sm.

1. By the road to Fulbourn from Hinton church; W. H. C.—4. By the gate of the north-west close by Madingley Wood; W. H. C.

7. **R. systýla** Woods.

Hedges and thickets. Sh. June, July.

First found by C. C. B. at Gamlingay in 1855.

1. In the lane, at about halfway, between Hinton and Teversham.—3. By White Wood, Gamlingay. Toft. Comberton; Hardwick; N.

8. **R. arvénsis** Huds.

M. Pl. 11. Lyons, 37. Relh. 200.

Hedges and thickets. Sh. June, July.

First found by Prof. J. Martyn.

1. Hinton. Stetchworth. Dullingham. Teversham. Shudy Camps; H.—2. Whittlesford. Shelford. Odsey; A. M. B.—3. Eltisley. Gamlingay. Toft. Eversden. Grantchester, by the footpath to Haslingfield, and by that from Cambridge. Western part of the District commonly; N.—4. By Madingley Wood and Moor Barns. Histon.

ROSACEÆ. 79

Graveley; N.—5. Horningsey. Upware. Bottisham; H. —6. Ely. Witcham.—7. Doddington.

CRATÆGUS Linn.

† 1. **C. Oxyacántha** Linn. *Whitethorn. Hawthorn.*

Oxyacanthus, R. C. 111. *Mespilus Apiifolio sylvestris sive Oxyacantho*, M. M. 118. *C. Oxyacantha*, M. Pl. 11. Relh. ed. 1, 189. *Mespilus Oxyacantha*, Relh. ed. 3, 197.

Hedges. Sh. May, June.

Abundant in hedges, but perhaps hardly a native plant in this county. The *C. monogyna* (Jacq.) is much the more common form.

PÝRUS Linn.

† 1. **P. commúnis** Linn. *Wild Pear Tree.*

P. sylvestris, R. Cat. Angl. ed. 1, 256. Lyons, 35. Relh. 197.

Woods and hedges. T. April, May.

1. In the Four-acre close, Fulbourn ; S. W. W. *Hinton, Mr C. Miller;* Lyons.—3. Crane's Lane, near Kingston !; Long Stow; (a single tree in each place) N. *Wimpole;* Relh.

2. **P. Málus** Linn. *Crab Tree.*

Malus sylvestris sive agrestis, R. C. 94. *M. sylvestris officinarum*, M. M. 117. *P. Malus*, M. Pl. 11. Relh. 198.

Woods and hedges. T. May.

Thinly scattered throughout the county; rarest in the Fens. I have no locality recorded in the (7) Chatteris District.

3. **P. Aucupária** Gaert. *Quicken Tree. Mountain Ash. Rowan Tree.*

Sorbus sylvestris foliis domesticæ similis, R. C. App. ii. 17; Cat. Angl. ed. 2, 278. M. M. 118. *S. Aucuparia,*

M. Pl. 11. Relh. ed. 1, 190. *P. Aucuparia*, Relh. ed. 3, 198.

Woods. T. May, June.

1. Wooded part of the Devil's Ditch.—3. White Wood, Gamlingay, but perhaps planted there.

4. **P. torminalis** Sm. *Service Tree.*

Sorbus torminalis, R. C. App. ii. 17; Cat. Angl. ed. 2, 279. *Cratægus folio laciniato*, M. M. 118. *C. torminalis*, M. Pl. 11. *P. torminalis*, Relh. 198.

Woods and hedges. T. April, May.

3. *Gransden;* P. Dent.

LYTHRACEÆ.

Lýthrum Linn.

1. **L. Salicária** Linn. *Purple Loosestrife.*

1. *Lysimachia purpurea* and *L. purpurea trifolia caule hexagono*, R. C. 93. *Salicaria trifolia, caule hexagono, flore purpureo*, M. M. 98. *L. Salicaria*, M. Pl. 11. Relh. 188.

Ditch-banks and damp places. P. July, August.

Common throughout the county.

2. **L. Hyssopifólia** Linn.

Gratiola angustifolia, R. C. 71. *Salicaria Hyssopifolio*, M. M. 98. *L. Hyssopifolia*, M. Pl. 11. Relh. 188.

Damp places where water has stagnated. A. June to October.

1. Stourbridge Fair Green, by footpath to Ditton. *Teversham Moor;* Relh. *Hinton Moor;* J. M.—4. Near the brick-kilns on the Chesterton road; H. By the Histon road, and near Cambridge Castle; J. C. *Abundant between Oakington and Histon;* Ray. *Madingley;* J. M.—5. Damp hollow by the Chippenham Avenue.—6. Ely!; W. M.

LYTHRACEÆ.

PÉPLIS Linn.

1. P. Pórtula Linn. *Water Purslane.*

Alsine rotundifolia seu Portulaca aquatica, R. C. App. ii. 2. *Portula*, M. M. 98. *P. Portula*, M. Pl. 8. Relh. 146.

Wet places. A. July, August.

3. Gamlingay.

ONAGRACEÆ.

EPILÓBIUM Linn.

1. E. hirsútum Linn. *Great Willow-herb.*

Lysimachia siliquosa hirsuta magno flore, R. C. 93. *Chamænerion villosum magno flore*, M. M. 83. *E. ramosum*, M. Pl. 9. *E. hirsutum*, Relh. 156.

Wet places by ditches and streams. P. July, August.
Common throughout the county.

2. E. parviflórum Schreb.

L. siliquosa hirsuta parvo flore, R. C. 94. *C. villosum parvo flore*, M. M. 83. *E. hirsutum*, M. Pl. 9. *E. parviflorum*, Relh. 156.

Damp places. P. July, August.
Common throughout the county.

3. E. montánum Linn.

L. siliquosa glabra major, R. C. 93. *C. glabrum majus*, M. M. 83. *E. montanum*, M. Pl. 9. Relh. 157.

Dry places. P. June, July.

1. Balsham, Borley and Westley Woods. Near Hinxton. —3. About Grantchester. Common in the western part of the District.—4. Madingley Park. Girton. Childerley; N. —6. Ely.—7. Doddington Wood.—8. Wisbech; A. P.

4. E. tetragónum Linn.

L. siliquosa glabra minor, R. C. 93. *C. glabrum minus*, M. M. 83. *E. tetragonum*, M. Pl. 9. Relh. 157.

Damp ditch-banks. P. July, August.

1. Balsham Wood. Burrough-Green.—3. Barton Road. Behind the Colleges, Cambridge. Comberton. Toft. Eversden Wood. Kingston. Caldecot; East Hatley; Hatley St George; Caxton; Croxton; N.—4. Girton. Dry Drayton. Madingley. By the Mare Way. Cottenham Fen. Elsworth; N.—5. Bottisham Fen.—7. Doddington. Chatteris.

5. E. obscúrum Schreb.

Damp places. P. July, August.

3. Near Tetworth House by Gamlingay, just in the county, first found in 1859!; N.

6. E. palústre Linn.

L. siliquosa glabra minor angustifolia, R. C. 93. *C. angustifolium glabrum*, M. M. 83. *E. palustre*, M. Pl. 9. Relh. 157.

Boggy places. P. July, August.

1. *Teversham Moor;* Ray. *Hinton Moor;* Relh.—2. Triplow; N.—3. Gamlingay; H. Gravel-pit between Caxton and Eltisley; N.—8. Wisbech!; J. B.

CIRCÆA Linn.

1. C. lutetiána Linn. *Enchanter's Nightshade*.

R. C. 38. M. M. 77. M. Pl. 1. Relh. 7.

Woods and hedge-banks. P. June to August.

1. Wood Ditton Park Wood. Westley Wood. Balsham Wood. Hinton; S. W. W. Fulbourn; W. H. C. Babraham; J. W.—2. Shelford; S. W. W. Bassingbourn; D. B.—3. Eversden, Kingston, Hardwick and other Woods.—5. Fen Ditton.

HALORAGACEÆ.

Myriophýllum Linn.

1. M. verticillátum Linn. *Small Water Yarrow or Milfoil.*

Millefolium aquaticum minus, R. C. 99. *Pentapterophyllon aquaticum flosculis ad foliorum nodos,* M. M. 83. *M. verticillatum,* M. Pl. 21. Relh. 393.

Ponds and ditches. P. July, August.

1. By the long drove in Wilbraham Fen. Sawston; G. S. G. *In the brook above the Paper Mills;* Ray. *By the path to Teversham;* Relh.—2. Peat-holes near Triplow.—3. Gamlingay. Barrington!; G. S. G.—4. Waterbeach Fen.—5. Wicken and Bottisham Fens. Baitsbite; H.—6. By Sandy's Cut; N.

2. M. spicátum Linn. *Feathered Water Milfoil.*

M. aquaticum pennatum spicatum, R. C. 99. *Potamogeton foliis pennatis,* M. M. 16. *M. spicatum,* M. Pl. 21. Relh. 393.

Ponds and ditches. P. June, July.

1. *By Coldham's Common;* Relh. — 3. Grantchester Fields; W. H. C.—4. Old Ouse river near Aldreth. Swavesey. Fen Drayton.—5. Upware; H. Fen Ditton.—6. Roswell Pits, Ely. *Stretham Ferry;* Ray.—7. Vermuden's Drain near Chatteris.—8. Newton. Wisbech.

3. M. alterniflórum Cand.

Ponds and ditches. P. May to August.

3. Brick-pit at Gamlingay, where it was first noticed in 1848 by C. C. B.

HIPPÚRIS Linn.

1. H. vulgáris Linn. *Mare's-tail.*

Equisetum palustre brevioribus foliis polyspermum, R. C. 49. *Limnopeuce*, M. M. 13. *Hippuris vulgaris*, M. Pl. 13. Relh. 2.

Ponds and slow streams. P. June, July.

1. On the common at one mile from Cambridge by the London road. Hinton. Quy Water, near Fulbourn. Trumpington. Near the Paper Mills; S. W. W.—2. Shepreth. Ickleton.—3. Barton. Wimpole. Grantchester. Harston; Kingston; Gravel-pit between Caxton and Eltisley; W. W. N. —4. Waterbeach Fen.—5. Wicken, Bottisham, and Reche Fens.—6. Roswell Pits, Ely. Sandy's Cut; N. Aldreth. —7. Chatteris.—8. Newton. Wisbech Canal.

CUCURBITACEÆ.

BRYÓNIA Linn.

1. B. dioïca Linn. *Red Bryony.*

B. alba, R. C. 23. M. Pl. 3. Relh. ed. 1, 375. *B. alba officinarum*, M. M. 61. *B. dioïca.* Relh. 413.

Hedges and thickets. P. May to September.

Common in the (1) Cambridge, (2) Royston, (3) Wimpole, and (4) Cottenham Districts.—5. Chippenham. Snailwell. Landwade. Horningsey. Quy. Fen Ditton. Upware.— 6. Ely. Haddenham. Sutton. Witcham. Witchford. Stuntney; N.—7. Doddington Wood.—8. Wisbech.

PORTULACEÆ.

MÓNTIA Linn.

1. M. fontána Linn. *Blinks.*

Alsine aquatica surrectior, R. C. App. i. 3. *A. palustris Portulaceæ aquaticæ similis*, R. C. App. ii. 1; Cat. Angl. ed.

1, 17. *Alsinoïdes palustris vulgaris*, M. M. 70. *M. fontana*, M. Pl. 3. Relh. 52.

Wet sandy places. A. April to August.

3. In the newly inclosed fields on Gamlingay Heath, very abundant there formerly.—4. *Hill of Health;* J. M. —6. Ely.

PARONYCHIACEÆ.

Herniária Linn.

1. H. glábra Linn.

Bab. Man. ed. 4, 123.

Sandy places. P. July.

1. Near an enclosure, half a mile to the south of Six-mile-Bottom Railway Station, first found in 1855 by Mr Newbould. Mr Hemsted's station near Newmarket (Bot. Guide, i. 48) is believed to have been at six miles from that town towards Bury, and therefore in Suffolk.

Lepigónum Wahl.

1. L. rúbrum Wahl. *Chickweed Spurry.*

Spergula alsineformis, R. C. 159. *S. purpurea*, M. M. 95. *Arenaria rubra*, M. Pl. 10. Relh. 179.

Sandy places. A. June to September.

1. Linton; R. B. S.—2. Sand-pits at Odsey; A. M. B. —3. Gamlingay Heath.—6. Ely; H.

2. L. médium Fries.

Damp sandy places near the sea. P. June to October.

8. On both sides of the river and town at Wisbech. On both sides of the river at Foul Anchour.

This is probably a distinct species.

3. L. marínum Wahl. *Sea Spurry.*

Spergula marina Dalechampii, R. C. 159. M. M. 95. *Arenaria marina*, M. Pl. 10. Relh. 179.

Near the sea. P. June to August.

1. *Near the windmill on the foot-way to Hinton, plentifully;* Relh. (that is, near the Cambridge Poor-house).—6. By a drove-way leading to the fen at about a mile on the way from Ely to Cambridge; Rev. L. Jenyns.—8. On both sides of the river at Foul Anchour. It is probable that both the former stations belong to *L. medium*, and that Foul Anchour is our only locality for this plant.

Spérgula Linn.

1. S. arvénsis Linn.

Spergula, R. C. 159. M. M. 95. *S. arvensis*, M. Pl. 10. Relh. 186.

Cultivated land. A. June to August.

1. Hinton. Gogmagog Hills. Linton; R. B. S.—3. Gamlingay. Toft.—4. *Madingley;* Ray.

β. *S. vulgaris* Boeningh.

3. Gamlingay.—5. Snailwell. Chippenham.

Scleránthus Linn.

1. S. ánnuus Linn. *Knawell.*

Polygonum exiguum, R. C. 121. *Knawell*, M. M. 19. *S. annuus*, M. Pl. 10. Relh. 172.

Sandy fields. A. June to August.

1. Furze Hills, Hildersham. Newmarket. Hinton. Cambridge by the Botanical garden. Linton; G. S. G.—3. Gamlingay.—4. *Oakington;* Ray. *Gravel Hill near the Observatory;* Relh. *Hill of Health;* J. M.—5. To the east of Chippenham.

CRASSULACEÆ.

Sédum Linn.

1. S. Teléphium Linn. *Orpine. Livelong.*

Telephium, R. C. 162. *Crassula officinarum*, M. M. 64. *S. Telephium*, M. Pl. 10. Relh. 180.

Hedge-banks. P. July, August.

1. *In a grove to the west of Burrough-Green church;* Ray.—2. *Shelford;* Ray.

‡2. S. álbum Linn. *Whiteflowered Stonecrop or Prickmadam.*

S. minus officinarum, R. C. 153. M. M. 64. *S. album*, M. Pl. 10. Relh. 181.

Rocks and walls. P. July, August.

7. *Plentiful on the thatched houses at Chatteris;* Ray.

‡3. S. dasyphýllum Linn.

Relh. ed. 1, 175; ed. 3, 180.

Walls. P. June, July.

1. On the wall of the Hall garden near to the church (and formerly on the church itself) at Fulbourn. On the roof of a house opposite the blacksmith's shop at Trumpington in 1857; soon after which date the house was pulled down.—2. *Royston;* J. F.

3. S. ácre Linn. *Wall Pepper. Stonecrop.*

S. sive sempervivum minimum acre, R. C. 153. *S. minimum officinarum*, M. M. 64. *S. acre*, M. Pl. 10. Relh. 181.

Walls and dry gravelly places. P. June, July.

1. Walls of Barnwell Abbey. Furze Hills, Hildersham. Allington Hill. Six-mile-Bottom. Babraham. Linton. Devil's Ditch; H.— 2. By the Baldock road, Royston; D. B.

Odsey; A. M. B.—3. Barrington (perhaps not a native there); N.—4. Swavesey. On wall at Castle End, Cambridge.—5. Chippenham.—6. Ely.—8. Wisbech; J. B.

4. **S. sexanguláre** Linn.

Lyons, 33. M. Pl. 10. Relh. 181.
Walls. P. July.

4. *Near Trinity Conduit Head*, where it was first noticed by Mr Lyons, but is now lost.—5. *Chippenham gravel-pit;* Relh.—6. *About the cathedral at Ely;* T. M.; but lost for many years; W. M.

[5. **S. refléxum** Linn.

Relh. ed. 1, 173; ed. 3, 182.
Walls. P. July, August.

1. *Cambridge;* J. F.—I know of no certain locality for this plant in the county.

β. *S. glaucum* Sm.

S. minus hematoïdes, R. C. 153. M. M. 64. *S. reflexum* β, Relh. 182.

These synonyms are now referred to the *S. glaucum*, but Ray says, "the ordinary Prickmadam or Stonecrop. In muris et tectis passim." *S. glaucum* grows near Mildenhall in Suffolk, but has not, I believe, been seen in this county.]

SEMPERVÍVUM Linn.

[1. **S. tectórum** Linn. *Houseleek*.

S. globuliferum, Lyons, 36. *S. tectorum*, Lyons, 35. M. Pl. 11. Relh. 192.

House-tops. P. July.

1. Cambridge.—5. Horningsey. Fen Ditton.—8. Wisbech; J. B.

This is not a native of the county.]

GROSSULARIACEÆ.

Ríbes Linn.

‡1. **R. Grossulária** Linn. *Gooseberry.*

R. Uva-crispa, Lyons, 24. M. Pl. 5. Relh. 101.

Hedges and thickets, doubtfully native. Sh. April, May.

1. Fulbourn; S. W. W. By the river below Cambridge!; H. Chalk-pit-close, Hinton; W. H. C. *Great Shelford Churchyard; Parker's Piece;* Lyons.—2. Triplow; N.—3. Gayne's Coppice, Comberton; S. W. W. Arrington; Harlton; Caldecot; Bourn; N. Grantchester; By the brook between Coton and Whitwell; W. H. C. *Harston; Triplow;* Relh.—4. Oakington; N. *Girton;* Lyons.—5. Swaffham!; H. Baitsbite; W. H. C.

‡2. **R. nígrum** Linn. *Black Currants. Squinancy Berries.*

R. nigrum vulgo dictum folio olente, R. C. 139. M. M. 118. *R. nigrum*, M. Pl. 5. Relh. 100.

In damp and swampy places. Sh. April, May.

1. *By the river side at Abington;* Ray. *Linton; Coldham's Common;* Relh.

‡3. **R. rúbrum** Linn. *Red Currants.*

Lyons, 23. M. Pl. 5. Relh. 100.

Damp woods and shady places. Sh. April, May.

1. *In an island between Bourn Bridge and Abington, a little above the sluice;* J. M. *Linton;* Relh.—3. Near the river at Grantchester; W. H. C.—5. Swaffham Prior.

SAXIFRAGACEÆ.

Saxifrága Linn.

1. **S. granuláta** Linn.

S. alba, R. C. 150. *S. alba officinarum*, M. M. 96. *S. granulata*, M. Pl. 10. Relh. 171.

Gravelly banks. P. May.

1. Furze Hills, Hildersham. Behind the house at the Gogmagogs; W. H. C. Balsham; R. B. S. *Linton;* Relh. *Coldham's Common;* J. F.—3. St John's College walks, probably brought in with turf formerly. Toft!; N.—4. Gravel Hill near the Observatory. Just below Chesterton; H. Histon road; W. H. C. *Hill of Health;* J. M.—6. Soham; W. M.

2. S. tridactylites Linn. *Rue Whitlow-grass.*

Paronychia rutaceo folio, R. C. 113. *S. verna annua humilior,* M. M. 96. *S. tridactylites,* M. Pl. 10. Relh. 172.

Walls. A. April to June.

1. Tennis-Court Road, Cambridge. Hinton. Fulbourn. Trumpington. West Wratting.—2. Litlington. Royston; D. B.—3. Gamlingay. Coton. Haslingfield; W. H. C. Barrington; Caldecot; Bourn; Great Eversden; N.—4. Chesterton. Madingley. Histon. Swavesey. *Hill of Health; Gravel Hill;* J. M.—5. Chippenham.—6. Sutton. Witchford.

UMBELLIFERÆ.

Hydrocótyle Linn.

1. H. vulgáris Linn. *Marsh Pennywort. White-rot.*

Cotyledon aquatica, R. C. 41. *H. vulgaris,* M. M. 43. M. Pl. 6. Relh. 109.

Boggy and marshy places. P. June to August.
Common throughout the county.

Sanícula Linn.

1. S. europæa Linn. *Sanicle.*

Sanicula sine Diapensia, R. C. 150. *S. officinarum,* M. M. 43. *S. europæa,* M. Pl. 6. Relh. 110.

Woods and thickets. P. June, July.

1. Hinton. Devil's Ditch. Wilbraham. Fulbourn. West

Wratting. Balsham, Yenhall, Wesley, and Wood Ditton Park Woods. Teversham; W. H. C.—2. Plantation by the Baldock Road, Royston; D. B. Odsey; H. F.—3. Whitwell. Hardwick. Eversden, Kingston, Gamlingay and Hayley Woods. Eltisley. Wimpole. Croxton; N.—4. Moor Barns Thicket. Madingley Wood and Park. Dry Drayton.

CICÚTA Linn.

1. **C. virósa** Linn. *Water Hemlock.*

Lyons, 28. M. Pl. 7. Relh. 122.
Ponds and ditches. P. July, August.
6. *Between Ely and Prickwillow, in a creek over against the tiled house upon Rimney Bank;* J. M. *In the river at a mile from Prickwillow Bridge;* Sir T. Cullum in Bot. Guide, i. 50. Mr W. Marshall has never been able to find it.

A'PIUM Linn.

1. **A. graveólens** Linn. *Celery. Smallage.*

A. palustre seu officinarum, R. C. 14. *A. officinarum*, M. M. 41. *A. graveolens*, M. Pl. 7. Relh. 128.
Marshes and ditches. P. June to August.
1. On the common at one mile from Cambridge towards London. Cow Fen. Quy Water; S. W. W. Marshy Meadows at Sawston; G. S. G.—3. Behind the Colleges, Cambridge. Great Eversden. Eltisley. Barton Road. Entrance of Coton by foot-way from Cambridge. Haslingfield. Barrington. Arrington; Longstow; Eltisley; N. Malton; D. B.—4. Baitsbite; H. Dry and Fen Drayton. Waterbeach. Oakington. Elsworth; N.—5. Horningsey. Upware. Swaffham Bulbeck. Wicken.—6. Abundant at Ely. —8. Wisbech. Sutton.

It is probably an escape from cultivation in the Cambridge and Wimpole districts and a native in those (4, 5, 6, 8) through which the tide formerly flowed.

Petroselínum Hoffm.

1. P. ségetum Koch. *Honewort*.

Selinum Sii foliis, R. C. App. i. 9. *Sium arvense seu segetum*, M. M. 39. *Sison segetum*, M. Pl. 7. Relh. 119.

Damp chalky fields. B. August, September.

2. Ickleton.—3. By the road behind the Colleges, Cambridge. Grantchester. By the foot-way to Coton. Barton Road. Kingston. Toft. Haslingfield; W. H. C. Bourn; Caxton; Near Croxton Windmill; N.—4. Madingley Road. Between Girton and Oakington, by the roadside.—6. Witchford. *Haddenham;* Relh.

[Lyons records *P. sativum*, Parsley, as found on the Gogmagog Hills. An accident.]

Helosciádium Koch.

1. H. nodiflórum Koch.

S. umbellatum repens, R. C. 156. M. M. 39. *Sium nodiflorum*, M. Pl. 6. Relh. 118.

Ditches and brooks. P. July, August.

Common throughout the county.

β. *H. repens* Koch.

Sium repens, Relh. ed. 2, 114; ed. 3, 119.

1. Stourbridge Fair Green! S. W. W. *Coldham's;* Relh. —5. Quarry at Upware.—6. Near the windmills overlooking West Fen, Ely.—8. *Wisbech!;* Relh.

2. H. inundátum Koch.

Sium inundatum, Lyons, 27. M. Pl. 7. Relh. ed. 1, 117. *Hydrocotyle inundatum*, Relh. ed. 2, 105; ed. 3, 117.

Ponds. P. June, July.

First found by Mr C. Miller.

1. Paper Mills; S. W. W. *In the village of Hinton;* Relh.—3. Gamlingay.—6. Roswell pits, Ely.

UMBELLIFERÆ.

Síson Linn.

1. S. Amómum Linn. *Stonewort.*

Sison sive officinarum Amomum, R. C. 156. *Sium quod Amomum vulgare officinarum*, M. M. 39. *S. Amomum*, M. Pl. 7. Relh. 119.

Dampish places on a chalky soil. B. August.

Rather plentiful in the (1) Cambridge, (2) Royston, (3) Wimpole, and (4) Cottenham Districts.—5. Quy road. Horningsey. Swaffham Bulbeck.—6. Ely. Sutton. Stuntney; N.

Ægopódium Linn.

1. Æ. Podagrária Linn. *Goutweed. Ashweed. Herb Gerard.*

Angelica sylvestris minor, R. C. 12. *Podagraria vulgaris*, M. M. 39. *Æ. Podagraria*, M. Pl. 7. Relh. 128.

Damp shady places. P. June, July.

1. Dullingham. Wilbraham. Balsham. West Wratting.—2. Royston; D. B.—3. Newnham, and grounds of St John's and Clare Colleges, Cambridge. Grantchester. Toft; Bourn; N.—5. Bottisham; H.—8. Near Elm.

Cárum Linn.

*****1. C. Cárui** Linn. *Caraway.*

Carum sive Careum, R. C. App. ii. 4; Cat. Angl. ed. 2, 55. *C. officinarum*, M. M. 40. *C. Carui*, M. Pl. 7. Relh. 126.

Pastures, hardly naturalized. P. June.

1. *Stourbridge Fair Green;* Ray. *Christ's College Piece;* Relh.—4. Madingley road; J. W.—5. *In the fields at Quy;* J. F.—8. Abundant in some pastures near Wisbech; A. P.

UMBELLIFERÆ.

Búnium Linn.

1. B. flexuósum Wither. *Pig or Earth Nut.*

Bulbocastanum, R. C. 24.   M. M. 39.   *B. Bulbocastanum*, M. Pl. 7.   Relh. ed. 1, 118.   *B. flexuosum*, Relh. ed. 2, 109; ed. 3, 114.

Sandy or gravelly woods and heaths.   P. May, June.

1. Wood Ditton Park Wood.—3. Gamlingay.   Eversden and Kingston Woods.   St John's College walks, but probably brought there with turf.   Long Stow; N.—4. Field by Moor Barns.

2. B. Bulbocástanum Linn.

Suppl. to Eng. Bot. fol. 2862.
First found by the Rev. W. H. Coleman in 1839.
Chalky fields.   P. June, July.

1. Fields above the chalk-pits at Hinton.   Between the Temple at Great Wilbraham and Streetway Hill.—2. Fields to the west of Melbourn; H. F.   Odsey; A. M. B. It occurs in several places just in Hertfordshire (Fl. Herts. 118).

Pimpinélla Linn.

1. P. mágna Linn.   *Great Burnet-Saxifrage.*

P. saxifraga hircina major, R. C. 118.   M. M. 40. *P. major*, M. Pl. 7.   *P. magna*, Relh. 127.

Woods.   P. July, August.

1. Wood Ditton Park, Borley, and Balsham Woods. *Linton;* Relh.—3. Gamlingay Wood; H.   Hayley, Eltisley, and Swavesey Woods; Gransden; Caxton; Hatley St George; N.—4. Honey Hill near Childerley; N.—7. Doddington Wood.

UMBELLIFERÆ.

2. **P. Saxifraga** Linn. *Small Burnet-Saxifrage.*

P. saxifraga major altera, R. C. 118. *P. s. m. minor*, R. C. 119. M. M. 40. *P. saxifraga officinarum*, M. M. 40. *P. Saxifraga*, M. Pl. 7. Relh. 126.

Dry pastures. P. July to September.

1. Hinton. Devil's Ditch. Balsham. The Links, Newmarket. Near Hinxton; N.—2. Litlington. Royston. Newton. Steeple Morden; Foxton; N. Odsey; A. M. B. —3. Gamlingay. Barrington. Kingston and elsewhere to the westward.—4. Near the Observatory. Childerley; both Papworths; Graveley; Elsworth; N.—5. Chippenham. Snailwell Heath. To the east of Newmarket; N. Horningsey.

Síum Linn.

1. **S. latifólium** Linn. *Great Water Parsnip.*

S. majus latifolium, R. C. 156. *S. officinarum*, M. M. 39. *S. latifolium*, M. Pl. 6. Relh. 118.

Ditches and streams. P. July, August.

3. Bourn Brook near Grantchester.—4. Middle Hill Drove, Waterbeach.—5. Wicken Fen. By the footway to the Quarry at Upware.—6. Roswell Pits, Ely. Aldreth.— 7. Doddington Turf-fen. Between the New Bedford River and Black Bank, by the railway; N.—8. Wisbech; A. P. Between Eastrey and Burnt House Drove, by the railway; N.

2. **S. angustifólium** Linn.

S. erectum umbellatum sive Pastinaca aquatica, R. C. 156. M. M. 39. *S. erectum*, M. Pl. 7. *S. angustifolium*, Relh. 118.

Ditches. P. August.

Rather common throughout the county in wet places. I have not heard of it in the (8) Wisbech District, where it is probably to be found.

UMBELLIFERÆ.

BUPLEÚRUM Linn.

1. B. tenuissimum Linn. *Least Hare's-ear.*

Auricula leporis minima, R. C. App. i. 3. *B. minimum,* M. M. 43. *B. tenuissimum,* M. Pl. 6. Relh. iii. 111.

Salt marshes. A. August, September.

1. *On that side of Hinton Moor which is next to the Hills Road;* J. M.—3. *Eltisley, towards St Neots;* Ray.—8. By the river-side below Wisbech. Foul Anchour.

† 2. B. rotundifólium Linn. *Thorough-wax.*

Perfoliata, R. C. 116. *P. officinarum,* M. M. 42. *B. rotundifolium,* M. Pl. 6. Relh. 111.

Cultivated land. A. July.

1. Abington; S. W. W. Between Red Cross Turnpike and the Gogmagog Hills; W. H. C. Hildersham. Balsham; R. B. S. Linton; G. S. G. *Stapleford;* Relh. *By the footway to Teversham;* J. M.—2. Odsey; H. F.—3. Caldecot. Near Kingston Wood. Barrington!; Long Stow; Comberton; Eltisley; Gamlingay; N.—4. Dry Drayton; Childerley; N. Knapwell!; J. C.—5. Corn-field at Upware. Reche; H. *Between Quy and Bottisham;* Relh.

ŒNÁNTHE Linn.

1. Œ. fistulósa Linn. *Water Drop-wort.*

Œ. aquatica, R. C. 104. M. M. 39. *Œ. fistulosa,* M. Pl. 7. Relh. 120.

By ponds and ditches. P. July to September.

1. Stourbridge Fair Green.—3. Gamlingay. Sheeps' Green, Cambridge. Grantchester fields. Gravel-pits between Caxton and Eltisley; Croxton; N.—4. Long Stanton. Waterbeach and Cottenham Fens. Mare Way. Fen Drayton. Swavesey. Histon road.—5. Horningsey. Bottisham

and Wicken Fens. Quarry at Upware.—6. Ely. Near Sandy's Cut; N.—7. Doddington. Chatteris.—8. Wisbech; A. P.

2. Œ. Lachenálii Gmel.

Œ. pimpinelloïdes, Relh. ed. 1, Suppl. i. 11; ed. 3, 120.
Marshes. P. July to September.

1. By the road to Hinton. Shelford Common; Teversham Moor; W. H. C. Sawston Moor; G. S. G.—2. Triplow.—3. Harlton; Kingston!; Great Eversden Clunch-pit; N. Between Barton and Comberton; Coton; W. H. C.—4. Histon Road; W. H. C. Papworth St Everard; N.—5. Wicken and Bottisham Fens.—6. Ely West Fen.—7. Doddington Turf-fen.—8. Newton. Wisbech.

3. Œ. silaifólia Bieb.

Œ. peucedanifolia, Relh. ed. 2, 116; ed. 3, 121.
Wet meadows. P. June.

3. At the Cambridge end of the Grantchester fields, in 1833. This is the original and only station. I have not again found it there.—4. Mr Newbould thinks that he found it in Over Fen: he certainly gathered it on the opposite side of the Ouse in Huntingdonshire. Being an early plant it is usually cut with the grass when just in flower.

4. Œ. Phellándrium Lam.

Cicutaria palustris, R. C. 34. *Phellandrium officinarum*, M. M. 41. *P. aquaticum*, M. Pl. 7. Relh. 121.
In water, ditches and ponds. B. July to October.

3. Sheeps' Green, Cambridge. Pond at Edge Hill by Kingston Wood.—4. Brick-pits near the Observatory. Waterbeach Fen. Swavesey.—5. Common in the fen-ditches.—6. Common. 7. Doddington Turf-fen. Chatteris.—8. Wisbech. Between Peterborough and Whittlesey; N.

5. Œ. fluviátilis Colem.

Coleman in Ann. Nat. Hist. xiii. 188 (1844); Eng. Bot. Suppl. fol. 2944. It is the *Millefolium aquaticum* of Dillenius in Ray's Syn. ed. 3, p. 216.

First recognised in the county by C. C. B. in 1843.

Ditches and streams. B. July to September.

1. Hinton Brook. Quy Water.—3. Sheeps' Green, Cambridge. Grantchester. Harston. Barrington. Burnt Mill, Haslingfield; N.—4. Waterbeach. Old Ouse near High Bridge.—5. Fen Ditton. Horningsey. Wicken Fen. —6. Ely. Aldreth.—7. Vermuden's Drain, Chatteris.

Æthúsa Linn.

1. Æ. Cynápium Linn. *Fool's-parsley.*

Cicutaria fatua, R. C. 34. *Cicuta minor Petroselino similis*, M. M. 41. *Æ. Cynapium*, M. Pl. 7. Relh. 122.

Cultivated land. A. July, August.

Found throughout the county.

Fœnículum Hoffm.

*1. F. officinále All. *Fennel.*

Anethum Fœniculum, Lyons 28. M. Pl. 7. Relh. 125.

Waste places near houses. An escape from cultivation. P. July, August.

1. *In a close by Parker's Piece, Cambridge;* Lyons.— 2. *Foulmire Mill; Triplow;* Relh.—3. *Caxton Churchyard;* J. M.—4. *Girton Churchyard;* Relh.—5. Burwell; Swaffham Bulbeck; Reche; H.

Séseli Linn.

1. S. Libanótis Koch.

Apium petræum seu montanum album, R. Syn. ed. 1, 70 (1690). *Athamanta Libanotis*, M. Pl. 6. Relh. 115.

Chalk-hills. P. July, August.

1. In old chalk-pits, and by hedgerows on both sides of the road from Hinton to the Gogmagog Hills.

Siláus Besser.

1. S. praténsis Bess. *Green Saxifrage.*

Saxifraga anglica facie Seseli pratensis, R. C. 151. *Seseli pratense nostras*, M. M. 41. *Seseli caruifolium*, M. Pl. 7. *Peucedanum Silaus*, Relh. 116.

Damp pastures, chiefly on a clay soil. P. June to September.

1. Wood Ditton. Dullingham. Balsham. Near Bottisham. Hildersham. West Wratting. Shelford. Fulbourn.—3, 4. Common.—5. Chippenham. Upware. Swaffham Bulbeck.—6. Stuntney; N.

Angélica Linn.

1. A. sylvéstris Linn.

R. C. 12. M. M. 40. M. Pl. 6. Relh. 117.

Damp places. P. July, August.

1. Dullingham. Wood Ditton. Little Linton. Sawston; N.—2. Triplow. Gatwell End; N.—3. Rather common.—4. Fen Drayton. Madingley. *Chesterton;* J. M.—5. Chippenham. Wicken Fen. Horningsey.—7. Between the Bedford rivers near the railway.

Peucédanum Linn.

1. P. palústre Moench.

Selinum palustre, Relh. ed. 1, Suppl. i. 11; ed. 3, 115.

Fens. P. July, August.

5. Burwell, Wicken, and Reche Fens. Anglesey Abbey; H.—6. *By the side of drain, eastward from Prickwillow;* Rev. Dr Goodenough (Bot. Guide, i. 47).

UMBELLIFERÆ.

PASTINÁCA Linn.

1. P. satíva Linn. *Wild Parsnep.*

Elaphoboscum, R. C. 48. *P. sylvestris latifolia*, M. M. 38. *P. sativa*, M. Pl. 7. Relh. 124.

Borders of fields. B. July.

Common throughout the county, except on the peat soil.

HERÁCLEUM Linn.

1. H. Sphondylium Linn. *Cow Parsnep.*

Sphondylium, R. C. 159. M. M. 37. *H. Sphondylium*, M. Pl. 6. Relh. 117.

Borders of fields. B. July.

Common throughout the county, except on the peat soil.

β. *angustifólium.*

Sphondylium majus aliud laciniatis foliis, R. C. App. ii. 17. M. M. 38. *H. Sphondylium* β. Relh. ed. 1, Suppl. ii. 10.

1. *In a close to the right hand of the lane leading from Hinton church to Teversham;* Relh.—3. *Kingston Wood;* Dent.

DAÚCUS Linn.

1. D. Caróta Linn. *Wild Carrot. Bird's-nest.*

D. officinarum, R. C. 44. *Staphylinus sylvestris*, M. M. 42. *D. Carota*, M. Pl. 6. Relh. 113.

Borders of fields and dry places. B. June to August.

Common in the (1) Cambridge, (2) Royston, (3) Wimpole, and (4) Cottenham Districts.—5. Swaffham Prior. Bottisham. Fen Ditton. Horningsey.—6. Between the Black Bank and Ely; N.—7. Doddington.—8. Wisbech. Foul Anchour. Whittlesey; N.

UMBELLIFERÆ. 101

CAÚCALIS Hoffm.

1. C. daucoïdes Linn.

C. tenuifolia flosculis subrubentibus, R. C. 30. M. M. 42. *C. leptophylla*, M. Pl. 6. *C. daucoïdes*, Relh. 112.

Corn-fields on a chalky soil. A. June.

1. Hinton. Gogmagog Hills. Shelford. Linton. — 2. Near Chrishall Grange; G. S. G. Known's Folly; N. Odsey; H. F. — 3. Near Fox-hole's-down farm, Barrington!; N. Near Kingston Wood. — 4. Dry Drayton; N. — 5. Swaffham Prior. Reche. *Anglesey Abbey;* Relh.

2. C. latifólia Linn.

Caucalis altera purpurascens foliis latioribus, et crassioribus, semine etiam majore, R. C. 29. M. M. 42. *Tordylium latifolium*, M. Pl. 6. *C. latifolia*, Relh. 112.

Corn-fields on a chalky soil. A. July.

Formerly abundant in the county, now exceedingly rare.

1. *Between Cambridge and the Gogmagog Hills;* Mr Woodward. *By the footpath to Hinton church;* Ray. *Teversham;* J. M.—3. *Near Madingley Wood!* in 1782; Rev. T. Gisborne. *Comberton;* Ray. *Near Kingston Wood;* J. M.—4. *By the road to Histon;* J. M.—5. Reche!, in 1833; Dr Lemann.

TORÍLIS Adans.

1. T. Anthríscus Gaert. *Hedge Parsley.*

Caucalis minor flosculis rubentibus, R. C. 30. M. M. 42. *C. Anthriscus*, M. Pl. 6. Relh. 112.

Hedges and dry banks. A. July, August.

Common in the (1) Cambridge, (2) Royston, (3) Wimpole, and (4) Cottenham Districts.—5. Swaffham Prior. Newmarket. Fen Ditton. Horningsey. Upware.—6. Stuntuey; N.—7. Doddington. Chatteris. Wimblington.—8. Wisbech. Near Whittlesey.

T. Anthriscus succeeds *Chærophyllum temulum* in the hedgerows, as that succeeds *Anthriscus sylvestris;* thus furnishing a continual succession of white and similar flowers from April to August.

2. **T. inféstа** Spr. *Small Corn Parsley.*

Caucalis ségetum minor Anthrisco hispido similis, R. C. App. ii. 5; Cat. Angl. ed. 1, 62. M. M. 42. *C. arvensis,* M. Pl. 6. Relh. ed. 1, 110. *C. infesta,* Relh. ed. 3, 113.

Corn-fields and other dry cultivated ground. A. July, August.

1. Cambridge. Hinton. Balsham. Six-mile-Bottom. Teversham. Hildersham; G. S. G.—2. Whittlesford. Hauxton. Royston; N.—4. By Coton Footpath. Orwell; Shepreth; N.—5. Waterbeach. Childerley; Oakington; Papworth St Agnes; N.—6. Newmarket. Chippenham.—7. Ely. Haddenham. Stuntney; N.—8. Wisbech.

3. **T. nodósa** Gaert. *Hedgehog Parsley.*

C. nodosa echinato semine, R. C. 30. M. M. 42. *C. nodosa,* M. Pl. 6. Relh. 113.

Banks and dry fields. A. May to July.

Not unfrequent in the (1) Cambridge, (2) Royston, (3) Wimpole, and (4) Cottenham Districts.—5. Chippenham. Horningsey. Upware.—6. Ely. Haddenham. Sutton.—7. Doddington. Chatteris.—8. Wisbech.

Scándix Linn.

1. **S. Pecten-Véneris** Linn. *Shepherd's-needle.*

Pecten-Veneris, R. C. 115. *S. semine rostrato vulgaris,* M. M. 38. *S. pecten,* M. Pl. 7. Relh. ed. 1, 120. *S. Pecten-Veneris,* Relh. ed. 3, 123.

Fields. A. June to September.

Common in the (1) Cambridge, (2) Royston, (3) Wimpole, and (4) Cottenham Districts.—5. Newmarket. Chippenham. Quy. Horningsey.—6. Ely. Sutton. Witcham.—8. Wisbech.

Anthríscus Hoffm.

1. **A. sylvéstris** Hoffm. *Wild Chervil.*

Cicutaria vulgaris, R. C. 35. *Chærophyllum sylvestre perenne Cicutæ folio*, M. M. 38. *Chærophyllum sylvestre*, M. Pl. 7. Relh. 123.

Hedges and banks. P. April to June.

Common in the (1) Cambridge, (2) Royston, (3) Wimpole, and (4) Cottenham Districts.—5. Newmarket. Chippenham. Horningsey.—6. Ely. Stuntney ; N.—7. Chatteris. Doddington.—8. Wisbech ; A. P.

2. **A. vulgáris** Pers. *Small Hemlock Chervil.*

Myrrhis sylvestris seminibus asperis, R. C. 102. *Caucalis sylvestris foliis Chærophylli*, M. M. 42. *Scandix Anthriscus*, M. Pl. 7. Relh. 123.

Gravelly banks. A. May, June.

1. Hinton. Near Allington Hill. Hildersham.—2. Odsey ; A. M. B.—3. Between Burrell's Walk and the Barton road, Cambridge. Gamlingay.—4. Gravel-hill near Cambridge. Histon road. Chesterton.—5. Horningsey.—8. North brink, Wisbech ; A. P.

Chærophýllum Linn.

1. **C. témulum** Linn. *Common Wild Chervil.*

Cerifolium sylvestre, R. C. 31. *Myrrhis annua semine striato lævi*, M. M. 38. *C. temulum*, M. Pl. 7. Relh. ed. 1, 122. *C. temulentum*, Relh. ed. 3, 124.

Hedge-banks. P. June, July.

Common in the (1) Cambridge, (2) Royston, (3) Wimpole, and (4) Cottenham Districts.—5. Upware. Horningsey. Newmarket. Chippenham. — 6. Ely. Sutton. Witcham. Witchford. Stuntney; N.—7. Doddington. Chatteris.—8. Wisbech.

Coníum Linn.

1. C. maculátum Linn. *Hemlock.*

Cicúta, R. C. 34. *Cicuta officinarum*, M. M. 41. *C. maculatum*, M. Pl. 6. Relh. 114.

Hedge-banks and waste places. B. June, July.

1. Hinton. Dullingham. Fulbourn. Balsham. Shudy Camps. Hildersham. Linton. Near Paper Mills; N.—2. Kneesworth. Whaddon. Newton.—3. Tolerably common from Toft and Haslingfield westwards.—4. Histon. Cottenham Fen. Mare Way. Westwick. Childerley; Elsworth; N. Graveley; T. Y.—5. Fen Ditton. Upware.—6. Ely. Witchford. Witcham. Sutton.—7. Doddington.—8. Wisbech.

Smýrnium Linn.

‡1. S. Olusátrum Linn. *Alexanders.*

Hipposelinum, R. C. 76. *S. officinarum*, M. M. 41. *S. Olusatrum*, M. Pl. 7. Relh. 125.

Waste places near old houses. B. May, June.

1. At the back of the chalk-pit close, Hinton. Great Wilbraham; N. Fulbourn, in the angle between the roads leading to Cambridge and to the church; W. H. C.—2. Foulmire; N. Marshy meadows at Sawston; G. S. G.—3. Croydon; N. *Haslingfield;* Relh.—4. Between Chesterton and the railway. Girton, in the angle between the brook and the lane leading to the Huntingdon road; W. H. C.

CORIÁNDRUM Linn.

*1. **C. satívum** Linn. *Coriander.*

Relh. ed. 1, Suppl. iii. 2; ed. 3, 121.
Fields. An escape from cultivation. A. June.
1. *Road to Hinton;* Relh.—6. Near Witchford.—7. Hundred-foot bridge; W. M.—8. Wisbech.

ARALIACEÆ.

ADÓXA Linn.

1. **A. Moschatellína** Linn. *Moschatel.*

Radix cava minima viridiflore, R. C. 130. *Moschatellina foliis Fumariæ bulbosæ,* M. M. 63. *A. Moschatellina,* M. Pl. 9. Relh. 165.
Shady hedges. P. April, May.
1. *Shelford;* Relh.—3. On the banks of the ditches surrounding St John's College grove.—4. Chesterton, in a lane near the railway bridge, and a lane between the east end of the village and the Ely road.

HÉDERA Linn.

1. **H. Hélix** Linn. *Ivy.*

H. Helix and *H. arborea,* R. C. 72. *H. arborea officinarum,* M. M. 119. *H. Helix,* M. Pl. 5. Relh. 101.
Woods, hedges, and old walls. Sh. October, November.
Common in the (1) Cambridge, (2) Royston, (3) Wimpole, and (4) Cottenham Districts.—5. Swaffham Prior. Chippenham.—8. Wisbech.

CORNACEÆ.

CÓRNUS Linn.

1. **C. sanguínea** Linn. *Dog-wood. Gatter-tree.*

C. fœmina, R. C. 39. M. M. 120. *C. sanguinea,* M. Pl. 4. Relh. 65.

Woods and hedges. Sh. June.

Common in the (1) Cambridge, (2) Royston, (3) Wimpole, and (4) Cottenham Districts.—5. Landwade. Chippenham. Upware. Swaffham Bulbeck. Fen Ditton. Wicken.— 7. Chatteris. Doddington.—8. Wisbech; J. B.

LORANTHACEÆ.

Víscum Linn.

1. V. álbum Linn. *Mistletoe.*

Viscum, R. C. 177. *Viscus officinarum*, M. M. 117. *V. album*, M. Pl. 22. Relh. 406.

Parasitical on trees. P. March, April.

1. On an old apple-tree at Grantchester; S. W. W.— 5. Bottisham!; H.

CAPRIFOLIACEÆ.

Sambúcus Linn.

1. S. E´bulus Linn. *Danewort. Wallwort. Dwarf Elder.*

Ebulus, R. C. 47. *S. humilis officinarum*, M. M. 119. *S. Ebulus*, M. Pl. 7. Relh. 129.

Banks on a chalky soil. P. August.

1. Linton; J. W. *By the horseway* [from Barnwell?] *to Hinton, and in the churchyard;* Ray.—2. In the village of Duxford; N.—3. Near a solitary tree on a hill above the road from Barton turnpike to Coton. Great Eversden; Caxton church-yard; N. *Barrington;* Relh.—4. Near the well at Madingley. *Near the road to Histon; Oakington;* Relh.— 5. Burwell; Swaffham Bulbeck; H.

2. S. nígra Linn. *Elder.*

Sambucus, R. C. 148. *S. officinarum*, M. M. 119. *S. nigra*, M. Pl. 7. Relh. 125.

Woods and hedges. T. June.

Tolerably abundant throughout the county. Found at Orwell in the (3) Wimpole District with white fruit by Relhan.

β. *laciniata.*
8. *At Wisbech;* Relh.

VIBÚRNUM Linn.

1. **V. Lantána** Linn. *Wayfaring Tree. Cotton Tree.*

Viburnum, R. C. 175.   M. M. 119.   *V. Lantana,* M. Pl. 7.   Relh. 129.

Hedges and copses on a chalky soil. T. May.

Rather common in the (1) Cambridge, (2) Royston, (3) Wimpole, and (4) Cottenham Districts.—5. Quy road. Fen Ditton. Bottisham!; H.—6. Sutton.

2. **V. O'pulus** Linn. *Guelder-Rose. Marsh or Water Elder.*

Sambucus aquaticus sive palustris, R. C. 147. *Opulus Ruelii,* M. M. 119.   *S. Opulus,* M. Pl. 7.   Relh. 129.

Damp hedges and copses. T. June, July.

1. Hinton. Wood Ditton. Devil's Ditch. Fulbourn. Balsham. West Wratting. Brinkley.—2. Newton. Steeple Morden; N.—3. Barton. Toft. Near Eltisley Wood. Grantchester road. Haslingfield. Comberton. Eversden. Kingston Wood. Near Hayley Wood; Caldecot; N.—4. Madingley Wood. Honey Hill, Childerley; N. King's Hedges.—8. Near Elm; A. P.

LONICÉRA Linn.

‡1. **L. Caprifólium** Linn.

Lyons, 22.   M. Pl. 5. and 35.   Relh. 98.

Figured in Engl. Bot. from a specimen gathered at Hinton in 1800, but it was found there by Mr J. Lyons before 1763.

Thickets. Sh. May, June.

1. Chalk-pit-close and a neighbouring coppice, Hinton.—2. Coppice near the Red Lion Inn at Triplow.—3. Coppices to the north of Haslingfield. By the footpath from Great to Little Eversden; In a hedge by Grantchester Mill; W. H. C.

2. L. Periclýmenum Linn. *Woodbine. Honeysuckle.*

Periclymenum, R. C. 116. *Caprifolium officinarum*, M. M. 119. *L. Periclymenum*, M. Pl. 5. Relh. 98.

Woods and hedges. Sh. June to September.

Common in the (1) Cambridge, (2) Royston, (3) Wimpole, and (4) Cottenham Districts.—5. Upware.—6. Ely.—7. Doddington Wood.—8. Wisbech; J. B.

RUBIACEÆ.

SHERÁRDIA Linn.

1. S. arvénsis Linn.

Rubeola arvensis, R. C. 140. *Asperula cærulea repens*, M. M. 45. *S. arvensis*, M. Pl. 3. Relh. 57.

Chalky and gravelly fields. A. May to July.

Common in the (1) Cambridge, (2) Royston, (3) Wimpole, and (4) Cottenham Districts.—5. Chippenham. Newmarket.—6. Ely. Witcham. Sutton.—7. Doddington.—8. Wisbech.

ASPÉRULA Linn.

1. A. cynánchica Linn. *Squinancy-wort.*

Rubia cynanchica, R. C. App. i. 8. *Rubeola vulgaris quadrifolia lævis floribus purpurantibus*, M. M. 47. *A. cynanchica*, M. Pl. 4. Relh. 57.

Chalky banks. P. June, July.

1. Hinton. Gogmagog Hills. Devil's Ditch. Allington Hill. Inclosure, a little south of Six-mile-Bottom. Hildersham. Brinkley. Fleam Dyke.—2. Newton. Near Sawston!; Morden Heath Farm; By the Ickneild Way, to the west of Royston; N. Between Whittlesford and Shelford. Odsey; A. M. B.—3. Old chalk-pits at Haslingfield, and near Fox-hole's-down Farm, Barrington. Harlton; N. —5. Snailwell Heath. Reche; N. High Ditch Lane, Fen Ditton.

2. A. odoráta Linn. *Woodruff.*

Asperula R. C. 17. *Mollugo sylvatica quæ Asperula officinarum*, M. M. 44. *A. odorata*, M. Pl. 3. Relh. 57.

Woods and copses. P. May, June.

1. *Fulbourn; Hall-wood, Wood Ditton;* Relh.—3. Gamlingay Wood. *Kingston;* Relh.

Gálium Linn.

1. G. cruciátum Wither. *Crosswort. Mugweed.*

Cruciata, R. C. 42. *C. officinarum*, M. M. 44. *Valantia cruciata*, M. Pl. 23. Relh. ed. 1, 377. *Galium cruciatum*, Relh. ed. 3, 58.

Waste ground, on a sandy or chalky soil. P. June, July.

1. Wood Ditton. Chalk-pit-close, Hinton. Balsham. Between Fulbourn and Great Wilbraham.—2. Near Ashwell; N.—3. Gamlingay. Arrington. Croydon. Comberton; Toft; Long Stow; Bourn; N.—4. Fen Drayton. Madingley. Chesterton. Moor Barns. Knapwell; N.—5. Fen Ditton.

2. G. tricórne Wither.

Aparine semine læviore, R. C. App. ii. 2; Cat. Angl. ed. 1, 25. *G. humilius folio hirsuto semine minus aspero*, M. M.

44. *G. spurium*, M. Pl. 4. Relh. ed. 1, 65. *G. tricorne*, Relh. ed. 3, 59.

Dry chalky fields. A. June to September.

Not uncommon amongst corn in the (1) Cambridge, (2) Royston, (3) Wimpole, and (4) Cottenham Districts.—5. Swaffham Bulbeck; Bottisham; H. Fen Ditton.

3. G. Aparine Linn. *Cleavers. Goose-grass.*

Aparine, R. C. 13. *G. vulgare asperum quod Aparine officinarum*, M. M. 44. *G. Aparine*, M. Pl. 4. Relh. 61.

Hedges, &c. A. June to August.

Common throughout the county.

[*G. spurium* is found near Chesterford, in Essex, close to the borders of Cambridgeshire.]

4. G. ánglicum Huds.

G. Parisiense, Relh. ed. 1, 67. *G. anglicum*, Relh. ed. 2, 57; ed. 3, 61.

Walls and dry sandy places. A. June, July.

5. Park-wall at Chippenham.—8. *Wisbech;* Relh. Outwell churchyard-wall (Florigr. Brit. i. 185), but that place is just in Norfolk.

5. G. eréctum Huds.

Relh. ed. 1, Suppl. iii. 1; ed. 3, 59.

Hedges and pastures. P. June, July.

1. Old entrance to the Great Chalk-pit, Hinton. Near Allington Hill, Hildersham. Stapleford; Babraham; S. W. W. Teversham!; H.—2. Pit at Meggot's Mount.—3. Croydon. Wimpole. Gamlingay. Barrington. Coton. Eversden. Harlton; Orwell; N.—4. King's Hedges.—5. By footpath to Baitsbite from Fen Ditton. Bottisham; S. W. W.

6. G. Mollúgo Linn. *Great Bastard Madder.*

Mollugo montana sive Gallium, R. C. 100. *G. album vulgare*, M. M. 44. *G. Mollugo*, M. Pl. 4. Relh. 60.

Hedges and banks. P. July, August.

Common in the (1) Cambridge, (2) Royston, (3) Wimpole, and (4) Cottenham Districts.—5. Chippenham.—8. Wisbech; J. B.

7. G. vérum Linn. *Lady's Bed-straw. Cheese-rening.*

G. luteum, R. C. 59. *G. officinarum*, M. M. 44. *G. verum*, M. Pl. 4. Relh. 60.

Dry and sandy places. P. July, August.

Tolerably common in the (1) Cambridge, (2) Royston, (3) Wimpole, and (4) Cottenham Districts.—5. Chippenham. Fen Ditton. To the east of Newmarket; N.—6. Ely; N.—7. Near Doddington.—8. Newton. Wisbech.

8. G. saxátile Linn.

Mollugo montana minor Gallio albo similis, R. C. App. ii. 13; Cat. Angl. ed. 1, 212. M. M. 44. *G. montanum*, M. Pl. 4. Relh. ed. 1, 66. *G. saxatile*, Relh. ed. 3, 59.

Sandy and gravelly places. P. July, August.

1. To the south of Six-mile-Bottom.—3. Gamlingay Heath.—5. Kennet Heath.

9. G. uliginósum Linn.

Lyons, 20. M. Pl. 4. Relh. 59.

First found by Prof. J. Martyn.

Wet places. P. July, August.

2. Triplow; Sawston; Gatwell End; N.—3. Gamlingay. Eversden Wood.—5. Bottisham, Snailwell, and Wicken Fens. Horningsey!; H. Brook below Chippenham. Quarry near Upware.

10. G. palústre Linn.

G. album, R. C. 59. *Mollugo minor palustris*, M. M. 44. *G. palustre*, M. Pl. 4. Relh. 58.

Wet places. P. July, August.

1. Balsham Wood. Deersley's Wood, near Newmarket. Long Pasture, near Hildersham. Linton.—2. Peat-holes, Triplow.—3. Grantchester Meadows. Gamlingay. Eltisley. Croydon. Barrington. In a pond near Kingston Wood. Eversden; Toft; Harlton; Caxton; Hayley Wood; Tadlow; Caldecot; Croxton; N.—4. Madingley chalk-pit. Between Waterbeach and Upware. Mare Way. Cuckoo Lane. Willingham. Histon. Near Two-pot House on St Neots road; Oakington; N.—5. Fen Ditton. Snailwell and Wicken Fens. By Bottisham Lode. By lake in Chippenham Park.—6. Witchford.—8. Wisbech.

11. G. elongátum Presl.

First recognized at Fen Ditton in 1848 by C. C. B.

Wet places. P. July, August.

2. Sawston Moor!; N.—3. By Bourn Brook; Haslingfield; Kingston; Gamlingay; Tadlow; N.—4. Mare Way. Cottenham Fen. Pit to the south-west of Madingley Park. Fen Drayton. Between Girton and Oakington.—5. Clayhythe. Wicken Fen. Near Sandy's Cut; N.—6. Roswell Pits, Ely. Between Haddenham and Witchford.—7. Doddington Turf-fen.—8. Newton.

[Relhan gives *Rubia peregrina* as a native of this county (ed. 3, 61) on the authority of Mr Skrimshire, who is stated (Bot. Guide, i. 45) to have found it "frequent, plentiful, and luxuriant in the quick-hedges at Crabmarsh, near Wisbech, in which neighbourhood it has not been cultivated above 50 years." It seems probable that this was *R. tinctoria* retaining its ground from former cultivation.]

VALERIANACEÆ.

CENTRÁNTHUS Cand.

‡1. **C. rúber** Cand. *Red Valerian.*

Valeriana rubra, Lyons, 2. M. Pl. 2. Relh. 17.
Old walls. P. June to September.
1. Walls about Trinity and St John's Colleges [now lost]; Wall opposite the church at Hildersham; W. H. C. *Babraham;* Lyons.—3. *Coton;* Relh.—6. *On the south side of Ely Cathedral;* J. M.

VALERIÁNA Linn.

1. **V. officinális** Linn. *Valerian.*

V. sylvestris major, R. C. 173. M. M. 36. *V. officinalis*, M. Pl. 2. Relh. 18.
Damp chalky ground. P. June, July.
1. Chalk-pit-close, Hinton. Balsham and Borley Woods. West Wratting. Fulbourn.—2. Steeple Morden; H. F.—3. Gamlingay, Eltisley, Eversden and Kingston Woods. Haslingfield. Caldecot.—4. Madingley Wood.—8. Woodhouse, Wisbech; Elm; J. B.

1. **V. sambucifólia** Mikan.

Bab. Man. ed. 3, 156.
Damp and peaty places. P. June, July.
First found by Mr James Carter in 1849.
1. Wood Ditton Park Wood.—2. Peat-holes, Triplow.—3. Comberton; Hayley Wood; Croxton; N. Hardwick and Eversden Woods.—5. Wicken Fen. Fens between Reche and Burwell!; J. C.

N.B. The *V. sylvestris major* of Ray's "Synopsis" (200) seems to be the *V. sambucifolia,* and the *V. s. m. montana* (added by Dillenius on the same page) the *V. officinalis* of modern authors. The plant of Ray's "Catalogus" (173) is certainly our *V. officinalis*. Apparently the typical plant of Linnæus is what we now call *V. sambucifolia*, and his var. β is our *V. officinalis*. This is unfortunate, but as the names are generally adopted both here and on the Continent, it cannot now be corrected.

1. **V. dioïca** Linn.

V. sylvestris minor, R. C. 173. *V. sylvestris sive palustris minor altera*, R. C. App. ii. 19. *Valerianastrum palustre vulgare*, M. M. 36. *V. dioica*, M. Pl. 2. Relh. 17.

Boggy places. P. May, June.

1. By the brook at Fulbourn. Wood Ditton Park Wood. Road to Hinton and by the brook. Cow Fen. Teversham. Quy Water.—2. Peat-holes, Triplow. Melbourn Common. Sawston Fen.—3. Harlton; Comberton; N. Grantchester Meadows.—5. Wicken and Bottisham Fens.

VALERIANÉLLA Moench.

1. **V. olitória** Moench. *Corn Salad. Lamb's Lettuce.*

Lactuca agnina, R. C. 82. *Valerianella arvensis, præcox humilis semine compresso*, M. M. 36. *Valeriana Locusta*, M. Pl. 2. Relh. 18.

Corn-fields and banks. A. May, June.

1. Abington.—3. Paradise, Cambridge; Rev. Dr Cookson. *Comberton;* J. M. *Grantchester;* J. F. *Near the church at Gamlingay;* Ray.

2. **V. Auricula** Cand.

Cultivated land. A. July, August.

First found by the Rev. W. W. Newbould before 1852.

3. Eversden!; Near a gravel-pit between Caxton and Eltisley; rare in both places; N.—8. Wisbech; J. B.

3. V. dentáta Deitr.

Sm. Eng. Fl. i. 45.
Cultivated land and banks. A. June, July.
Discovered by the Rev. J. Holme before 1824.
1. Hinton; H. By Borley Wood. Hildersham. Linton; G. S. G. Near Hinxton; N.—2. To the east of Royston; N. Odsey; H. F.—3. In fields occasionally throughout the (3) Wimpole and (4) Cottenham Districts.

DIPSACACEÆ.

Dípsacus Linn.

1. D. sylvéstris Linn. *Teasel.*

Dipsacus, R. C. 44. *D. sylvestris officinarum*, M. M. 35. *D. sylvestris*, M. Pl. 3. Relh. 55.
Hedge-banks and road-sides. B. August, September.
Common in the (1) Cambridge, (2) Royston, (3) Wimpole, and (4) Cottenham Districts.—5. Snailwell. Wicken. Horningsey. Fen Ditton.—6. Ely. Witchford. Witcham. Sutton. Stuntney; N.—8. Wisbech.

2. D. pilósus Linn. *Shepherd's Rod.*

D. minor, R. C. 46.   M. M. 35.   *D. pilosus*, M. Pl. 3. Relh. 55.
Moist shady places. B. August.
1. Chalk-pit-close, Hinton; S. W. W. *By the side of one of the closes near the grove, Barnwell;* J. M.—3. Trumpington Spinney; W. H. C. By and in Kingston Wood. Kingston Stones; Toft; N.—5. Between Upware and Wicken. Quy road. *Fen Ditton;* J. M.

KNAÚTIA Coult.

1. K. arvénsis Coult.

Scabiosa major vulgaris, R. C. 151. *S. officinarum*, M. M. 35. *S. arvensis*, M. Pl. 3. Relh. 56.

Fields and banks. P. July to September.

Not uncommon in the (1) Cambridge, (2) Royston, (3) Wimpole, and (4) Cottenham Districts.—5. Newmarket. Chippenham. Quy road. Horningsey. Wicken Fen.— 8. Wisbech. Between Peterborough and Whittlesey; N.

SCABIÓSA Linn.

1. S. succísa Linn. *Devil's-bit.*

Morsus Diaboli, R. C. 100. *S. folio integro*, M. M. 35. *S. succisa*, M. Pl. 3. Relh. 56.

Meadows and pastures. P. July.

1. Fulbourn. Stetchworth. Dullingham. Chalk-pit-close, Hinton; Shelford; W. H. C. Great Wilbraham.— 2. Sawston; Gatwell End, near Steeple Morden; N. Royston; D. B.—3. Cambridge. Gamlingay. Kingston. Eversden. Comberton. Hardwick, Pincote, and Kingston Woods; Croxton; N.—4. Madingley Wood; W. H. C.— 5. Bottisham Fen.

2. S. Columbária Linn.

S. minor sive columbaria, R. C. 152. *S. minor vulgaris*, M. M. 35. *S. Columbaria*, M. Pl. 3. Relh. 56.

Dry, chalky, and gravelly places. P. July, August.

1. Gogmagog Hills. Dullingham. Hinton. Hills Road, Cambridge. Devil's Ditch. Hildersham. Near Hinxton; N.—2. Litlington. Kneesworth. Triplow Heath. Royston; Steeple Morden; N.—3. Common.—4. Cambridge. By Madingley Wood. By Huntingdon road. Near Ar-

bury. Dry Drayton; Near the church at Papworth St Everard; N. *Hill of Health;* T. M.—5. Snailwell Heath. Chippenham. To the east of Newmarket; N.

COMPOSITÆ.

EUPATÓRIUM Linn.

1. **E. cannabinum** Linn. *Hemp-Agrimony.*

R. C. 51. M. Pl. 17. Relh. 334. *E. cannabinum officinarum,* M. M. 28.

By streams. P. August, September.

1. Cow Fen. Hinton Brook. Temple, Wilbraham. Wood Ditton. Newmarket. Fulbourn Moor; W. H. C. Catley Park; J. W. Sawston; N.— 2. Whittlesford. Hauxton. Triplow. Gatwell End, near Steeple Morden; N. Bury Lane, Meldreth; D. B.—3. Eversden; Triplow; Comberton; Harlton; The Moat, Caxton; Barrington; N.—4. Waterbeach. Brick-pits near the Observatory. Cottenham; S. W. W.—5. Chippenham. Bottisham and Wicken Fens. Horningsey. Quy !; H.—6. Near Sandy's Cut; N. *Aldreth Causeway;* Relh.—7. Doddington.— 8. Wisbech; J. B.

PETASÍTES Gaert.

1. **P. vulgáris** Desf. *Butterbur.*

R. C. 117. *P. officinarum,* M. M. 28. *Tussilago Petasites,* M. Pl. 19. Relh. 340.

Banks of rivers. P. April.

1. Little Linton.—2. By the railway near Sawston; N. Shepreth; N. W. *Hauxton Mills;* Relh.— 3. Paradise, Cambridge, and on the opposite side of the river. Malton. Barrington. Harston. By the brook at Gransden; N. *Grantchester;* J. M.—5. Swaffham Prior !; H. Snailwell Fen. Swaffham Bulbeck.—8. Wisbech; J. B.

COMPOSITÆ.

Tussilágo Linn.

1. T. Fárfara Linn. *Coltsfoot.*

Tussilago R. C. 172. *T. officinarum*, M. M. 25. *T. Farfara*, M. Pl. 19. Relh. 340.

In wet, chalky, and clayey fields. P. March, April.

Common in the (1) Cambridge, (2) Royston, (3) Wimpole, and (4) Cottenham Districts.—5. Horningsey. Upware. Chippenham.—6. Ely. Witcham.—7. March. Maney; N. —8. Wisbech. Whittlesey.

Áster Linn.

1. A. Tripólium Linn. *Sea Starwort.*

Tripolium majus et minus, R. C. 169. *Aster maritimus purpureus Tripolium dictus*, M. M. 27. *A. Tripolium*, M. Pl. 19. Relh. 344.

Salt-marshes. P. August, September.

8. County Drain, near Tidd Gout. On the river-banks, both above and below Wisbech.

Erígeron Linn.

1. E. ácris Linn. *Fleabane.*

Conyza cærulea acris, R. C. 36. *Aster arvensis cæruleus acris*, M. M. 26. *E. acris*, M. Pl. 19. Relh. 340.

Dry, gravelly places and walls. P. July, August.

1. Wall next Cow Fen, and Tennis-court road, Cambridge. Abington; W. H. C. Fulbourn. Deserted railway near Hinxton; N. *Linton;* J. M.—2. Whittlesford. Known's Folly, near Royston; Foxton; N. Near Melbourn Heath Farm; H. F.—3. Barton; N.—5. Park-wall at Chippenham. Exning. Churchyard-wall, Fen Ditton. —8. Tidd Gout. Wisbech.

BÉLLIS Linn.

1. B. perénnis Linn. *Daisy.*

B. minor sylvestris simplex, R. C. 20. *B. m. officinarum*, M. M. 29. *B. perennis*, M. Pl. 19. Relh. 348.

Open pastures and banks. P. March to October.

Common throughout the county, except on the peat soil.

SOLIDÁGO Linn.

1. S. Virgaúrea Linn. *Golden Rod.*

Virga aurea, R. C. App. i. 9. *V. a. officinarum*, M. M. 27. *S. Virgaurea*, M. Pl. 19. Relh. 344.

Woods and thickets. P. July to September.

3. Gamlingay.

I'NULA Linn.

‡1. I. Helénium Linn. *Elecampane.*

Enula campana, R. C. 48. *Aster omnium maximum Helenium dictus*, M. M. 26. *I. Helenium*, M. Pl. 19. Relh. 345.

Almost certainly an escape from ancient cultivation. P. July, August.

3. Eversden; N. *Eversden Wood.*—Relh. *Barton and other places;* Ray.—4. *Moor Barns; In a close to the north-west of the Church at Lolworth; In a close near the road from Madingley to Drayton;* Relh. *In great plenty about Madingley;* Ray.

2. I. Conýza Cand. *Ploughman's Spikenard.*

Conyza major, R. C. 37. M. M. 28. *C. squarrosa*, M. Pl. 19. Relh. 339.

Hedges and bushy places on chalk. P. July to September.

1. Camois Hall, Wood Ditton. By the back road from Little Abington to Linton. Stapleford; S. W. W. Deserted railway near Hinxton; N.—2. Whittlesford; S. W. W. Hinxton; N. *Shelford;* Relh.—5. Landwade. *Chippenham;* Relh.

PULICÁRIA Gaert.

1. P. vulgáris Gaert. *Dwarf Fleabane.*

Conyza minor, R. C. 37. *Aster palustris parvo flore globosa*, M. M. 27. *Inula Pulicaria*, M. Pl. 19. Relh. 346.

Moist, sandy, and gravelly places. A. August, September.

1. *Hinton;* Relh.—2. *Hauxton;* Relh.—4. In a lane leading from the eastern end of Chesterton to the Ely road; W. H. C. *Near Cambridge by road to Histon;* Ray. *By the river, Chesterton;* Relh.—6. By the road-side near a windmill to the west of Ely; W. M.

2. P. dysentérica Gaert.

Conyza media, R. C. 36. *Aster pratensis autumnalis Conyzæ folio*, M. M. 27. *Inula dysenterica*, M. Pl. 19. Relh. 345.

Damp places. P. August, September.

Common in the (1) Cambridge, (2) Royston, (3) Wimpole, and (4) Cottenham Districts. — 5. Chippenham. Wicken. Upware. Horningsey. — 6. Ely. Witcham.— 7. Doddington.—8. Wisbech.

BÍDENS Linn.

1. B. tripartíta Linn.

Eupatorium cannabinum fœmina, R. C. 51. *B. foliis tripartito divisis*, M. M. 30. *B. tripartita*, M. Pl. 18. Relh. 333.

Marshy places and by ponds. A. August, September.

3. Comberton; Great Eversden; Caxton; Long Stow; Tadlow; Childerley; N. Barton. Queens'-Green, Cambridge. —4. Dry Drayton. Waterbeach Fen. Between Baitsbite and Clayhythe; W. H. C.—5. Not unfrequent in the Fens.— 6. Near Sandy's Cut; N.—7. Chatteris.—8. Wisbech; J. B.

2. B. cérnua Linn.

Eupatorium aquaticum folio integro, R. C. 51. *B. folio non dissecto*, M. M. 30. *B. cernua*, M. Pl. 19. Relh. 333. *B. minima*, M. Pl. 19. Lyons, 48.

Marshy places and by ponds. A. August, September.

1. Cow Fen, Cambridge. *Hinton Moor;* J. F.—3. Eltisley.—4. Waterbeach. Cottenham; S. W. W.—5. Swaffham Bulbeck!; H.—6. Near Sandy's Cut; N. *Aldreth Causeway;* J. M.

A'NTHEMIS Linn.

1. A. arvénsis Linn. *Dog's or Corn Chamomile.*

Cotula non fœtida, R. C. 41. *Chamæmelum inodorum*, M. M. 30. *A. arvensis*, M. Pl. 19. Relh. 351.

Dry cultivated fields. A. June, July.

1. Gogmagog Hills. Stetchworth. By the Cambridge road over Newmarket Heath. Streetway Hill.—2. Bassingbourn. Triplow. Newton. Royston.—3. By the footway to Coton. Barton. Barrington. Caldecot; N.—4. Dry Drayton. Madingley.—5. Chippenham.

2. A. Cótula Linn. *Maithes. Mayweed.*

Cotula fœtida, R. C. 40. *Chamæmelum fœtidum*, M. M. 30. *A. Cotula*, M. Pl. 19. Relh. 351.

Fields and waste places, on a damp and clayey soil. A. July to September.

Common in the (1) Cambridge, (3) Wimpole, and (4) Cottenham Districts.—2. Royston, and probably elsewhere. —5. Chippenham. Newmarket; N. Quy!; H. Horningsey. —6. Haddenham.—7. Doddington.

3. A. nobilis Linn. *Chamomile.*

Damp, gravelly, and sandy places. P. July, August.
8. Crab Marsh, Wisbech!; J. B.

ACHILLÉA Linn.

1. A. Ptármaca Linn. *Sneezewort.*

Ptarmaca, R. C. 127. *Millefolium pratense folio serrato,* M. M. 30. *A. Ptarmaca,* M. Pl. 19. Relh. 351.

Wet pastures. P. July, August.

1. Quy; S. W. W. *Hinton;* J. M.—2. *Triplow;* Relh. —3. Gamlingay. Near Pincote Wood; The Moats, Caxton; N.—4. Cottenham Fen. *Girton;* Relh.—5. Quy and Bottisham Fens. Horningsey!; H.—6. Mepal; H.—7. Chatteris. Doddington.—8. Near Whittlesey; N.—*All over the Fens, abundantly;* Ray.

2. A. Millefólium Linn. *Yarrow. Millefoil.*

Millefolium vulgare album, and *M. v. flore diluti ruboris,* R. C. 100. *M. officinarum,* M. M. 30. *A. Millefolium,* M. Pl. 19. Relh. 353.

Pastures and waste ground. P. June to August.
Common throughout the county.

CHRYSÁNTHEMUM Linn.

1. C. Leucánthemum Linn. *Ox-eye.*

Bellis major, R. C. 20. *Matricaria vulgaris flore amplo folio serrato,* M. M. 29. *C. Leucanthemum,* M. Pl. 29. Relh. 348.

COMPOSITÆ. 123

Meadows and pastures. P. June to August.

Common in the (1) Cambridge, (3) Wimpole, and (4) Cottenham Districts.—2. Probably common. Royston. Odscy; A. M. B.—5. Chippenham. Horningsey. Upware.—6. Witchford.—7. Doddington.—8. Wisbech; J. B.

2. C. ségetum Linn. *Corn Marigold.*

R. C. 33. M. Pl. 19. Relh. 349. *Matricaria folio minus secto glauco flore luteo,* M. M. 29.

Corn-fields. A. June to August.

2. Fields by the Newmarket road from Royston, in small quantity; D. B.—3. Gamlingay.—4. Between Histon and Rampton. Near Cottenham. *Oakington, plentifully;* Ray. *Madingley;* J. M.—6. *Wilburton;* Relh.—8. Wisbech; J. B.

MATRICÁRIA Linn.

†1. M. Parthénium Linn. *Feverfew.*

Parthenium, R. C. 114. *M. officinarum,* M. M. 29. *M. Parthenium,* M. Pl. 19. Relh. ed. 1, 322. *Pyrethrum Parthenium,* Relh. ed. 3, 349.

Waste places, near houses. P. July, August.

2. Royston; D. B.—3. Gamlingay. Comberton; Toft; Little Eversden; Harlton; N.—4. Dry Drayton. Huntingdon road. Childerley; Elsworth; N.—5. Chippenham.—7. Doddington. Chatteris.

2. M. inodóra Linn.

Chrysanthemum inodorum, Relh. ed. 1, Suppl. i. 14. *Pyrethrum inodorum,* Relh. ed. 3, 350.

Fields. A. July, August.

1. Balsham. Near Hinxton; N.—2. Ickleton. Newton. Royston; Steeple Morden; Foxton; N.—3. Gamlingay. Croydon. Wimpole. Barrington. Comberton. Mal-

ton; Great Eversden; Caxton; Croxton; Toft; N.— 4. Near the Observatory. Huntingdon road. Fen Drayton. Oakington. Papworth St Everard; Graveley; Elsworth; N.—6. Ely. Stuntney; N.—7. Doddington.—8. Wisbech. Newton.

3. M. Chamomílla Linn.

M. Pl. 19. Relh. 350.

Gravelly land. A. June, July.

1. Cambridge!; H. Hildersham. Fulbourn.—2. Royston; D. B.—3. Gamlingay. Harston; Caxton; N.— 4. Cuckoo Lane, Histon. Willingham. Fen Drayton. Oakington. Elsworth; N.—5. Horningsey.—6. Haddenham. Ely. Witchford. Witcham. Sutton.—7. Chatteris.—8. Wisbech; J. B.

Artemísia Linn.

1. A. Absínthium Linn. *Wormwood.*

A. vulgare, R. C. 1. M. M. 31. *A. Absinthium*, M. Pl. 19. Relh. 336.

Waste ground. P. July, August.

1. Babraham; Abington Park; W. H. C. Cambridge!; H. Linton; J. W. *Chalk-pit close, Hinton;* J. M.—2. Shelford. Foxton; N. Royston; D. B.— 3. Gamlingay. Toft. Coton; W. H. C.—4. Chesterton; S. W. W. Madingley; W. H. C. Kesby's Hut, Willingham; N.—5. Wicken.—6. Haddenham. Witchford. —8. Newton. Near Whittlesey; N.

A variety with larger and broader floral leaves was sent to Smith from Gamlingay as *A. cœrulescens* (Eng. Fl. iii. 409).

2. A. vulgáris Linn. *Mugwort.*

Artemisia, R. C. 15. *A. officinarum*, M. M. 31. *A. vulgaris*, M. Pl. 19. Relh. 336.

Gravelly banks. P. July to September.

1. Hinton. Stetchworth. Shudy Camps. Hildersham. Linton; J. W. Newmarket. Brinkley.—2. Near Morden Heath Farm!; Royston; Newton; N. Whittlesford. Odsey; A. M. B.—3. Gamlingay. Bourn; Steeple Morden; N.—4. Huntingdon road. Oakington.—5. Landwade. Quy!; H. Chippenham. Wicken. Swaffham Bulbeck.—6. Near Roswell Pits, Ely. Witchford.—7. Chatteris.—8. Wisbech. Near Peterborough; N.

3. A. maritima Linn.

Relh. ed. 1, Suppl. ii. 14; ed. 3, 336.

Salt marshes. P. August, September.

8. Horseshoe-corner, Wisbech. On both sides of the river at Foul Anchour.

[*A. campestris* Linn. (*Abrotanum campestre* and *A. inodorum*, R. C. 1) is "said to be found on Newmarket Heath by Mr Sare, in Howe's *Phytographia Britannica*." Ray says, "We have searched diligently but can as yet find neither there." It is not unlikely to have grown near Chippenham, for the country there is like, and not far distant from, that in which the plant is found.]

TANACÉTUM Linn.

1. T. vulgare Linn. *Tansy*.

Tanacetum, R. C. 161. *T. officinarum*, M. M. 30. *T. vulgare*, M. Pl. 19. Relh. 335.

Way-sides. P. August.

1. *Gogmagog Hills;* J. M.—2. Whittlesford; S. W. W.—3. Gamlingay. Rarely at Toft; N. Eversden, an escape; T. Y. *Coton; Barton;* J. M.—4. Near Madingley windmill; W. H. C. *Cottenham;* Relh.—8. *Elm;* J. M.

FILÁGO Linn.

1. F. germánica Linn. *Cudweed.*

Gnaphalium minus sive Herba impia, R. C. 63. *Filago seu Impia,* M. M. 27. *F. germanica,* M. Pl. 20. Relh. ed. 1, 327. *Gn. germanicum,* Relh. ed. 3, 339.

Dry fields. A. July, August.

Common except in the Fens, from whence it is only recorded at—6. Haddenham Sand-pit. Ely.—7. Doddington.—8. Wisbech. By Shire Drain, Tydd.

2. F. apiculáta G. E. Sm.

Sandy places. A. July, August.

3. Gamlingay, where it was first found in 1849 by C. C. B.

3. F. spathuláta Presl.

Phytol. iii. 216. Ann. Nat. Hist. ser. 2, ii. 293.

First found in the county by Mr G. S. Gibson at Hildersham in July, 1848.

Dry fields and waste land. A. July, August.

1. By the Hills Road at three miles from Cambridge, and thence all the way to Linton. By the Cambridge road on Newmarket Heath. Hildersham.—2. Triplow Heath. Melbourn; Royston; N.—3. Caldecot. Kingston. Bourn; N.—5. In plenty to the south of Chippenham Park. About Newmarket.—7. By Wimblington Railway-station.

4. F. minima Fries.

Gn. minimum, R. C. 64. *F. minor,* M. M. 28. *F. montana,* M. Pl. 20. Relh. ed. 1, 327. *Gn. minimum,* Relh. ed. 3, 338.

Dry, sandy, and gravelly places. A. June to September.

1. Hildersham; W. H. C.—2. Odsey; A. M. B.—3. Gamlingay. Near the Observatory.—5. Chippenham.

COMPOSITÆ.

GNAPHÁLIUM Linn.

‡ 1. **G. luteo-álbum** Linn.

Relh. ed. 2, 323; ed. 3, 337. Eng. Bot. 1002.

Sandy and gravelly places. A. July, August.

Between Hauxton and Little Shelford, by the road-side, and in a gravel-pit to the right of the road; Relh. He sent a specimen from thence in 1802, which was figured in English Botany. I cannot find the plant by the road-side, and the gravel-pit I believe to be levelled and cultivated.

2. **G. uliginósum** Linn. *Common Cudweed.*

G. vulgare, R. C. 64. *G. uliginosum*, M. Pl. 19. Relh. 338.

Damp, sandy, and gravelly places. A. July, August.

2. Near Ashwell; N.—3. Gamlingay. Grantchester foot-path. Comberton; Barton; Toft;. Near Hayley Wood; N.—6. Sutton; Mepal!; Ely!; H. Sand-pit, Haddenham. —7. Doddington Wood.—8. Wisbech. Newton.

3. **G. sylváticum** Linn.

G. anglicum, R. C. App. i. 5. M. Pl. 19. *Filago altera*, M. M. 27. *G. rectum*, Relh. 338.

Heaths. P. July to September.

3. Gamlingay.

ANTENNÁRIA R. Br.

1. **A. dioïca** Gaert. *Cat's-foot.*

G. montanum album, R. C. 64. *Elichrysum montanum flore rotundiore*, M. M. 28. *Gn. dioïcum*, M. Pl. 19. Relh. 337.

Heaths. P. June, July.

1. Gogmagog Hills in 1828!; H. *Shelford Common;* Relh.—3. *Gamlingay;* Relh.—4. *By the road to Histon;*

J. M.—5. Newmarket Heath, to the right of the road from Cambridge. Kennet Heath.

Senécio Linn.

1. S. vulgáris Linn. *Groundsel.*

R. C. 154.   M. Pl. 19.   Relh. 341.   *S. officinarum,* M. M. 29.

Cultivated and waste ground.   A. The whole year.
Common throughout the county.

2. S. viscósus Linn.

S. hirsutus viscidus major odoratus, R. C. 154. *Jacobœa incana viscosa annua Senecionis folio,* M. M. 26. *S. viscosus,* M. Pl. 19. Relh. 341.

Waste ground.   A. July to September.

3. *Gamlingay;* Relh.—6. *Mepal;* Relh.   *On almost all the fen banks in the Isle of Ely;* Ray.—7. *Chatteris;* Relh.

3. S. sylváticus Linn.

Relh. ed. 2, 327; ed. 3, 342.

Dry and gravelly places.   A. July to September.

1. Furze Hills, Hildersham.—3. Gamlingay.

4. S. erucifólius Linn.

Jacobœa Senecionis folio incana perennis, R. C. App. ii. 10. M. M. 26. *S. erucifolius,* M. Pl. 19. Relh. ed. 1, 316. *S. tenuifolius,* Relh. ed. 3, 342.

Banks and waste places, chiefly on a chalky soil.   P. July, August.

Common in the (1) Cambridge, (2) Royston, (3) Wimpole, and (4) Cottenham Districts.—5. Horningsey. Teversham.—6. Ely. Haddenham.—7. Doddington. Chatteris. —8. Wisbech. Newton.

5. **S. Jacobæa** Linn. *Ragwort.*

Jacobæa vulgaris, R. C. 80. M. M. 26. *S. Jacobæa*, M. Pl. 19. Relh. 342.

Waste ground. P. July to September.
Common throughout the county.

6. **S. aquáticus** Huds.

Jacobæa latifolia, R. C. 80. *J. l. palustris sive aquatica*, M. M. 26. *S. aquaticus*, M. Pl. 19. Relh. 343.

Wet meadows and by streams. P. July, August.
Tolerably common throughout the county.

7. **S. paludósus** Linn. *Marsh Fleabane.*

Conyza palustris, R. C. 37. *Jacobæa palustris altissima foliis serratis*, M. M. 26. *S. paludosus*, M. Pl. 19. Relh. 343.

Fen ditches. P. June, July.

5. Wicken Fen, in 1857.—6. Padnal Fen; W. M. About three miles below Ely, in 1833!; H. *We have found it in many places about the Fens, as by a great ditch side near Stretham Ferry;* Ray.—7. Chatteris; J. M.

8. **S. palústris** Cand. *Fleabane-Mullet.*

Conyza foliis laciniatis, R. C. 37. *Aster palustris laciniatus*, M. M. 27. *Othonna palustris*, M. Pl. 20. *Cineraria palustris*, Relh. 346.

Fen ditches. P. June, July.

5. *In a ditch at the edge of the moor near the Park at Chippenham;* Relh.—6. Found a few years since in West Fen, Ely! by W. M.—7. *About March and Chatteris;* Ray.

9. **S. campéstris** Cand.

Jacobæa montana lanuginosa angustifolia non laciniata, R. C. 80. M. M. 26. *Othonna integrifolia*, M. Pl. 20.

Cineraria alpina, Relh. ed. 1, 320.   *C. integrifolia,* Relh. ed. 3, 347.

Chalk Downs.   P. June.

1. Gogmagog Hills, in and beyond the Park, and on the Wool-street. Devil's Ditch. Fleam Dyke. Newmarket Heath; Dullingham; W. H. C. Balsham Heath; R. B. S. —5. Newmarket Heath, near Swaffham!; H.

Carlína Linn.

1. **C. vulgáris** Linn.

C. sylvestris quibusdam, aliis Atractylis, R. C. 29. M. M. 34. *C. vulgaris,* M. Pl. 18. Relh. 333.

Dry, sandy, and gravelly places.   B. July to October.

1. Devil's Ditch. Gogmagog Hills. Shudy Camps. Chalk-pit opposite Babraham House; S. W. W. Hinton; Shelford!; H. Furze Hills, Hildersham.—2. Heydon Ditch. Litlington Pit. Steeple Morden; N. Royston; D. B. Melbourn Heath Farm; H. F.—3. Orwell; Kingston; Near Hayley Wood; Eversden Quarry; N. By the bridle-way between the Barton Road and Coton Cross; W. H. C.— 4. Madingley, outside the Park.—5. Devil's Ditch; Swaffham Prior; H. Snailwell Heath. Swaffham Bulbeck. East of Newmarket.—8. By Shire Drain, near Tydd.

A'rctium Linn.

1. **A. tomentósum** Pers.

Lappa major altera, R. C. App. ii. 3. (*Bardana major altera,* R. Cat. Angl. ed. 1, 38.) *L. m. montana capitulis tomentosis sive Arctium Dioscoridis,* M. M. 32. *A. Lappa* δ, M. Pl. 18. Relh. ed. 1, 303.

Waste places.   B. August.

3. *Kingston Wood;* P. Dent.—4. Moor Barns. Dry Drayton. Willingham.—7. Doddington.—8. Wisbech.

2. A. május Schk. *Great Burdock.*

A. Lappa β, M. P. 18. Relh. ed. 1, 302. (*L. major capitulo glabro maxima*, Dill. in R. Syn. ed. 3, 196.)

Waste places. B. August.

1. Devil's Ditch, near the railway.—2. Near Arrington Bridge. Whittlesford. Shelford. Duxford; Hinxton; N.—3. and 4. Common.—5. Chippenham. Snailwell. Fen Ditton. Upware. Swaffham Bulbeck. Bottisham.—6. By West Fen, Ely. Stuntney!; Sandy's Cut; N.—7. Doddington Wood. Chatteris.

3. A. minus Schk.

A. Lappa γ and ε, M. Pl. 18. Relh. ed. 1, 303. (*Lappa major montana, capitulis minoribus, rotundioribus et magis tomentosis*, R. Syn. ed. 3, 197. *L. m. capitulis parvis glabris*, Dill. in R. Syn. ed. 3, 197.)

Waste places. B. August.

1. Dullingham. Stetchworth. Shudy Camps. Balsham. Wilbraham.—2. Near Hauxton Bridge. Newton; near Ashwell; W. W. N.—3. In the lane between Trumpington church and the river. Toft. Croydon. Kingston Wood. Wimpole; Hayley Wood; Caldecot; Great Eversden; Eltisley; N.—4. Madingley. Long Stanton. Waterbeach. Childerley; Papworth St Everard; N.—5. Chippenham. Upware. To the east of Newmarket; N.—6. Ely.—7. Doddington Wood.—8. Newton.

4. A. pubens Bab.

Bardana major, R. C. 19. *Lappa major Arctium Dioscoridis*, M. M. 32. *A. Lappa* α, M. Pl. 18. Relh. ed. 1, 303.

Waste places. B. August.

1. Near Hinxton; N.—2. Near Hauxton bridge. Steeple Morden; N.—3. Barton. By Barton road. Toft.

Eversden. Trumpington. Arrington; Kingston; Caxton; N.—4. Middle Hill Drove, in Waterbeach Fen. Upware Ferry. King's Hedges. Impington. Huntingdon road. Oakington.—5. Chippenham.—6. Stuntney!; N.—8. By Whittlesey Railway-station; N.

N.B. All these species of *Arctium* are included under *A. Lappa*, without distinction as varieties, in Relh. ed. 2, 314. In his ed. 3, 327, *A. Lappa* and *A. Bardana* are adopted from Smith. It seems tolerably certain that they were known to Ray, although not until long after his last publication relative to the plants of this county.

SERRÁTULA Linn.

1. **S. tinctória** Linn. *Sawwort.*

Serratula, R. C. 154. M. M. 34. *S. tinctoria*, M. Pl. 18. Relh. 328.

Woods, thickets, and hedge-banks. P. July.

1. Borley Wood. Balsham. Thickets near the brook at Fulbourn. *Hinton Moor;* J. M.—3. Gamlingay. Hardwick. Hardwick Wood. Eversden Wood. Harlton; Comberton; Eltisley; Hayley, Kingston, and Swavesey Woods; N.—4. Madingley, behind the park.

CENTAURÉA Linn.

1. **C. nígra** Linn. *Black Matfellon or Knapweed.*

Jacea nigra vulgaris, R. C. 79. M. M. 34. *C. nigra*, M. Pl. 20. Relh. 353.

Banks and pastures. P. August, September.

Common throughout the county, but apparently less so in the Fens.

2. **C. Cýanus** Linn. *Blue-bottle.*

Cyanus, R. C. 42. *C. minor officinarum*, M. M. 34. *C. Cyanus*, M. Pl. 19. Relh. 353.

Corn-fields. A. June to August.

1. Cambridge, near the railway-bridge. Gogmagog Hills. Six-mile-Bottom. Hildersham.—2. Newton. Melbourn. Royston; D. B. Odsey; A. M. B.—3. Gamlingay. Barrington. Caldecot. Toft. Comberton. Orwell; N.—4. Cottenham. Histon. Oakington.—5. Chippenham. Upware. Exning; E. S.—8. Wisbech; J. B.

3. C. Scabiósa Linn. *Great Matfellon or Knapweed.*

Jacea segetum major purpurea, R. C. 79. *J. major*, M. M. 34. *C. Scabiosa*, M. Pl. 19. Relh. 353.

Fields and hedges. P. July to September.

Common in the (1) Cambridge, (2) Royston, and (3) Wimpole Districts.—4. Dry Drayton; N.—5. Chippenham. Upware. Newmarket. — 6. Haddenham. — 8. Wisbech. Near Whittlesey; N.

‡ 4. C. solstitiális Linn. *Yellow Star-thistle.*

Hensl. Cat. ed. 1, 14.
First found by Prof. Henslow.
Cultivated land. A. July to September.
4. Swaffham Prior in 1828.

5. C. Calcitrapa Linn. *Common Star-thistle.*

Carduus stellatus, R. C. 28. *Calcitrapa vulgaris foliis laciniatis*, M. M. 33. *C. Calcitrapa*, M. Pl. 20. Relh. 354.

Gravelly and sandy places. A. July, August.

1. Cow Fen; H. *Parker's Piece;* T. M. *Newmarket*, 1792, Rev. J. Hemsted in Sowerby's MSS.—3. By the road behind the Colleges in 1818; J. W.—6. Ely; Grunty Fen!; H.—8. Richmond Manor, Wisbech!; A. P.

ONOPÓRDUM Linn.

1. O. Acánthium Linn. *Cotton Thistle.*

Acanthium, R. C. 2. *Carduus tomentosus Acanthium dictus vulgaris*, M. M. 31. *O. Acanthium*, M. Pl. 18. Relh. 332.

Waste ground. B. August.

1. Hills road, Cambridge. Hildersham. Behind Gogmagogs Park. Fulbourn.—2. Triplow.—3. Toft. Barton. Cambridge. Barrington. Gamlingay. Harston; N.— 3. Near the Observatory. Back of Madingley Park, and at the chalk-pit.—4. Snailwell. Wicken. Chippenham. Newmarket.—6. Haddenham. Witchford. Ely.

CÁRDUUS Linn.

1. C. nútans Linn. *Musk Thistle.*

R. C. 28. M. M. 31. M. Pl. 18. Relh. 329.
Waste ground. B. May to August.
Common throughout the county.

2. C. críspus Linn. *Welted Thistle.*

C. polyanthi secunda species, R. C. App. i. 4. *C. spinosissimus capitulis minoribus*, M. M. 32. *C. acanthoïdes*, M. Pl. 18. Relh. 329.

Banks and waste places. B. June to August.

Rather common, especially near Cambridge, and in the western part of the county. Less frequent in the Fens. Not noticed in (8) Wisbech District.

Ray names the *C. crispus* β. *acanthoïdes* as a native of the county in his Cat. App. i. 4, under the name of *C. p. prima species.* I have not seen or heard of its being found here.

3. C. lanceolátus Linn. *Spear Thistle.*

C. lanceatus and *C. l. angustifolius,* R. C. 28. *Eriocephalus vulgaris capite turbinato flore purpureo,* M. M. 33. *C. lanceolatus,* M. Pl. 18. Relh. 329.

Waste ground. B. July, August.

Common throughout the county.

4. C. erióphorus Linn. *Woolly-headed Thistle.*

C. capite tomentoso, R. C. 29. *Eriocephalus capite rotundo maximo,* M. M. 33. *C. eriophorus,* M. Pl. 18. Relh. 331.

Waste ground. B. August.

1. Chalk-pit-close, Hinton. Arrington Hill, Linton; W. H. C. Between Bartlow and Shudy Camps in plenty. Balsham. Shuckburgh Castle, Newmarket Heath. Hildersham; G. S. G.—2. Heydon Ditch. Odsey (but supposed to be now lost); A. M. B.—3. Toft. Fox's Bridge, Comberton. Eversden and Kingston Woods. Coton!; H. Croydon. Caldecot. Harlton; Caxton; N.—4. Childerley; N. Closes by Madingley Wood; W. H. C. To the north of Chesterton. 5. Burwell; H.—6. By Roswell pits, Ely.

5. C. arvénsis Curt. *Creeping Thistle.*

Cirsium vulgatissimum viarum, R. C. 29. *Cir. arvense Sonchifolio radice repente,* M. M. 33. *Serratula arvensis,* M. Pl. 18. Relh. ed. 1, 303. *Car. arvensis,* Relh. ed. 3, 330.

Fields and waste ground. P. July.

Common throughout the county.

6. C. palústris Linn.

Cirsium caule crispo, R. C. 29. *Cir. pratense polycephalon vulgare,* M. M. 32. *C. palustris,* M. Pl. 18. Relh. 330.

Wet meadows. A. July, August.

1. Shelford Common and Hinton !; H. Great Wilbraham. West Wratting. Linton. Westley Wood.—2. Triplow. Newton. Hauxton. Steeple Morden; N.—3. Gamlingay. Eltisley. Haslingfield. Sawston; Comberton; Bourn; Hardwick; Caxton; Near Hayley and Eversden Woods; N.—4. Waterbeach. Dry Drayton.—5. Chippenham. Snailwell. Wicken Fen.—6. Witcham.—7. Doddington Turf-fen.

7. C. praténsis Huds.

Cirsium Anglicum primum, R. C. 35. *Cir. Britannicum Clusii repens*, M. M. 32. *C. heterophyllus*, M. Pl. 18. Relh. ed. 1, 306. *C. pratensis*, Relh. ed. 3, 331.

Fens and marshy meadows. P. June to August.

1. Shelford Moor (now drained); H. *Hinton Moor;* J. M. By the brook near Fulbourn.—2. Peat-holes, near Triplow. Sawston Fen!; N.—4. *Cottenham and Willingham Fens;* Relh.—5. Horningsey, Burwell, Bottisham, and Wicken Fens. Anglesey Abbey; H.—6. *Littleport; Willingham;* Relh.

This is the *Melancholy Thistle* of Ray.

[*Carduus palustris mitior, Bardanæ capitulo summo caule singulari*, Pluk. Almag. 82. He found it in the Isle of Ely. It was probably *C. pratensis.*]

8. C. acaúlis Linn. *Ground Thistle.*

R. C. 27. M. Pl. 18. Relh. 331. *Cir. acaulos flore purpureo*, M. M. 32.

Dry chalky pastures and banks. P. July to September.

Common in the (1) Cambridge, (2) Royston, (3) Wimpole, and (4) Cottenham Districts.—5. Chippenham. Snailwell. Exning. Upware. Swaffham Bulbeck. Newmarket.—6. Haddenham.—8. By the river at Wisbech; A. P.

COMPOSITÆ. 137

SÍLYBUM Gaert.

1. S. mariánum Gaert. *Milk Thistle. Lady's Thistle.*

Carduus lacteus, R. C. 28. *Silybum albis maculis notatum,* M. M. 33. *Carduus marianus,* M. Pl. 18. Relh. 330.
Waste places. B. June, July.

This plant seems to have been tolerably abundant formerly by waysides. It is now very rare.—1. Fulbourn; W. H. C. Great Abington in 1819; J. W.—4. On the bank below the Ely road just out of Cambridge. North of Chesterton; H.—8. In a few places about Wisbech; A. P.

LÁPSANA Linn.

1. L. commúnis Linn. *Nipplewort. Tetterwort. Dock-cresses.*

Lampsana, R. C. 84. M. M. 25. *L. communis,* M. Pl. 18. Relh. 326.
Waste and cultivated ground. A. July, August.
Common throughout the county.

ARNÓSERIS Gaert.

1. A. pusilla Gaert.

Hieracium minimum, R. C. 74. *Lampsana minor aphyllocaulos,* M. M. 25. *Hyoseris minima,* M. Pl. 18. Relh. 324.
Sandy fields. A. June to August.
3. Fields on the old heath at Gamlingay.

CICHÓRIUM Linn.

1. C. I'ntybus Linn. *Wild Succory.*

Cichoreum sylvestre, R. C. 33. *Cichorium sylvestre officinarum,* M. M. 25. *C. Intybus,* M. Pl. 18. Relh. 327.
Waste ground on a gravelly and chalky soil. P. July, August.

Common in the (1) Cambridge, (2) Royston, (3) Wimpole, and (4) Cottenham Districts.—5. Chippenham. Horningsey. Upware.—6. Stuntney; N.—8. Wisbech; J. B.

HYPOCHŒRIS Linn.

1. **H. glábra** Linn.

Relh. ed. 1, Suppl. iii. 6; ed. 3, 325.
Sandy and gravelly places. A. July, August.
3. *Gamlingay, by the road to White Wood!;* Relh.—5. Chippenham; W. H. C.

2. **H. radicáta** Linn. *Cat's-ear.*

Hieracium longius radicatum, R. C. 74. *H. vulgaris major*, M. M. 24. *H. radicata*, M. Pl. 18. Relh. 326.
Pastures and waste ground. P. July.
1. Stetchworth. Balsham. Furze-hills, Hildersham. Shuckburgh Castle, Newmarket Heath.—3. Cambridge. Comberton. Haslingfield. Toft. Caxton; Harlton; N. Gamlingay.—4. Long Stanton. Cottenham. Fen Drayton.—5. Bottisham!; H. Newmarket; N.—6. Haddenham.

3. **H. maculáta** Linn.

Hieracium montanum caule aphyllo non ramoso flore pallidiore, R. C. App. i. 6. *Hier. primum latifolium*, M. M. 22. *H. maculata*, M. Pl. 18. Relh. 325.
Chalky and sandy hills. P. July, August.
1. Devil's Ditch. Old chalk-pit on Little-trees Hill, Gogmagogs. Furze-hill (next Linton), Hildersham.—2. *On a hillock* (probably one of the Crowley Hills) *in the open field between Triplow Heath and Foulmire;* T. M.—5. Devil's Ditch near Reche; H.

THRÍNCIA Roth.

1. T. hirta Cand.

Hedypnoïs hirta, Relh. ed. 2, 337; ed. 3, 321.
Gravelly places and fields. P. July to September.
1. Cambridge.—2. Sawston; G. S. G. Royston; N.—
3. St John's College walks and elsewhere at Cambridge. Coton. Gamlingay. Caxton; Old Brick-fields between Bourn and Kingston Wood; Eltisley; Croxton; N.—4. Madingley. Fen Drayton. Swavesey.—5. Chippenham. Bottisham and Wicken Fens. To the east of Newmarket.—6. Witcham.—
7. Doddington Turf-fen.

APÁRGIA Schreb.

1. A. hispida Willd. *Hawkbit.*

Hieracium caule aphyllo hirsuto, R. C. 74. *Taraxaconoïdes perennis et vulgaris*, M. M. 24. *Leontodon hispidum*, M. Pl. 18. Relh. ed. 1, 295. *Hedypnoïs hispida*, Relh. ed. 3, 320.
Meadows and pastures. P. June to September.
Common in the (1) Cambridge, (2) Royston, (3) Wimpole, and (4) Cottenham Districts.—5. Chippenham. Horningsey.—6. Haddenham. Stuntney; N.—8. Newton.

2. A. autumnalis Willd.

Hieracium minus præmorsa radice, R. C. 74. *Scorzonera folio laciniato, radice succisa*, M. M. 24. *Leontodon autumnale*, M. Pl. 18. Relh. ed. 1, 295. *Hedypnoïs autumnalis*, Relh. ed. 3, 321.
Meadows and pastures. P. August.
Common throughout the county.

Tragopógon Linn.

1. T. minor Fries. *Goat's-beard.*

T. luteum, R. C. 164. M. M. 24. *T. pratense*, M. Pl. 17. Relh. 315.

Hedge-banks. B. June, July.

Common in the (1) Cambridge, (2) Royston, (3) Wimpole, and (4) Cottenham Districts.—5. Fen Ditton. Horningsey. Snailwell.—6. Ely. Witchford. Sutton. Stuntney; N.—8. By old drain between Four Gouts and Foul Anchour. Crab Marsh; J. B.

Pícris Linn.

1. P. hieracioïdes Linn.

Hieracium asperum majore flore in agrorum limitibus, R. C. 75. M. M. 22. *P. hieracioïdes*, M. Pl. 17. Relh. 316.

Dry banks. P. July to September.

1. Chalk-pit-close, Hinton. Balsham. Near Hinxton; N. *Gogmagog Hills;* J. M.—2. About Royston and Guilden Morden; N.—3. Common.—4. Common.—8. South Brink, Wisbech; J. B.

Helmínthia Juss.

1. H. echioïdes Gaert. *Ox-tongue.*

Buglossum luteum, R. C. 24. *Helminthotheca hispidosa vulgaris annua*, M. M. 24. *Picris echioïdes*, M. Pl. 17. Relh. 316.

Dry banks. P. July to September.

1. Devil's Ditch. Balsham. Hildersham.—2. Newton. Triplow. Guilden Morden; N.—3. Common.—4. Common; as by Madingley road at Cambridge.—5. Horningsey.—6. Ely.—7. Doddington.—8. South Brink, Wisbech; J. B.

COMPOSITÆ.

LACTÚCA Linn.

1. L. saligna Linn.

L. sylvestris laciniata minima, R. C. 83. M. M. 21. *L. saligna,* M. Pl. 18. Relh. 318.

Chalky places. B. July, August.

1. Between the London road near Cambridge and Cow Fen; Ray.—4. About a mile from Histon on the way to Cottenham; Relh.—5. River bank a little below Clayhythe; Relh.

Figured in Eng. Bot. from a Cambridge specimen sent by Relhan in the year 1800.

2. L. virósa Linn. *Acrid Lettuce.*

L. sylvestris, R. C. 83. *L. s. major odore Opii,* M. M. 21. *L. virosa,* M. Pl. 18. Relh. 318.

Dry banks. B. July, August.

1. Chalk-pit-close, Hinton.—4. *Ditch near Denny Abbey; By the second bridge beyond Histon on the way to Cottenham;* Relh.—5. *Burwell Pit!;* Relh.—6. *By the Cambridge road at a mile or two from Ely;* Ray.

3. L. Scariola Linn. *Prickly Lettuce.*

L. sylvestris costa spinosa, R. C. 82. M. M. 21. *L. virosa* β, M. Pl. 12. *L. Scariola,* Relh. 318.

Waste places. B. July, August.

1. Shuckburgh Castle, Newmarket Heath.—3. *Burrell's Walk, Cambridge; By the lane leading out of Haslingfield towards Cambridge;* Relh.—4. *Denny Abbey;* Rev. J. Hemsted, who sent, in 1795, the specimen figured in Eng. Bot. *Lane between Long Stanton and Swavesey; Between Histon and Rampton;* Relh.—5. *Burwell Pit!;* Relh.—6. Grunty Fen; W. M. Ely; H.

4. L. muralis Cand.

Prenanthes muralis, Relh. ed. 1, 293; ed. 3, 319.

Dry places. A. July.

1. On willows by the old sluice at Grantchester; S. W. W. Between Trumpington Church and the bridge.—2. By the road-side between Whittlesford and Shelford.—5. *Wall of Chippenham Park;* Relh. He sent the specimen figured in Eng. Bot. from thence in the year 1798.

Leóntodon Linn.

1. **L. Taráxacum** Linn. *Dandelion.*

Dens leonis, R. C. 44. M. M. 23. *L. Taraxacum*, M. Pl. 18. Relh. 319.

Everywhere. P. March to October.

Common throughout the county.

γ. *erythrospermum.*

Dens leonis angustioribus foliis, R. Cat. Angl. ed. 1, 92 (in part). In the *Synopsis* Ray restricts that term to the *Tar. erythrospermum* (Cand.).

3. Gamlingay.

δ. *L. palustre*, Sm.

L. palustre, Lyons, 48. Relh. ed. 3, 320. *L. Taraxacum* γ. *palustre*, M. Pl. 18. Relh. ed. 1, 294.

First noticed by Mr C. Miller on Hinton Moor.

1. Shelford Common in 1829. Coldham's Common; H.—2. Sawston Fen; G. S. G.—4. Madingley Chalk-pit.—5. Bottisham Fen. Anglesey Abbey; H.—6. Stretham Fen.

Sónchus Linn.

1. **S. oleráceus** Linn. *Sowthistle. Hare's Lettuce.*

S. lævis, R. C. 158. *S. l. officinarum*, M. M. 21. *S. oleraceus* α and β, M. Pl. 17. Relh. 317.

Cultivated and waste ground. A. June to August.

Common throughout the county.

2. S. ásper Hoffm. *Sowthistle.*

S. asper laciniatus et non laciniatus, R. C. 158. *S. a. officinarum,* M. M. 21. *S. oleraceus* γ and δ, M. Pl. 17. Relh. 317.

Cultivated and waste ground. A. June to August.
Common throughout the county.

3. S. arvénsis Linn. *Corn Sowthistle.*

S. repens, multis Hieracium majus, R. C. 158. M. M. 21. *S. arvensis,* M. Pl. 17. Relh. 317.

Fields. P. August, September.
Common throughout the county.

4. S. palústris Linn.

Lyons, 47. M. Pl. 18. Relh. 316.
Marshes. P. July, August.
6. *Near Stretham Ferry;* Mr J. Lyons. Not found there for many years; Relh. in 1820.

CRÉPIS Linn.

1. C. fœtida Linn.

Hieracium minus Cichorei vel potius Stœbes folio hirsutum, R. C. 75. *Hieracioïdes vulgaris fœtida,* M. M. 23. *C. fœtida,* M. Pl. 18. Relh. 323.

Chalky places. B. June, July.
1. Devil's Ditch; Hooker in New Bot. Guide. Banks of deserted railway between Chesterford and the Woolstreet!; G. S. G. *About Cambridge;* Withering. *Gravel-pits by the lower road to the Gogmagog Hills;* J. M.—2. *Between Shelford and Whittlesford;* J. M.

2. C. virens Linn.

Hieracium luteum glabrum sive minus hirsutum, R. C. 75. *Hieracioïdes vulgatissima, pene glabra, annua, folio longo dentato*, M. M. 23. *C. tectorum*, M. Pl. 18. Relh. 323.

Waste ground. A. June to September.
Common throughout the county.

3. C. biénnis Linn.

Relh. ed. 1, 296; ed. 3, 324.

Road sides and banks. B. June, July.

1. Gogmagog Hills above Hinton. Road-side near the Furze-hills, Hildersham. *Linton;* Relh.—5. Road-side between Ditton and Horningsey, and by the footpath near White Hall. *By the horse-road from Bottisham to Newmarket,* Relh.

HIERÁCIUM Linn.

1. H. Piloséllá Linn. *Creeping Mouse-ear.*

Pilosella repens, R. C. 117. M. M. 23. *H. Pilosella,* M. Pl. 18. Relh. 321.

Dry banks and pastures. P. May to August.

Common in the (1) Cambridge, (2) Royston, (3) Wimpole, and (4) Cottenham Districts.—5. Snailwell Heath. Chippenham. Horningsey. Quy road. Newmarket.— 6. Ely.—8. Wisbech; J. B.

2. H. murórum Linn. ?

Pulmonaria gallica sive aurea latifolia, R. C. App. i. 8. *H. murorum folio pilosissimo,* M. M. 22. *H. murorum,* M. Pl. 18. Relh. 322.

Heaths. P. June, July.

1. Walls of the Botanical Garden (old); Relh. It is *H. cœsium,* and only an escape from the garden.—3. *Gamlingay;* Ray.

3. H. umbellátum Linn.

H. fruticosum angustifolium majus, R. C. App. ii. 9. M. M. 22. *H. umbellatum,* M. Pl. 18. Relh. 323.

Gravelly and sandy places. P. July to September.
1. *Furze-hills, Hildersham;* J. M.—3. *Gamlingay;* P. Dent.

4. H. boreale Fries.

H. fruticosum latifolium hirsutum, R. C. App. ii. 9. M. M. 22. *H. f. l. glabrum*, R. C. App. ii. 9 (but the plant so named in Ray's *Synopsis* is different). *H. sabaudum*, M. Pl. 18. Relh. 322.

Banks. P. August, September.
1. *Linton;* Relh.—3. Gamlingay.

CAMPANULACEÆ.

JASÍONE Linn.

1. J. montána Linn. *Sheep's Scabious.*

Rapunculus Scabiosæ capitulo cœruleo, R. C. 138. M. M. 68. *J. montana*, M. Pl. 20. Relh. 92.

Dry places. P. July.
1. Newmarket Heath. *Furze-hills, Hildersham;* J. M. —3. Gamlingay.

CAMPÁNULA Linn.

1. C. glomeráta Linn. *Clustered Bell-flower.*

Trachelium minus, R. C. 164. *C. pratensis flore conglomerato*, M. M. 68. *C. glomerata*, M. Pl. 5. Relh. 90.

Chalky banks. P. July, August.

Common in the (1) Cambridge, (2) Royston, and (3) Wimpole Districts.—4. Madingley, near the chalk-pit and outside the park. Papworth St Everard; T. Y.—5. Exning; E. S. Newmarket.

2. C. latifólia Linn. *Giant Bell-flower.*

Relh. ed. 1, Suppl. 2, 10; ed. 3, 90.
Woods. P. July, August.

1. Wood Ditton. *Cheveley;* Rev. G. Crabbe in Bot. Guide.—3. Gaine's Coppice, Comberton; where it was first noticed by the Rev. Mr Newton. The coppice was grubbed up in 1852.—6. In a plantation by Bentham's Monument, Ely, in 1822; Rev. L. Jenyns.

3. C. Trachélium Linn. *Throatwort.*

Trachelium majus, R. C. 164. *C. major et asperior folio Urticæ,* M. M. 67. *C. Trachelium,* M. Pl. 5. Relh. 90.

Hedges and thickets. P. July, August.

1. Wooded part of the Devil's Ditch. By the footway from West Wratting to Weston Colville.—3. Eversden and Kingston Woods. Gamlingay. Caldecot; N.—4. Madingley.

[C. rapunculoïdes Linn.

Hedges. P. July, August.

1. Chalk-pit-close, Hinton!; H.—3. Hedge recently (1859) removed in St John's College Cricket-field.—4. Plantation adjoining Madingley Wood·!; Mr W. M. Frost. An escape from cultivation in gardens.]

4. C. rotundifólia Linn. *Harebell.*

R. C. 26. M. M. 67. M. Pl. 5. Relh. 98.

Dry places. P. July, August.

1. Hinton. Stetchworth. Ascent of Newmarket Heath from Bottisham. Allington Hill. Shelford. Furze-hills, Hildersham. Near Hinxton; N. *Gogmagog Hills;* J. M. —2. Sawston!; Hauxton; Royston; Whittlesford; Near Ashwell; N. Shepreth; N. W.—3. Gamlingay. Chalk-hills, near Barrington. Orwell.—4. Road-side near Arbury Camp.—5. Chippenham. Newmarket. Upware. Swaffham Bulbeck.

[**C. pátula** Linn.

Mr W. Walton, M.A. of Trinity College, informs me that he found one fine plant of this species in the chalk-pit at Haslingfield in 1857; and Mr Job Watson believes that he found plants of it in Barnwell gravel-pits in 1818. I suspect them to have been of accidental occurrence in each place.]

Specularia Heist.

1. S. hybrida Cand.

Speculum Veneris minus, R. C. 159. M. M. 68. *Campanula hybrida*, M. Pl. 5. Relh. 91.

Corn-fields. A. June to September.

Generally distributed through the (1) Cambridge, (2) Royston, (3) Wimpole, (4) Cottenham, and (5) Burwell Districts. — 6. Ely. Witcham. Sutton. — 8. Wisbech; J. B.

ERICACEÆ.

Callúna Salisb.

1. C. vulgáris Salisb. *Ling.*

Erica vulgaris, R. C. 50. M. Pl. 9. Relh. 159. *Scopa prior*, M. M. 123.

Heaths. Sh. June to August.

1. Newmarket Heath. Plantation near Six-mile-Bottom. Balsham Heath, only one bush left in 1856; R. B. S. *Linton Heath* in 1818; J. W. *Gogmagog Hills;* Ray.— 2. Triplow Heath; W. H. C. In small quantity near Ickleton Grange; G. S. G. — 3. Gamlingay. — 5. Newmarket Heath. Kennet Heath.

Erica Linn.

1. E. Tétralix Linn.

E. pumila altera Belgarum, R. C. App. ii. 6; Cat. Angl. ed. 2, 97. *E. Tetralix*, M. Pl. 9. Relh. 159.

Boggy heaths. Sh. July, August.
3. Gamlingay.

2. E. cinérea Linn.

Relh. ed. 1, 156; ed. 3, 159.
Heaths. Sh. July, August.
3. Gamlingay.

VACCÍNIUM Linn.

1. V. Oxycóccos Linn. *Cranberry.*

Vaccinia palustria, R. C. App. ii. 18; Cat. Angl. ed. 2, 298. *Oxycoccus sive Vaccinia palustria*, M. M. 63. *V. Oxycoccos*, M. Pl. 9. Relh. 158.
Bogs. Sh. June, July.
3. Bogs on Gamlingay Heath; formerly abundant; confined in 1859 to one small spot by the stream flowing to the site of the old pond.

MONÓTROPA Linn.

1. M. Hypópitys Linn. *Yellow Bird's-nest.*

Relh. ed. 2, 164; ed. 3, 171.
Woods and thickets. P. July, August.
3. Fir plantation at the south west of Eversden Wood!; Plantation by the house on the hill at Wimpole!; N.—
4. Plantation by the St Neots road at about two miles from Cambridge. *Madingley Plantations;* Relh.

AQUIFOLIACEÆ.

I'LEX Linn.

1. I. Aquifólium Linn. *Holly.*

Agrifolium, R. C. 5. *Aquifolium vulgare*, M. Pl. 22. *I. Aquifolium*, Relh. 66.

Woods and hedges. T. June to August.

1. Stetchworth. Wood Ditton Park Wood.—4. Madingley Wood.

OLEACEÆ.

Ligústrum Linn.

1. L. vulgáre Linn. *Privet.*

Ligustrum, R. C. 87. *L. officinarum,* M. M. 121. *L. vulgare,* M. Pl. 1. Relh. 6.

Hedges and thickets. Sh. June, July.

Not uncommon, but probably often planted, in the (1) Cambridge, (2) Royston, (3) Wimpole, and (4) Cottenham Districts.—5. Swaffham Bulbeck.—8. Wisbech ; J. B.

Fráxinus Linn.

1. F. excélsior Linn. *Ash.*

F. vulgaris, R. C. 55. *F. officinarum,* M. M. 123. *F. excelsior,* M. Pl. 23. Relh. 6.

Woods and hedges, often planted. T. April.

1. Hinton.—3. Comberton. Toft. Eltisley. Gamlingay. Cambridge. Harlton ; Long Stow ; N.—4. King's Hedges. Childerley ; N.—5. Wicken.—8. Common on the silt near Wisbech ; J. B.

The Weeping Ash was "discovered, about 1750, at Gamlingay" (Loudon's Enc. of Trees and Shrubs, 640), "by Prof. Martyn" (Sm. Fl. Brit. 1, 13).

APOCYNACEÆ.

Vínca Linn.

† 1. V. minor Linn. *Lesser Periwinkle.*

Clematis daphnoïdes minor R. C. 35. *Pervinca vulgaris angustifolia,* M. M. 65. *V. minor,* M. Pl. 5. Relh. 103.

Woods and thickets. P. May, June.

1. Near the north end of the back lane at Hinton, and in the field next to the church in which is the foot-path to Cambridge. Thickets near the brook at Fulbourn and at the north-west end of the village. Hildersham; S. W. W. In the second close from the road opposite to Teversham church; Between Shelford and Stapleford; W. H. C. Rose-Green, Balsham; Streetly End; R. B. S.—2. Near Litlington Mill. Shelford!; H. *Triplow;* Relh.—3. Harlton; N. *White Wood, Gamlingay;* Relh.—4. Plantations round Madingley Park; W. H. C.—5. Swaffham; H. Bottisham; S. W. W.—8. Wisbech; J. B.

* 2. **V. májor** Linn. *Greater Periwinkle.*

Lyons, 24. M. Pl. 5. Relh. 103.

Hedges near houses. An escape from gardens. P. April, May.

1. In the back lane at Hinton. Bank of St Peter's College garden, next Cow Fen. *Whittlesford;* Lyons.— 2. Triplow; S. W. W. Royston; D. B.—3. Harlton; N. Grantchester churchyard; W. H. C. *By the summer-house* (which has lately, 1858, been removed) *at the south-west corner of St John's Fellows' Walks,* 1730; R. Jackson MS. in M. M. *Coton;* Relh.—4. Observatory; S. W. W. *Girton; Madingley; Rampton; Histon;* Relh. *Between Madingley and Drayton;* Lyons.—5. *Bottisham,* Relh.

GENTIANACEÆ.

CHLÓRA Linn.

1. **C. perfoliáta** Linn. *Yellow-wort.*

Centaurium luteum perfoliatum, R. C. 31. M. M. 73. *Blackstonia perfoliata,* M. Pl. 9. *Chlora perfoliata,* Relh. 158.

Damp chalky places. A. July to September.

1. Shelford Common. Hinton. Teversham Moor; W. H. C.—2. Sawston Fen!; N. Royston; D. B.—3. Old chalk-pit, Haslingfield. To the north of Hardwick Wood. Eversden Wood. Comberton; Barton; Long Stow; Caxton; Toft; Croxton; By Hayley Wood; N. *By Trumpington Road;* Relh.—4. Madingley chalk-pit and near St Neots Road. Histon Road!; H. Dry Drayton; Elsworth; N. —5. Between Quy and Bottisham; H.

ERYTHRÆA Renealm.

1. **E. pulchélla** Fries.

Chironia pulchella, Relh. ed. 2, 93; ed. 3, 97.
Gravelley places. A. July to September.
1. Shelford Common; W. H. C. *Teversham;* Relh.— 2. *Hauxton;* Relh.—3. Eversden Wood!; By Swansley Wood, St Neots Road; N.—4. Anglesey Abbey; Cowbridge, Swaffham Bulbeck!; H.—5. Chippenham.—8. Salt marshes, near Wisbech.

2. **E. Centaúrium** Pers. *Centaury.*

Centaurium minus, R. C. 31. *C. m. officinarum,* M. M. 73. *Gentiana Centaurium,* M. Pl. 5. Relh. ed. 1, 100. *E. Centaurium,* Relh. ed. 3, 97.

Barren pastures and newly cleared woods. A. July, August.
1. Trumpington. Shelford. Balsham. Between Fulbourn and Wilbraham. Hildersham; W. H. C. Linton; S. W. W. *Teversham; Hinton;* Relh.—2. Whittlesford. Duxford; S. W. W. Shepreth; N. F.—3. Gamlingay. Cambridge. Comberton. Eversden and Kingston Woods. Mare Way. Caxton; Croxton; Shepreth; Near Hayley Wood; N.—4. Near the chalk-pit, Madingley. Impington Road; W. H. C. Dry Drayton; Elsworth; N. *Hill of*

Health; J. M.—5. Bottisham; Newmarket Heath!; H. Chippenham.—8. Wisbech; J. B.

Gentiána Linn.

1. G. Amarélla Linn. *Fellwort.*

Gentianella fugax autumnalis minor, R. C. 60. *G. pratensis flore lanuginoso,* M. M. 67. *G. campestris,* M. Pl. 5. *G. Amarella,* Relh. 109.

Chalky fields. A. August, September.

1. Gogmagog Hills. Chalk-pit-close at Hinton. Devil's Ditch and Shuckburgh Castle, Newmarket Heath. Coldham's Common; S. W. W. Shelford Common; W. H. C. Balsham; R. B. S. Sawston; N.—2. Sawston Moor!; N. Near Melbourn Heath Farm; H. F. Odsey; A. M. B. Ickleton; C. B. C.—3. Old chalk-pit, Haslingfield. Eversden; By Hayley Wood near Long Stow!; Kingston Wood; Eltisley; Croxton; N.—4. Madingley chalk-pit; W. H. C.—5. Devil's Ditch; H. Exning; E. S. To the east of Newmarket; N.

[*G. campestris* grows on the river-bank at about two miles below Wisbech; J. B. That spot is in Norfolk, but close on the borders of Cambridgeshire.]

Villársia Vent.

1. V. nymphæoïdes Vent.

Nymphæa lutea minor flore fimbriato, R. C. 104. *Nymphoïdes aquis innatans,* M. M. 63. *Menyanthes nymphoïdes,* M. Pl. 4. *M. nymphæoïdes,* Relh. 85.

On water. P. July, August.

1. *In the Cam below Cambridge;* Relh.—4. In a pond to the left of the Ely Road at about two miles from Cambridge; S. W. W. Swavesey. Fen Drayton.—6. Roswell Pits, Ely. Littleport!; H. Sandy's Cut; N. *Stretham Ferry;* J. M.—7. In the Old Bedford River near Mepal;

H. In Vermuden's Drain near Chatteris.—8. Upwell; J. B.

MENYÁNTHES Linn.

1. M. trifoliáta Linn. *Buckbean.*

Trifolium palustre, R. C. 168. *Menianthes palustre Theophrasti*, M. M. 72. *M. trifoliata*, M. Pl. 4. Relh. 85.
Bogs and fens. P. May to July.
1. Cow Fen, Cambridge. *Hinton Moor;* J. M. —
2. Ickleton; Sawston Fen; G. S. G. Peat-holes near Triplow. — 3. Sheep's-Green, Cambridge. Gamlingay. — 5. Quy Fen; S. W. W. Horningsey. Bottisham and Wicken Fens.

CONVOLVULACEÆ.

CONVÓLVULUS Linn.

1. C. arvénsis Linn. *Field Bindweed.*

C. minor arvensis, R. C. 38. M. M. 67. *C. arvensis*, M. Pl. 5. Relh. 88.
Fields and hedge-banks. P. June to August.
Common throughout the county.
β. *C. minimus*, R. C. App. ii. 6; Syn. ed. 1, 102.
Between Harlton and Little Eversden.

2. C. sépium Linn. *Great Bindweed.*

C. major, R. C. 38. M. M. 67. *C. sepium*, M. Pl. 5. Relh. 89.
Damp hedges and thickets. P. July, August.
Generally distributed, but not very abundant, throughout the county; growing even on very wet fen-land, as in Wicken Fen.

CUSCÚTA Linn.

1. C. europæa Linn. *Great Dodder.*

C. major, R. C. 42. *C. officinarum,* M. M. 70. *C. europæa,* M. Pl. 6. Relh. 108.

Parasitical upon herbaceous plants. A. August, September.

3. On potatoes at Whitwell; On beans at Coton; S.W.W. Bourn (Mr Haycock); N. *On Torilis Anthiscus at Barton;* Relh.—4. *On thistles at Madingley; On the hop at Oakington; On beans at Swavesey;* Relh. On beans at Oakington; S. W. W.—5. Swaffham Prior!; Rev. J. Downes. *Newmarket;* Rev. J. Hemsted (Bot. Guide, i. 48). —6. Haddenham; N. Love Lane, and near the bridge, Ely; H.

2. C. Epithýmum Murr. *Small Dodder.*

C. europæa β, Relh. ed. 1, 102. *C. Epithymum,* Relh. ed. 3, 108.

Parasitical on small shrubby plants. A. July to September.

3. Gamlingay Heath.

† 3. C. Trifólii Bab. *Clover Dodder.*

Phytol. i. 467. Ann. Nat. Hist. xiii. 252.

First noticed by C. C. B. in 1842.

Clover-fields; rarely on other herbaceous plants. A. July to September.

1. Balsham. Cambridge.—2. Whittlesford; Mr C. Thurnall.—3. Coton. Barrington. Barton. Toft!; Eversden; Kingston; Eltisley; upon *Plantago lanceolata* at Long Stow (Rev. J. Rushton); N. Malton; D. B.—4. Dry Drayton; N.

[*C. Epilinum* was introduced with the seed of flax at Ely in 1853, but disappeared with that crop.]

BORAGINACEÆ.

ASPERÚGO Linn.

1. A. procúmbens Linn. *Madwort.*

Alysson germanicum echioïdes, R. C. 9. *A vulgaris,* M. M. 47. *A. procumbens,* M. Pl. 4. Relh. 83.

Rich waste ground. A. June, July.

1. *Newmarket;* Ray. It has been lost for many years.

CYNOGLÓSSUM Linn.

1. C. officinále Linn. *Hound's-tongue.*

R. C. 43. M. Pl. 4. Relh. 8. *C. officinarum,* M. M. 45. Waste ground. B. June, July.

1. Hinton. Inclosure south of Six-mile-Bottom. Fulbourn; N. Teversham; W. H. C.—3. By the Wimpole Road near Barton; S. W. W. Between Barton and Comberton; Little Eversden; N. — 4. By the river near Waterbeach; W. H. C.—5. By the river between Swaffham and Bottisham Lodes.—6. Aldreth. Witcham. Sutton.

BORÁGO Linn.

[1. B. officinális Linn. *Borage.*

Lyons, 21. M. Pl. 4. Relh. 82.

On rubbish. An escape from gardens. B. June, July.

1. *About the outskirts of Cambridge;* Relh.—2. By the Newmarket Road, Royston; D. B.—3. Barton Road, Cambridge. Toft; N.—4. Chesterton; H.—8. Common about Leverington; J. B.]

ANCHÚSA Linn.

[1. A. sempervírens Linn.

Lyons, 21. M. Pl. 4. Relh. 81.

Waste ground. An escape from gardens. P. May to August.

1. About Christ's and Emmanuel Colleges!; H.]

Lycópsis Linn.

1. L. arvénsis Linn. *Bugloss.*

Buglossa sylvestris minor, R. C. 24. *Echioïdes flore albo*, M. M. 46. *L. arvensis*, M. Pl. 4. Relh. 83.

Fields and hedges. A. June, July.

1. Great Wilbraham. Furze Hills, Hildersham. Six-mile-Bottom.—3. Gamlingay. Near the House-in-the-fields, Cambridge; J. W.—4. Chesterton; H. Near the Observatory; W. H. C.—5. Chippenham.

Sýmphytum Linn.

1. S. officinále Linn. *Comfrey.*

Consolida major, R. C. 37. *S. officinarum Consolida major*, M. M. 47. *S. officinale*, M. Pl. 4. Relh. 81.

Damp banks. P. May, June.

1. Teversham; N. On the common at one mile on the London Road, Cambridge. Cow Fen.—2. Meldreth Road, Royston; D. B. Shepreth; N. W.—3. Gamlingay. Cambridge. Between the foot-path to Grantchester and the river. By the river at Barrington. Toft; Bourn; Kingston; Near Eversden Wood; N.—4. Baitsbite. Waterbeach. Mare Way. Cottenham. Swavesey. Chesterton. Madingley Wood; W. H. C.—5. Horningsey. Upware. Wicken. Reche. Swaffham Bulbeck.—6. Turbutsey, Ely. —7. Doddington. Chatteris.—8. Wisbech; J. B.

β. *S. pátens*, Sibth.

Relh. ed. 2, 77; ed. 3, 82.

8. *By Deadman's Pond, Wisbech;* Mr Skrimshire.

[Mr Woodward says (in Wither. Bot. Arr. ed. 2, 230), that *S. tuberosum* is found on Fen-banks mixed with *S. officinale*. This is almost certainly a mistake.]

[*S. tauricum* has established itself on a hedge-bank near the Observatory.]

E′CHIUM Linn.

1. E. vulgáre Linn. *Viper's Bugloss.*

R. C. 47. M. Pl. 4. Relh. 83. *E. officinarum*, M. M. 46. *E. alterum*, R. C. 47. M. M. 46. *E. anglicum*, M. Pl. 4. *E. vulgare* β, Relh. 84.

Gravelly and chalky places. B. June, July.

Common in the (1) Cambridge, (2) Royston, and (3) Wimpole Districts.—4. Gravel Hill, near the Observatory. Dry Drayton; N. Papworth St Everard; T. Y.—5. Chippenham.—6. Thetford.

E. alterum of Ray is only a state of the plant with smaller flowers and longer stamens. He omitted it in his later works, but Dillenius restored it in the Synopsis, ed. 3.

LITHOSPÉRMUM Linn.

1. L. officinále Linn. *Gromwell.*

L. sive Milium solis, R. C. 90. *L. officinarum*, M. M. 46. *L. officinale*, M. Pl. 4. Relh. 80.

Gravelly woods, thickets, and hedge-banks. P. June to August.

1. Chalk-pit-close, and about the upper end of the brook and the last field by the footpath, and by the road to Hinton. By the brook between Shardlow's Well and the railway, Fulbourn. Linton; G. S. G. *Teversham;* J. M.—2. Steeple Morden; H. F. *Triplow;* Relh.—3. Not unfrequent in the western part; N.—4. Moor Barns. King's Hedges.—5. Bottisham!; H. *Chippenham;* Relh.

2. **L. arvénse** Linn. *Salfern. Corn Gromwell.*

Anchusa degener facie milii solis, R. C. 12. *Buglossum arvense annuum Lithospermi folio*, M. M. 46. *L. arvense*, M. Pl. 4. Relh. 80.

Arable land. A. May to July.

Generally distributed through the (1) Cambridge, (2) Royston, (3) Wimpole, and (4) Cottenham Districts.—5. Chippenham. Reche !; Swaffham Bulbeck!; H.—8. Near Four Gouts, Wisbech; J. B.

Myosótis Linn.

1. **M. palústris** Wither. *Forget-me-not.*

M. scorpioïdes palustris, R. C. 102. *Scorpiurus palustris perennis viridioribus foliis*, M. M. 47. *M. scorpioïdes β*, M. Pl. 4. *M. palustris*, Relh. 79.

Banks of rivers and wet ditches. P. June to August.
Generally distributed throughout the county.

2. **M. cæspitósa** Schultz.

First observed by Mr Wanton shortly before 1851.
Wet places. P. June to August.
3. Gamlingay. Near Kingston Wood. Comberton; N. —4. Pits near the Observatory; S. W. W. Between Girton and Oakington. Waterbeach.—5. Wicken and Bottisham Fens.—6. Ely. Barraway. Witcham.—8. Wisbech.

3. **M. arvénsis** Hoffm.

M. scorpioïdes hirsuta, R. C. 101. *Scorpioïdes annuus arvensis hirsutus cœruleus*, M. M. 47. *M. scorpioïdes α and β*, Relh. ed. 1, 75. *M. arvensis*, Relh. ed. 3, 79.

Fields, also in thickets. A. June to August.
Common throughout the county.

Prof. Henslow marks (Cat. ed. 1, 16) the *M. sylvatica* as a native of this county. I have only seen the large-flowered form of *M. arvensis*, which occurs not unfrequently in thickets. He tells me that he found it in a copse near Linton by Barrington Hill; that is to say, probably, the Rivey Wood. He has no specimen.

4. **M. collína** Hoffm.

First found by C. C. B. in 1835.

Dry banks. A. April, May.

1. Linton. Shuckburgh Castle, Newmarket Heath. Six-mile-Bottom. Furze-hills, Hildersham. — 3. Gamlingay. Grantchester, on walls. Barton.—4. By the Huntingdon road beyond Howe's House; S. W. W. Gravel Hill near the Observatory.

6. **M. versicolor** Ehrh.

M. scorpioïdes minor flosculis luteis, R. Cat. Angl. ed. 1, 217. *M. scorpioides* β, Relh. ed. 1, 75. *M. versicolor*, Relh. ed. 3, 79.

Sandy and gravelly places. A. May, June.

1. Furze-hills, Hildersham; S. W. W. Newmarket Heath!; H.—3. Gamlingay.—8. Wisbech; J. B.

SOLANACEÆ.

Solánum Linn.

1. **S. nígrum** Linn. *Black Nightshade.*

Solanum sive Solatrum vulgare, R. C. 157. *S. vulgare officinarum*, M. M. 62. *S. nigrum*, M. Pl. 5. Relh. 96.

Waste places. A. July to October.

1. Cambridge. Stetchworth.—3. Cambridge. Gamlingay. —4. Cambridge. Chesterton. Histon. Impington. Rampton. Oakington; N.—5. Newmarket, with both black and green fruit.—6. Stretham. Ely, with both black and green fruit. —8. Wisbech. Newton.

2. S. Dulcamára Linn. *Bitter-sweet.*

S. lignosum seu Dulcamara, R. C. 157. *S. l. officinarum,* M. M. 63. *S. Dulcamara,* M. Pl. 5. Relh. 96.

Woods and hedges. Sh. June, July.

Common in the (1) Cambridge, (2) Royston, (3) Wimpole, and (4) Cottenham Districts.—5. Snailwell. Quy Road. Horningsey. Upware. Chippenham.—6. Ely. Witcham. —7. Doddington. Chatteris.—8. Wisbech.

A'TROPA Linn.

1. A. Belladónna Linn. *Dwale. Deadly Nightshade.*

Solanum lethale, R. C. 157. *Belladonna,* M. M. 62. *A. Belladonna,* M. Pl. 5. Relh. 95.

Waste places. Old gravel and chalk pits. P. June to August.

1. In Mr Townley's closes at Fulbourn; S. W. W. *Gravel-pits, Barnwell,* in 1818; J. W. *In the lanes about Fulbourn, plentifully;* Ray. *Dovecote Close near Jesus Green;* Relh.—2. *Triplow;* Relh. — 4. *Cottenham;* Relh. [Established for a few years in the gravel-pits near the Observatory from cultivation, but now extirpated].—5. In a wood at Bottisham Hall. *Old chalk-pits at Swaffham Bulbeck;* Relh. *Reche;* J. M.—6. Soham; W. M. *Isleham;* Relh.—8. Near Newton, abundantly; A. P. Wisbech; J. B.

[*Lycium vulgare* (Dun) is frequent in hedges near houses.]

HYOSCÝAMUS Linn.

1. H. niger Linn. *Henbane.*

R. C. 78. M. Pl. 5. Relh. 95. *H. n. officinarum,* M. M. 66.

Waste places. B. May to July.

Not unfrequent by road-sides, but of uncertain locality, in the (1) Cambridge, (2) Royston, (3) Wimpole, and (4) Cottenham Districts; but rarely, if at all, on the peat soil. —5. Quy Road. Baitsbite. Between Wicken and Fordey. Upware.—6. Witcham. Witchford. Sutton.—8. Wisbech.

DATÚRA Linn.

‡1. D. Stramónium Linn. *Thorn-apple.*

Relh. ed. 2, 90; ed. 3, 94.

Waste ground. A. June, July.

1. *Barnwell Gravel-pits;* Relh.—3. Sprang from soil brought into St John's College walks from excavations in Trinity Street in July, 1857, but did not continue there.—6. On dunghills at Ely!; H.—8. Wisbech.

OROBANCHACEÆ.

OROBÁNCHE Linn.

1. O. Rápum Thuill. *Broom-rape.*

O. sive Rapum Genistæ, R. C. 110 (in part). M. M. 77. *O. major*, M. Pl. 14. Relh. 256.

Parasitical on Broom, Furze, and other shrubby leguminous plants. P. June, July.

3. Gamlingay. *Grantchester;* J. M.—4. *Histon;* Relh.

2. O. elátior Sutt.

O. sive Rapum Genistæ, R. C. 110 (in part). *O. elatior*, Relh. ed. 2, 248; ed. 3, 257.

Parasitical on *Centaurea Scabiosa*, on balks and banks in chalky places. P. June, July.

1. Above Hinton Chalk-pit. Balsham. Hildersham. Gogmagog Hills. Near Hinxton; N. *Between Shelford and Stapleford;* Relh.—2. Chalk-pit, Newton; S. W. W.

Meggot's Mount; Sawston!; Chalk-pit, Foxton; N. *Hauxton;* Relh.—3. Comberton. In the old chalk-pit, Haslingfield. Above Fox-hole's-down Farm, Barrington. Harlton; N. *Between Cambridge and Grantchester;* Ray.

3. O. Pícridis F. W. Schultz.

Ann. Nat. Hist. Ser. 2, ii. 149.
First found by Mr Newbould in 1848.
Parasitical on *Picris hieracioïdes.* P. July.

3. Comberton. To the right of the road a little to the north of Caldecot; A little to the south of Hardwick Wood; Between Caxton and Eltisley; N.

4. O. minor Sutt.

Hensl. Cat. ed. 1, 17.
Parasitical chiefly on clover; probably often introduced with the seed. A. ? June, July.

Formerly abundant in clover-fields; H.—1. Balsham; R. B. S.—2. Royston; D. B.—5. Chippenham.

[**O. ramósa** (Linn.), Relh. ed. 2, 248; ed. 3, 257. This plant was found amongst Flax near Wisbech and Upware formerly. It has no claim to be included in our Flora.]

SCROPHULARIACEÆ.

Verbáscum Linn.

1. V. Thápsus Linn. *High-taper. Great Mullen.*

Tapsus barbatus, R. C. 161. *Verbascum officinarum,* M. M. 73. *V. Thapsus,* M. Pl. 5. Relh. 93.
Waste, gravelly, and chalky ground. B. July, August.
1. Stetchworth. Bartlow. Hinton!; H. Cow Fen; Between Shelford and the Gogmagog Hills; Fulbourn; W. H. C. Hildersham.—2. Ickleton. Foxton; N. Vicar-

age grounds, Royston; D. B.—3. Gamlingay. Harston. Near Grantchester church. Barton. Comberton; Toft; Barrington; Caxton; Eltisley; Great Eversden; N.— 4. Chalk-pit, Madingley; W. H. C. Rectory wall, Cottenham. Graveley; N.—5. Chippenham. Landwade. Upware. Between Fen Ditton and Baitsbite; W. H. C. Newmarket; N.

2. **V. nigrum** Linn. *Black Mullen.*

R. C. 173.   M. M. 73.   M. Pl. 5.   Relh. 94.

Chalky and gravelly banks, and waste places. P. July, August.

1. Wilbraham. Dullingham. Fulbourn; Linton; Abington; Bartlow; H. Between Shelford and Stapleford; Between Abington and Hildersham; Babraham; W. H. C. Gogmagog Hills; J. W. Near Hinxton; N.—2. Triplow; S. W. W.—3. Harston; N.—5. Badlingham, near Chippenham.

[Lyons introduced *V. Lychnitis* (Linn.) into the list of Cambridgeshire plants, as found " in ruderatis." Relhan inserted it in his *Flora* on the sole authority of Lyons. It is now nearly a century since Lyons wrote, and no botanist is known to have found the plant in this county. It seems therefore reasonable to suppose that some mistake caused its name to appear in the *Fasciculus.*]

ANTIRRHÍNUM Linn.

* 1. **A. majus** Linn. *Snapdragon.*

Lyons, 41.   M. Pl. 14.   Relh. 255.

Old walls. P. July to September.

1. Walls at Jesus College. About the Gas-works, Barnwell. Hildersham; W. H. C.—4. Histon; W. H. C. —5. Bottisham !; H.—6. On the Cathedral and old walls, Ely.

2. A. Oróntium Linn.

Gravelly fields. A. July, August.

2. Not uncommon about Odsey; A. M. B. and H. F.

Lináría Mill.

* 1. L. Cymbalária Mill.

Lyons, 40. M. Pl. 14. Relh. 253.

First found by Prof. J. Martyn at Drayton.

Old walls. P. May to October.

1. On the walls and bridges at Cambridge. Fulbourn. Brinkley. *Bartlow church;* Relh.—2. Royston; D. B.—4. *On the great house at [Dry] Drayton;* J. M.—5. Churchyard wall, Fen Ditton.—6. Churchyard wall, Sutton. On the cathedral and old walls, Ely.—8. Wisbech; J. B.

2. L. Elátine Mill. *Sharp-pointed Fluellin.*

Elatine folio acuminato, R. C. 48. M. M. 75. *Antirrhinum Elatine,* M. Pl. 14. Relh. 253.

Arable fields. A. July to September.

1. Near Hinxton; N. *Barnwell; Hinton;* Relh. *Teversham;* J. M.—2. Odsey; H. F.—3. Coton; H. Grantchester; J. W. Caldecot; Long Stow; Near Hayley Wood; Croxton; Eversden; Toft; Kingston; N.—4. By Huntingdon road. Closes by Madingley Wood. Dry Drayton. Papworth St Everard; Elsworth; N.—5. Upware. Exning; E. S.

3. L. spúria Mill. *Round-leaved Fluellin.*

Elatine folio subrotundo, R. C. 48. *E. officinarum,* M. M. 75. *Antirrh. spurium,* M. Pl. 14. Relh. 253.

Arable fields. A. July to September.

1. Cambridge. Stetchworth. By Borley Wood. Weston Colville. Hildersham; G. S. G. Near Hinxton; N.

Barnwell; Hinton; Relh. *Teversham;* J. M.—2. Known's Folly, Royston; D. B. Odsey; A. M. B.—3. Cambridge· Coton. Grantchester. Barton. Hardwick; Comberton; Toft; Caldecot; Kingston; Caxton; Long Stow; Near Hayley Wood; Eversden; Eltisley; Croxton; N.—4. Madingley. Dry Drayton; Papworth; Elsworth; N. *By Histon road;* J. M. — 5. Bottisham; H. Wicken; W. M. Upware. Exning; E. S.

4. L. minor Desf. *Calves-snout.*

Antirrhinum minimum, R. C. 13. *L. Antirrhinum dicta,* M. M. 75. *Antirrh. minus,* M. Pl. 14. Relh. 254.

Arable fields. A. June to August.

1. Hinton. Gogmagog Hills. Stetchworth. By Borley Wood. Six-mile-Bottom. Hildersham. Linton; R. B. S. Near Hinxton; N. Brinkley. Fulbourn. *Babraham; Newmarket Heath;* Relh.—2. Between Known's Folly and Melbourn; Near Ashwell; Steeple Morden; Foxton; N. Odsey; H. F.—3. Not rare in the western part of this district; N.—4. Dry Drayton; Papworth St Everard; N. —5. Bottisham; H. Exning; E. S. Snailwell, Chippenham.

5. L. vulgaris Mill. *Toad-flax.*

L. v. nostras, R. C. 88. *L. officinarum,* M. M. 74. *Antirrh. Linaria,* M. Pl. 14. Relh. 254.

Banks and hedges. P. June, July.

1. Wilbraham. Six-mile-Bottom. Allington Hill. Newmarket. Brinkley. Hildersham. Near Hinxton; N. Cow Fen; Hinton; Gogmagog Hills; W. H. C.—2. Shepreth. Triplow. Whittlesford. Morden Heath Farm!; N. Royston; D. B. Odsey; A. M. B.—3. Gamlingay. By Barton road, near Cambridge. Shepreth; Barrington; N.—4. Near the railway-bridge, Chesterton.—5. Newmarket.

Chippenham. Upware; N. Bottisham. *By the foot-path by the river to Fen Ditton;* Relh.—8. Near Whittlesey; N. Wisbech; J. B.

SCROPHULÁRIA Linn.

1. S. nodósa Linn. *Knotted Figwort.*

S. major, R. C. 152. *S. officinarum,* M. M. 74. *S. nodosa,* M. Pl. 14. Relh. 255.

Moist hedge-banks and thickets. P. June, July.

1. Wood Ditton. Devil's Ditch, near Stetchworth. Balsham. Deersley's Wood, Newmarket. West Wratting. Linton Wood; S. W. W. The Rivey, Linton. Between Little Abington and Hildersham; W. H. C.—3. Eversden and Kingston Woods. Cambridge. Gamlingay; H. Toft. Caldecot. Long Stow; Hayley Wood; Bourn; Eltisley; N. —4. Madingley Wood. Oakington. Long Stanton. Waterbeach. *Impington;* Histon; Relh.—5. To the east of Newmarket; N.—7. Doddington Wood.

2. S. aquática Linn. *Water Betony.*

Betonica aquatica, R. C. 20. *S. a. officinarum,* M. M. 74. *S. aquatica,* M. Pl. 14. Relh. 255.

Banks of streams and wet ditches. P. July, August.

Not unfrequent in the (1) Cambridge, (2) Royston, (3) Wimpole, and (4) Cottenham Districts.—5. Snailwell Fen. Wicken Fen. Horningsey. Chippenham.—7. Doddington Turf-fen.—8. By the canal, Wisbech.

LIMOSÉLLA Linn.

1. L. aquática Linn. *Mudwort.*

Plantaginella palustris, R. C. App. i. 7. *P. aquatica,* M. M. 72. *L. aquatica,* M. Pl. 14. Relh. 256.

Muddy places where water has stagnated. A. July to September.

3. Gamlingay in 1827!; H.—4. *In a cart-rut just beyond Milton, on the way to Ely;* J. M.

MELAMPÝRUM Linn.

1. **M. cristátum** Linn. *Crested Cow-wheat.*

M. c. flore purpureo, R. C. 95. M. M. 76. *M. cristatum,* M. Pl. 14. Relh. 250.

Woods and thickets, rarely in fields. A. July.

1. Devil's Ditch, near Stetchworth. Wood Ditton. Yenhall Wood, West Wratting. Balsham and Borley Woods. Hildersham Wood; S. W. W.—2. Near Royston; D. B.—3. Eversden, Hardwick, Kingston, and Gamlingay Woods. Caldecot; Hayley Wood; White Wood, Gamlingay; S. W. W.—4. Closes by Madingley Wood; W. H. C. Knapwell; J. C. *Very abundant between Madingley and Dry Drayton;* Ray.

2. **M. praténse** Linn. *Common Cow-wheat.*

M. sylvaticum flore luteo, sive Satureia lutea sylvestris, R. C. 95. *M. luteum latifolium,* M. M. 76. *M. sylvaticum,* M. Pl. 14. Relh. ed. 1, 240. *M. pratense,* Relh. ed. 3, 251.

Woods and thickets. A. July.

1. *In a wood at Stetchworth;* Ray.—3. White Wood, Gamlingay; H. and S. W. W. *In a Wood at Hatley St George, in great plenty;* Ray.

PEDICULÁRIS Linn.

1. **P. palústris** Linn. *Louse-wort.*

P. p. rubra elatior, R. Cat. Angl. ed. 1, 235. Lyons, 40. M. Pl. 14. Relh. 252.

Marshy places. A. May to July.

1. Sawston Moor; S. W. W. Shelford Common; W. H. C.—2. Peat-holes, Triplow. Foulmire Common.—3. Grantchester Meadows. Sheep's-Green, Cambridge. Gamlingay.—5. Bottisham and Wicken Fens.

2. P. sylvática Linn. *Red Rattle.*

Pedicularis, R. C. 115. *P. pratensis rubra vulgaris*, M. M. 75. *P. sylvatica*, M. Pl. 14. Relh. 252.

Wet heathy pastures. A. May to August.

1. Wood Ditton.—2. Sawston Moor; N.—3. Gamlingay.

RHINÁNTHUS Linn.

1. R. Crista-galli Linn. *Yellow Rattle. Cock's-comb.*

Pedicularis sive Crista galli lutea, R. C. 115. M. M. 75. *R. Crista galli*, M. Pl. 14. Relh. 249.

Clayey pastures. Parasitical on the roots of grasses (see Josh. Clarke in *Linn. Proceed.* ii. 255). A. June.

Common in the (1) Cambridge, (2) Royston, (3) Wimpole, (4) Cottenham, and (5) Burwell Districts.—6. Barraway.—8. Wisbech; J. B.

EUPHRÁSIA Linn.

1. E. officinális Linn. *Eye-bright.*

Euphrasia, R. C. 52. *E. officinarum*, M. M. 75. *E. officinalis*, M. Pl. 14. Relh. 250.

1. Devil's Ditch. Hinton. Balsham. Shelford. Wood Ditton. *Gogmagog Hills;* Ray.—2. Sawston!; N. Baldock Road, Royston; D. B. Odsey; A. M. B.—Shepreth; N. W.—3. Comberton; Harlton; Orwell; Long Stow!; Near Hayley Wood; Caxton; N. Gamlingay. Eltisley. Kingston.—4. Madingley. Waterbeach.—5. Chippenham. Wicken.—6. Barraway.—8. Newton.

β. *E. nemorósa*, Pers.

1. Near Ashwell; N.—3. Eversden; Caxton; Long Stow; Gamlingay; Kingston; Bourn; Caldecot; Eltisley; Croxton; N.—4. Elsworth; N.—5. Newmarket; N.

It will be seen that I am indebted to Mr Newbould for most of the information relative to these plants, taken separately. He has paid much attention to them.

2. E. Odontites Linn.

Cratæogonon Euphrosynes facie, R. C. 41. *Odontites*, M. M. 76. *E. Odontites*, M. Pl. 14. Relh. ed. 1, 239. *Bartsia Odontites*, Relh. ed. 3, 249.

Cultivated ground and waste spots. A. July, August.

1. Cambridge. Stetchworth. Balsham. Hildersham. Near Hinxton; N.—2. Eastwards from Royston; Steeple Morden; N.—3. and 4. Rather common in these districts.—5. Chippenham.—7. Doddington.—8. Wisbech. Near Tydd Gout.

E. rotundata (Ball) is recorded in the Annals of Nat. Hist. (ser. 2. iv. 30), as having been found in Cambridgeshire. I know nothing of the plant but what can be learned from the description there given.

VERONÍCA Linn.

1. V. scutelláta Linn.

Anagallis aquatica angustifolia, R. C. 11. *Beccabunga minima angustifolia*, M. M. 69. *V. scutellata*, M. Pl. 1. Relh. 9.

Boggy places. P. June to August.

1. Pits on Stourbridge Fair Green. *Teversham Moor;* J. M. *Hinton Moor;* J. F.—2. Sawston Moor; G. S. G.—3. Comberton; S. W. W. Gamlingay. At the eastern end of Eversden Wood; W. H. C.—4. Gravel-pits by the Histon Road!; H; a place that is now drained and cultivated.

Hill of Health; Relh.—5. Anglesey Abbey!; H. Pit by the Chippenham avenue.—6. Roswell pits, Ely.—7. Doddington.

2. V. Anagallis Linn.

Anagallis aquatica III., *sive major folio oblongo;* and *A. a.* IV., *sive minor folio oblongo,* R. C. 10. *Beccabunga minor angustifolia,* M. M. 69. *V. Anagallis,* M. Pl. 1. Relh. 8.

In nearly or quite stagnant water. P. June to August. Appears to be abundant throughout the county.

3. V. Beccabúnga Linn. *Brooklime.*

Anagallis aquatica I., *sive major folio subrotundo;* and *A. a.* II., *sive aquatica minor folio subrotundo,* R. C. 10. *Beccabunga officinarum,* M. M. 69. *V. Beccabunga,* M. Pl. 1. Relh. 8.

Ditches and streams. P. May to August.
Common throughout the county.

4. V. Chamædrys Linn. *Germander Speedwell.*

Chamædrys sylvestris, R. C. 32. *Beccabunga foliis subrotundis sessilibus profunde crenatis,* M. M. 69. *V. Chamædrys,* M. Pl. 1. Relh. 8.

Hedge-banks. P. May, June.

Common in the (1) Cambridge, (2) Royston, and (3) Wimpole Districts.—4. Cambridge. Fen Drayton. Swavesey. Over. Chesterton.—5. Chippenham.—6. Ely.—8. Wisbech; A. P.

5. V. montána Linn.

Chamædrys spuria foliis pediculis oblongis insidentibus, R. C. App. ii. 5. *Beccabunga foliis subrotundis crenatis caudis longis donatis*, M. M. 69. *V. montana*, M. Pl. 1. Relh. 9.

Woods and thickets. P. May, June.

1. Wooded part of the Devil's Ditch near Camois Hall. Wood Ditton Park Wood. *Fulbourn; J. F. Linton; J. M.*—3. Gamlingay. *Kingston Wood; J. M.*

6. V. officinalis Linn.

V. mas supina et vulgatissima, R. C. 174. *Beccabunga repens foliis subrotundis serratis*, M. M. 69. *V. officinalis*, M. Pl. 1. Relh. 8.

Dry banks and heaths. P. June to August.

1. Near Allington Hill. Westley Wood. Wood Ditton. Linton Wood; S. W. W. *Catledge Wood;* Relh. *Chalk-pit-close, Hinton; J. F.*—2. Royston; D. B. Odsey; *A. M. B.*—3. White Wood, Gamlingay.—4. *Madingley Wood; J. M.*—5. Newmarket Heath. Chippenham. Six-mile-Bottom.

7. V. spicáta Linn.

V. spicata recta minor, R. C. 174. M. M. 68. *V. spicata*, M. Pl. 1. Relh. 7.

Chalky heaths. P. July, August.

1. *About Horseheath; About Hare Park, Newmarket Heath;* Relh.—5. Beacon Course, Newmarket Heath!; H. *In several closes on Newmarket Heath; as in a close near the Beacon, on the left-hand of the way from Cambridge to Newmarket, in great plenty;* Ray.

8. V. serpyllifólia Linn.

V. pratensis, R. C. 174. M. M. 68. *V. serpillifolia*, M. Pl. 1. Relh. 8.

Damp places. P. May to July.

Apparently common throughout the county.

9. V. arvénsis Linn.

Alsine foliis Veronicæ, R. C. 8. *V. flosculis singularibus cauliculis adhærentibus*, M. M. 68. *V. arvensis*, M. Pl. 1. Relh. 10.

Gravelly and sandy places, wall-tops. A. April to July.

Common, except in the Fens, and there generally to be found on the gravelly mounds.

10. V. agréstis Linn.

Alsine foliis Trissaginis, R. C. 7. *V. floribus singularibus in oblongis pediculis Chamædryfolia*, M. M. 68. *V. agrestis*, M. Pl. 1. Relh. 10.

Cultivated ground. A. April to September.

1. Stetchworth. Brinkley.—3. St John's College Walks. Bourn; Caxton; Gamlingay; N. Little Eversden; T. Y.—4. Swavesey. Fen Drayton. Elsworth; N.—5. Chippenham. Newmarket. Swaffham Prior!; H.—6. Stretham!; Rev. H. Baker. Stuntney; N.—7. Doddington.—8. Wisbech.

11. V. políta Fries.

Hensl. Cat. ed. 2, 42.

Cultivated ground. A. April to September.

Common in the (1) Cambridge, (2) Royston, (3) Wimpole, and (4) Cottenham Districts.—5. Common about Chippenham. Upware. Horningsey.—6. Stretham!; Rev. H. Baker. Wentworth. Witchford.—8. Newton. Wisbech.

β. *grandiflóra*.

Bab. Man. ed. 1, 225.

3. Fields above Coton church in 1830. Toft.

*12. V. Buxbaúmii Ten.

Hensl. Cat. ed. 2, 42.

First noticed by Prof. Henslow in 1826.

Cultivated ground. A. April to September.

1. Hinton. Gogmagog Hills. Cambridge. Balsham; R. B. S.—2. Litlington. Foxton; N.—3. Barton; S. W. W. Malton. Kingston; Harlton; Shepreth; Eversden; N.— 8. Wisbech; J. B.

13. V. hederifólia Linn.

Alsine hederacea, R. C. 8. *Morsus gallinæ folio hederulæ*, M. M. 69. *V. hederifolia*, M. Pl. 1. Relh. 10.

Fields, banks, and cultivated land. A. April to June.

Probably common throughout the county. Not recorded as in the (7) Chatteris district.

LABIATÆ.

MÉNTHA Linn.

1. M. rotundifólia Linn.

Menthastrum folio rugoso rotundiore spontaneum flore spicato odore gravi, R. C. App. ii. 12; Cat. Angl. ed. 2, 198. *M. sylvestris rotundiore folio*, M. M. 48. *M. rotundifolia*, M. Pl. 13. Relh. 234.

Damp waste ground. P. August, September.

1. *Chalk-pit-close, Hinton;* Mr Dent.—4. *At the western end of Cottenham;* Relh.—5. *Exning;* Relh.

2. M. sylvéstris Linn.

Menthastrum spicatum folio longiore candicante, R. C. App. ii. 12; Cat. Angl. ed. 2, 198. *M. sylvestris folio longiore*, M. M. 49. *M. sylvestris*, M. Pl. 13. Relh. 233.

Wet waste places. P. July, August.

1. Cow Fen; W. H. C. *Near the first bridge on the road from Linton to Hildersham;* Relh.—3. *By the stump of a cross, Coton;* W. H. C. *Behind Grantchester, Harston,*

and Barrington Mills; Relh. No such plant is now to be found at Barrington.—5. *In a bushy close near Exning;* Mr Dent. *A quarter of a mile from the mill at Bottisham Lode;* Relh.

*3. M. víridis Linn. *Spear Mint.*

Hensl. Cat. ed. 2, 42.
Wet places. P. July, August.
5. Naturalized in Bottisham Park!; H.

4. M. piperita Linn. *Peppermint.*

Relh. ed. 1, 223; ed. 3, 234.
Wet places. P. July, August.
2. *By running water in the village of Hauxton;* Relh. He sent it to Smith from thence in the year 1800; Eng. Bot. fol. 687.—3. Barrington, but probably an escape. Near the Fox at Long Stow, an escape, and from thence down the water to Bourn; N.—4. Wet hollow in Moor Barns Thicket.—5. Bottisham; H.

5. M. aquática *Headed Mint.*

M. aquatica sive Sisymbrium, R. C. 97. *M. a. officinarum,* M. M. 48. *M. aquatica,* M. Pl. 13. Relh. ed. 1, 223. *M. hirsuta* a and β, Relh. ed. 3, 235.

Wet places. P. July, August.

Probably common throughout the county, although but few stations are recorded in the Fens.

6. M. praténsis Sole.

M. gracilis, Relh. ed. 2, 228; ed. 3, 236.
Wet places. P. August, September.
2. *Shelford; In the village and on the common at Hauxton;* Relh.—4. *By the river-side nearly opposite to Horningsey;* Relh.

7. M. satíva Linn.

M. hirsuta γ, Relh. ed. 2, 227. *M. h.* γ and δ, Relh. ed. 3, 236. *M. paludosa*, Relh. ed. 2, 228.

Wet places. P. July, August.

1. *Hinton;* Relh. West Wratting.—3. Cambridge. Long Stow. Croxton; N.—4. By the way from Long Stanton to Rampton. Dry Drayton. Mare Way. Waterbeach. Impington road.—5. Wicken. *Fordham;* Relh.—6. *Aldreth Causeway; Holt Fen, Stretham;* Relh.

8. M. arvénsis Linn.

Calamintha aquatica, R. C. 25. *M. arvensis verticillata hirsuta*, M. M. 48. *M. arvensis*, M. Pl. 13. Relh. 236.

Arable land. P. July to September.

1. Stetchworth. Dullingham. West Wratting.—2. Shelford.—3. Cambridge. Barton. By footpath to Comberton. Barrington. Gamlingay. Bourn; Long Stow; East Hatley; Croxton; N.—4. Closes by Madingley Wood. Near King's Hedges.—5. Chippenham. Exning. Wicken. Horningsey. —6. Barraway.—7. Doddington.—8. Newton.

β. *M. agréstis* Sm.
Relh. ed. 2, 229; ed. 3, 237.

1. *Pampisford;* Relh.—3. *Hauxton; Harlton;* Relh.— 4. *Cottenham;* Relh.

It is not clear what the *M. arvensis verticillata folio rotundiore odore aromatico* of Ray (Syn. ed. 2, 123) was. Lyons believed it to be the *M. gentilis*, and Smith appears to follow him. It was "found by Mr Wigmores at Shelford."

9. M. Pulégium Linn. *Penny Royal.*

Pulegium, R. C. 128. *P. officinarum*, M. M. 19. *M. Pulegium*, M. Pl. 13. Relh. 237.

Edges of ponds. P. August, September.

3. *On the common by the road-side at Harlton;* Relh.—
4. *In a bottom* [the Roman ditch] *over against the furthest house at Castle End, Cambridge;* Ray. *Denny Abbey;* J. M. [—7. Just out of this county at Earith; N.]—8. *Wisbech;* Relh.

Lýcopus Linn.

1. **L. europæus** Linn. *Water Horehound. Gipsy-wort.*

Marrubium aquaticum, R. C. 95. *Pseudo-marrubium palustre,* M. M. 49. *L. europæus,* M. Pl. 1. Relh. 12.

Banks of streams and ditches. P. July, August.

Apparently common throughout the county, although I have no recorded station for it in (7) Chatteris District.

Sálvia Linn.

1. **S. verbenáca** Linn. *English Clary.*

Horminum sylvestre, R. C. 78. *H. s. officinarum,* M. M. 49. *S. verbenaca,* M. Pl. 1. Relh. 13.

Dry gravelly banks. P. May, June.

1. Hinton. Fulbourn. Wilbraham. Paper Mills road, Cambridge. Shudy Camps. Bartlow. Abington. Hildersham. —2. Triplow. Ickleton. Great Shelford. Hinxton; N.— 3. Burrell's Walk and other banks, Cambridge. Little Gransden; N.—4. Willingham. By Ely road and on Castle Hill, Cambridge.—5. Quy road. Horningsey. Between Ditton and Baitsbite. Chippenham. Newmarket; N.

Oríganum Linn.

1. **O. vulgáre** Linn. *Wild Marjoram.*

R. C. 109. M. Pl. 14. Relh. 245.

Dry uncultivated places. P. August.

1. Gogmagog Hills. Hinton. Fulbourn. Near Hinxton; N. Between Hildersham and Linton!; H. Barnwell; J.W.

To the east of Linton near the river in plenty; Ray.—
2. Foulmire. Triplow. Bassingbourn; D. B. Odsey; A. M. B.
—3. Cambridge. *Gamlingay;* R. J.—4. *Histon;* Relh.

THÝMUS Linn.

1. T. Serpíllum Linn. *Common Thyme.*

Serpillum vulgare, R. C. 154. *S. officinarum,* M. M. 84.
S. hirsutum, R. C. 154. M. M. 84. *T. Serpillum,* M. Pl.
14. Relh. 246.

Dry heaths and chalky banks. P. June to August.

Common in the (1) Cambridge, (2) Royston, and (3) Wimpole Districts, chiefly in the chalky tracts.—4. Dry Drayton. Childerley; N. — 5. Snailwell Heath. Swaffham Bulbeck!; H. Newmarket.

2. T. Chamædrys Fries.

Ann. Nat. Hist. Ser. 2, xi. 433.

First noticed in 1852 by C. C. B.

Dry heaths, and chalky and gravelly banks. P. June to August.

1. On the lower side of the Devil's Ditch between the railway and Stetchworth. Furze-hills, Hildersham. Near Hinxton; N.—2. Newton!; N.—3. Very abundant at Gamlingay. Clunch-pit to the west of Harlton; Eversden Quarry; Barrington; N.—5. Gravel-pit, Chippenham.

CALAMÍNTHA Moench.

1. C. Népeta Clairv.

Calamintha flore minore odore Pulegii, R. C. App. i. 4.
C. odore Pulegii, M. M. 53. *Melissa Nepeta,* M. Pl. 14.
Relh. ed. 1, 235. *Thymus Nepeta,* Relh. ed. 3, 247.

Dry banks. P. July, August.

1. Hildersham. Linton. Between Babraham and Bartlow, plentifully; H. Chalk-pit opposite to Babraham House; S. W. W. Newmarket Heath.—2. Royston; D. B. Odsey; H. F.—3. *Grantchester;* Relh.—5. Chippenham. Snailwell. To the east and west of Newmarket.

2. C. officinalis Moench. *Common Calamint.*

C. vulgaris, R. C. 24. *C. montana officinarum,* M. M. 53. *M. Calamintha,* M. Pl. 14. Relh. ed. 1, 235. *T. Calamintha,* Relh. ed. 3, 247.

Dry banks. P. July to September.

1. Hinton road; Babraham; Abington; Fulbourn; W. H. C. Hildersham. Roman road near Balsham; R. B. S. *Linton;* Ray.—2. Shelford; S. W. W.—3. Gamlingay. Between Bourn and Caxton; N.—4. By Chesterton road, abundantly. *Churchyard, Girton;* Relh. — 5. Burwell. Swaffham Bulbeck and Swaffham Prior; H. Badlingham. Newmarket; N.

3. C. A'cinos Clairv. *Basil Thyme. Stone Basil.*

Acinos anglica, R. C. 14. M. M. 53. *Thymus Acinos,* M. Pl. 14. Relh. 246.

Dry, gravelly, and chalky places. A. June, July.

1. Gogmagog Hills. By the Roman road near Balsham; R. B. S. Fulbourn; Near Hinxton; N. Between Newmarket and Wood Ditton. Shuckburgh Castle, Newmarket Heath. Allington Hill. Abington!; H.—2. Near Ashwell; Steeple Morden; N. Near Royston; W. H. C. and H. F. Odsey; A. M. B.—5. Newmarket Heath. Chippenham.

4. C. Clinopódium Spenn. *Wild Basil.*

Clinopodium majus, R. C. 36. M. M. 52. *Cl. vulgare,* M. Pl. 14. Relh. 245.

Dry, bushy, exposed places. P. July, August.

1. Chalk-pit-close, Hinton. Devil's Ditch. Balsham. Westley Wood. Near Hinxton; N. *Bowyer's Close, near Hinton church;* Ray.—2. Guilden and Steeple Morden; Hinxton; N. Royston; D. B. Odsey; A. M. B.— 3. Coton. Eversden. Caldecot. Wimpole. Comberton. Malton; Bourn; Tadlow; Caxton; Near Hayley Wood; Croxton; N.—4. Cuckoo Lane, near Histon. Madingley. Dry Drayton. Graveley; Elsworth; Papworth St Everard; N.—5. Landwade. Chippenham. To the east of Newmarket. Upware. Bottisham!; H.

[*Melissa officinalis*, Linn., was found by Lyons (39) in Garrett Hostel Lane, beyond the bridge, Cambridge. I found it on the left-hand side of the Huntingdon road, a little before arriving at Howe's House, 1853. It is an escape from cultivation.]

SCUTELLÁRIA Linn.

1. S. galericuláta Linn. *Skull-cap.*

Lysimachia galericulata, R. C. 92. *S. palustris repens cærulea*, M. M. 54. *S. galericulata*, M. Pl. 14. Relh. 248.

Banks of rivers and ditches. P. July, August.

1. *Hinton and Teversham Moors;* Relh.—2. Hinxton; C. B. C. Shepreth; N. W. — 3. Malton. Comberton; Toft; Gamlingay; N.—4. Waterbeach Fen. Pits near the Observatory.—5. Bottisham and Wicken Fens.—6. Ely. —8. Wisbech; J. B.

PRUNÉLLA Linn.

1. P. vulgáris Linn. *Self-heal.*

Prunella, R. C. 126. *P. officinarum*, M. M. 50. *P. vulgaris*, M. Pl. 14. Relh. 248.

Damp pastures. P. July, August.

Appears to be common throughout the county.

NÉPETA Linn.

1. N. Cataria Linn. *Cat-mint. Neppe.*

Mentha Cattaria, R. C. 97. *M. officinarum,* M. M. 50. *N. Cataria,* M. Pl. 13. Relh. 232.

Dry banks by road-sides. P. July, August.

1. Hinton. Teversham. Dullingham. Bottisham. Hills Road, Cambridge. Stapleford; J. C. — 2. Triplow; Towards Arrington, and towards Known's Folly, Royston; Steeple Morden; N.—3. Bridle-way from Grantchester to Barton. Between Eversden and the quarry; Near Spring Hall, Haslingfield; Kingston; N.—4. Cuckoo Lane, Histon. Chesterton; W. H. C.—5. Chippenham. Snailwell. Newmarket. Upware. Landwade. Bottisham !; H. Horningsey. *In great plenty at Swaffham;* Ray.—8. Near Oxburgh Hall, Elm. *Wisbech;* J. M.

2. N. Glechoma Benth. *Ground Ivy. Alehoof. Tunhoof.*

Hedera terrestris, R. C. 72. *Chamæcissus officinarum,* M. M. 53. *Glechoma hederacea,* M. Pl. 13. Relh. 238.

Hedge-banks. P. April to June.

Common in the (1) Cambridge, (2) Royston, (3) Wimpole, and (4) Cottenham Districts.—5. Horningsey. Upware. Newmarket. Chippenham.—6. West Fen, Ely. Aldreth. Witchford. Stuntney; N.—7. Doddington Wood. —8. Wisbech; A. P.

LÁMIUM Linn.

1. L. amplexicaule Linn. *Henbit.*

Alsine hederula altera, R. C. 8. *L. folio caulem ambiente majus et minus,* M. M. 51. *L. amplexicaule,* M. Pl. 13. Relh. 239.

Gravelly and chalky fields. A. May to August.

1. Cambridge. Gogmagog Hills. Hinton. Fulbourn. Hildersham. Dullingham. Six-mile-Bottom. Newmarket. Near Hinxton; N.—2. Ickleton. Royston. Steeple Morden; Near Ashwell; Foxton; N. Odsey; A. M. B.—3. Barton. Haslingfield. Cambridge. Gamlingay.—4. Madingley. Dry Drayton. Near the Observatory.—5. Chippenham. Newmarket. Wicken.

2. L. incisum Willd.

L. purpureum β, Relh. ed. 2, 231; ed. 3, 238.

Dry fields and banks. A. April to June.

3. Barringtou!; N.—6. About Witchford, Witcham, and Sutton.—8. A little below Wisbech, perhaps just out of this county.

3. L. purpúreum Linn. *Red Archangel.*

L. rubrum, R. C. 84. M. Pl. 13. *L. r. officinarum*, M. M. 51. *L. purpureum*, Relh. 238.

Waste and cultivated ground. A. May to August.

Common throughout the county.

4. L. album Linn. *White Archangel or Dead-Nettle.*

R. C. 84. M. Pl. 13. Relh. 238. *L. a. officinarum*, M. M. 51.

Hedges. P. May, June.

Common in the (1) Cambridge, (2) Royston, (3) Wimpole, and (4) Cottenham Districts.—5. Horningsey. Newmarket. Chippenham.—6. Ely. Stuntney; N.—7. Doddington.—8 Wisbech.

5. L. Galeóbdolon Crantz. *Yellow Archangel.*

L. luteum, R. C. 84. *Galeobdolon*, M. M. 51. *Galeopsis Galeobdolon*, M. Pl. 13. *Galeobdolon Galeopsis*, Relh. ed. 1, 228. *G. luteum*, Relh. ed. 3, 240.

Woods and thickets. P. May, June.

1. Dullingham. Wooded part of Devil's Ditch. Wood Ditton Park Wood. Deersley's Wood, Newmarket. *Hinton;* Relh.—3. Gamlingay. Kingston, Eversden, and Hardwick Woods. Hayley Wood; N. *Whitwell;* Relh.—4. Madingley Wood.

Leonúrus Linn.

‡1. L. Cardíaca Linn. *Mother-wort.*

Cardiaca, R. C. App. ii. 3; Cat. Angl. ed. 1, 54. *C. officinarum,* M. M. 53. *L. Cardiaca,* M. Pl. 14. Relh. 244.

Hedges and waste places. P. August.

1. Barnwell Gravel-pits, 1818; J. W. *Between Trumpington and Shelford;* Relh.—3. By the Grantchester footpath at Newnham, Cambridge; J. C.—5. *Newmarket road beyond the Paper Mills;* T. M.—8. *Elm;* Relh. *Tydd St Giles;* J. F.

Galeópsis Linn.

1. G. Ládanum Linn. *Ironwort.*

Ladanum segetum quorundam flore rubro, R. C. 83. M. M. 51. *G. Ladanum,* M. Pl. 13. Relh. 239. *G. villosa,* Relh. ed. 1, 227 ?

Gravelly and chalky fields. A. August, September.

Common in the (1) Cambridge, (2) Royston, (3) Wimpole, and (4) Cottenham Districts.—5. Quy !; H. Chippenham.

2. G. Tétrahit Linn. *Wild Hemp.*

Cannabis spuria, R. C. 27. *Ladanum verticillis crebrioribus flore purpureo caule fulcrato,* M. M. 51. *G. Tetrahit,* M. Pl. 13. Relh. 239.

Waste and arable land. A. July to August.

1. Balsham. Hildersham. Newmarket. Shuckburgh Castle. Six-mile-Bottom.—3 and 4. Common.—5. Chippen-

ham. Upware. Horningsey.—7. Doddington. Chatteris.—
8. Newton. Wisbech; J. B. *Elm;* J. F.

3. G. versícolor Curt.

G. Tetrahit β, M. Pl. 13. Relh. ed. 1, 228. *G. versicolor*, Relh. ed. 3, 240.

Sandy and peaty land. A. July, August.

1. Hinton Moor; S. W. W. *Entrance to Fulbourn from Hinton;* Relh.—2. Sawston Moor; G. S. G. One plant in a barley-field, Royston; D. B.—4. Waterbeach Fen. Upware Ferry. Dry Drayton; N.—5. Bottisham Fen.—6. Ely; S. W. W. Littleport!; H. Near Sandy's Cut; Stuntney; N.—7. Doddington Fen.—8. Wisbech; J. B.

STÁCHYS Linn.

1. S. Betónica Benth. *Wood Betony.*

Betonica, R. C. 21. *B. officinarum*, M. M. 50. *B. officinalis*, M. Pl. 13. Relh. 241.

Woods and thickets. P. July, August.

1. Temple, Wilbraham. Abington Park; Linton Wood; S. W. W.—3. Gamlingay. Caldecot. Eversden Wood. Kingston, Hayley, Hardwick, and White Woods; N. Malton; D. B.—4. King's Hedges. Madingley Wood.

2. S. sylvática Linn. *Hemp-nettle.*

Galeopsis legitima Dioscoridis, R. C. 59. M. M. 50. *S. sylvatica*, M. Pl. 13. Relh. 241.

Hedges. ·P. July, August.

Rather frequent throughout the county, but least so in the Fens.

LABIATÆ.

3. S. palústris Linn. *Clown's All-heal.*

Sideritis anglica strumosa radice, R. C. 155. *Galeopsis angustifolia fœtida*, M. M. 51. *S. palustris*, M. Pl. 13. Relh. 242.

Banks of rivers and ditches. P. July, August.

1. Newmarket Heath!; H. Teversham.—2. Mill-stream, Meldreth; D. B.—3. Cambridge. Barrington. Toft; Great Eversden; Comberton; N. *Grantchester;* J. F.—4. Madingley. Waterbeach. Cottenham.—5. Wicken. Ditton!; H.— 7. Doddington. Chatteris.—8. Newton. Wisbech. Near Whittlesey and Eastrey; N.

β. *S. ambigua* Sm.

3. Coton; N. *Fen Ditton;* Relh. in a letter to Sowerby, A. D. 1795.

4. S. arvénsis Linn. *Petty All-heal.*

Sideritis humilis lato obtuso folio, R. C. App. ii. 17. *Glechoma arvensis*, M. Pl. 13. *S. arvensis*, Relh. 242.

Arable land. A. August, September.

1. Linton; G. S. G. *Gains near Teversham;* Dent. *Gogmagog Hills;* Relh.—3. Near Eversden Wood; Near Hayley Wood; N.—5. Upware.

BALLÓTA Linn.

1. B. fœtida Lam. *Stinking Horehound.*

Ballote, R. C. 19. M. M. 52. *B. nigra*, M. Pl. 13. Relh. 243.

Dry hedge-banks and road-sides. P. July, August.

Common throughout the county, but only on islands in the Fens.

MARRÚBIUM Linn.

1. M. vulgáre Linn. *White Horehound.*

M. album, R. C. 95. *M. vulgare*, M. M. 52. M. Pl. 14. Relh. 243.

By road-sides. P. August, September.

1. Wort's Causeway and Wool-street. Near Mr Townley's house, Fulbourn; S. W. W.—2. By footpath from Royston to Bassingbourn; D. B.—3. Gamlingay.—4. By the road to Chesterton; W. H. C. *Hill of Health;* J. M. —6. Ely!; H.—7. Doddington.—8. Wisbech; J. B.

Teúcrium Linn.

1. **T. Scorodónia** Linn. *Wood Sage.*

Relh. ed. 1, 220; ed. 3, 232.
Woods and shady places. P. July, August.
3. White Wood, Gamlingay.—6. Ely!; H.

2. **T. Scórdium** Linn. *Water Germander.*

Scordium majus, Ger. Herb. 535. *Scordium,* R. C. 152. *S. officinarum,* M. M. 54. *T. Scordium,* M. Pl. 13. Relh. 232.

Wet places. P. July, August.

4. On the left of the Histon Road, near the first pond. King's Hedges. In ditches by the road to Ely, abundantly; H. *Waterbeach; Cottenham;* Relh.—6. Roswell Pits, Ely. Mepal!; H.

[Ray, or rather Dent, mentions *T. Chamædrys*(?) under the name of *Chamædrys vulgo vera existimata* (Cat. App. ii. 5) as growing "in a little island nigh the road to Quey Water, on the right hand from Cambridge." There is every reason to consider this as a mistake.]

A'juga Linn.

1. **A. réptans** Linn. *Bugle.*

Bugula, R. C. 24. M. M. 54. *A. reptans,* M. Pl. 13. Relh. 231.

Damp shady places. P. May, June.

Common in the (1) Cambridge, (2) Royston, and (3) Wimpole Districts.—4. Madingley Wood. Chesterton.—5. Bottisham Fen.—8. Wisbech; J. B.

2. A. Chamæpitys Schreb. *Ground Pine.*

Chamæpitys vulgaris, R. C. 32. *C. officinarum*, M. M. 54. *Teucrium Chamæpitys*, Relh. 231.

Chalky fields. A. May to July.

2. Odsey; H. F. and A. M. B. *On the layers about the borders of Triplow Heath;* Ray.

VERBENACEÆ.

Verbéna Linn.

1. V. officinális Linn. *Vervain.*

V. vulgaris, R. C. 174. *V. officinarum*, M. M. 49. *V. officinalis*, M. Pl. 1. Relh. 233.

Waste ground. P. July, August.

1. Hinton. Cow Fen, Cambridge. Shudy Camps. Hildersham. Newmarket.—2. Triplow. Newton. Guilden Morden; Near Ashwell; Foxton; N. Odsey; A. M. B. By the drift-way, Royston; D. B.—3. Comberton. Haslingfield. Barrington. Toft. Kingston. Barton. Cambridge. Orwell; Harlton; Eltisley; Croxton; N. — 4. Histon. Waterbeach. Graveley; Elsworth; T. Y.—5. Chippenham. Newmarket. Reche. Wicken. Swaffham Bulbeck. Upware. Swaffham Prior!; H.—6. Barraway.—8. Wisbech.

LENTIBULARIACEÆ.

Pinguícula Linn.

1. P. vulgáris Linn. *Butterwort.*

Pinguicula vel Liparis, R. C. 119. *P. Gesneri*, M. M. 75. *P. vulgaris*, M. Pl. 1. Relh. 11.

Bogs and fens. P. May, June.

1. Shelford Common (as late as 1852); J. C. *Teversham; Hinton Moor;* J. M. *Fulbourn;* Relh.—2. Peat-holes, Triplow. Melbourn Common. Sawston Moor; S.W.W. —3. Gamlingay (recently).—5. Bottisham Fen.—6. *Fens near Ely;* Relh.: but W. M. has never seen it.

UTRICULÁRIA Linn.

1. U. vulgáris Linn. *Bladder-wort. Hooded Water Milfoil.*

Millefolium palustre galericulatum, R. C. 99. *Lentibularia*, M. M. 76. *U. vulgaris*, M. Pl. 1. Relh. 12.

Deep ditches and pits. P. June to August.

1. *Paper Mills;* Ray. *Teversham;* J. M.—2. Peat-holes, Triplow.—4. Pits near the Observatory. Waterbeach Fen. Over; N.—5. Baitsbite; H. Wicken and Bottisham Fens. —6. Ely. Downham; H. Aldreth. Near Sandy's Cut; West Moor Fen; N.—8. Wisbech; J. B.

2. U. mínor Linn.

Millefolium palustre galericulatum minus flore minore, R. C. App. ii. 13. *Lentibularia minor*, M. M. 76. *U. minor*, M. Pl. 1. Relh. 12.

Deep ditches and pits. P. June to August.

1. *Teversham;* Dent. *Fulbourn; Hinton;* Relh.— 2. *Sawston Moor;* Relh.—3. Bogs on Gamlingay Heath (now lost).—5. Bottisham Fen. *Chippenham;* Relh.— 6. *Fordham;* Rev. J. Hemsted.

PRIMULACEÆ.

Prímula Linn.

1. P. vulgáris Huds. *Primrose.*

P. veris vulgaris, R. C. 126. *P. v. officinarum*, M. M. 71. *P. vulgaris*, M. Pl. 4. Relh. 84.

Woods and shady places. P. April, May.

1. Hinton. Stetchworth.—2. Triplow. Steeple Morden; D. B. Shepreth; N. W.—3. St John's College Walks, Cambridge. Hardwick. Harston; Haslingfield; Croxton; N. Gamlingay Wood.—4. Madingley Wood. Papworth St Everard; N.

β. *cauléscens.*

P. veris β, M. Pl. 4. *P. inodora*, Relh. ed. 1, 81. *P. elatior*, Relh. ed. 3, 84.

Occasionally at the edges of woods.

2. P. véris Linn. *Paigle. Cowslip.*

P. veris major, R. C. 125. M. M. 71. *P. veris*, M. Pl. 4. Relh. 85.

Pastures. P. April, May.

Common in the (1) Cambridge, (2) Royston, (3) Wimpole, and (4) Cottenham Districts.—5. Chippenham.—8. Wisbech; A. P.

3. P. elátior Jacq. *Ox-lip.*

P. veris elatior pallido flore, R. C. 126. M. M. 71.

Woods on clay. P. April, May.

1. Wood Ditton Park Wood. Wood near Bartlow; H. Yenhall Wood, West Wratting. Balsham, Borley, and Westley Woods.—3. Eversden, Kingston, and Hardwick Woods. Hayley Wood. Between Long Stow and Bourn!; N.—4. Knapwell Wood!; J. C. *Madingley Wood;* Ray.

PRIMULACEÆ. 189

Hottónia Linn.

1. **H. palústris** Linn.   *Water-violet.*

Millefolium aquaticum dictum Viola aquatica, R. C. 98. *Myriophyllon*, M. M. 72. *H. palustris*, M. Pl. 4. Relh. 86.
Ponds and ditches.   P. May, June.
Common throughout the county, especially in the Fens.

Lysimáchia Linn.

1. **L. vulgáris** Linn.   *Common Yellow Loose-strife.*

L. lutea, R. C. 92. *L. officinarum*, M. M. 71. *L. vulgaris*, M. Pl. 4. Relh. 86.

By rivers and ponds.   P. July.

1. Quy Bridge; S. W. W. Copse by the footway to the old mill, Fulbourn; W. H. C. *Paper Mills;* Ray. *Hinton;* J. M.—2. Peat-holes, Triplow. Sawston Moor; G. S. G.—3. Harlton; N. *Grantchester;* J. M.—4. Pits near the Observatory. Waterbeach; J. W. *Rampton;* Relh.—5. Wicken Fen. Fen Ditton; S. W. W.—6. Ely. Near Sandy's Cut; N.—7. Doddington Fen.—8. Wisbech; J. B.

2. **L. Nummulária** Linn.   *Golden Money-wort. Herb Twopence.*

Nummularia, R. C. 103. M. M. 71. *L. Nummularia*, M. Pl. 5. Relh. 87.

Damp places.   P. June, July.

1. Wood Ditton Park Wood. Dullingham. Balsham. Stourbridge-Fair Green; W. H. C. Abington!; H. Trumpington; J. W.—3. St John's College Walks. Sheep's-Green, Cambridge. By Barton road, Haslingfield. Malton. Barrington. Hauxton. Kingston Wood. Harlton; Eversden; Caldecot; Toft; Bourn; Caxton; Croxton; N. Grantchester fields; W. H. C.—4. Girton road. Oakington. Girton. Long

Stanton. Chesterton. Histon road. Waterbeach. Mare Way. Cottenham Fen. Fen Drayton. Papworth; N.—5. Horningsey. Wicken Fen. Quarry at Upware.—6. Ely. Witchford.—7. Chatteris.—8. Elm; A. P. Wisbech; J. B. Whittlesey; N.

3. L. némorum Linn.

Relh. ed. 1, Suppl. 3, 1; ed. 3, 87.

Woods. P. June to August.

1. *Hall Wood, Wood Ditton;* Relh. That place is no longer a wood.

ANAGÁLLIS Linn.

1. A. arvénsis Linn. *Red Pimpernel.*

A. mas, R. C. 11. *A. terrestris mas officinarum,* M. M. 71. *A. arvensis,* M. Pl. 5. Relh. 87.

Arable land. A. June, July.

Generally distributed throughout the county.

β. *cœrulea. Blue Pimpernel.*

A. fœmina, R. C. 11. *A. terrestris fœmina,* M. M. 71. *A. cœrulea,* Relh. 88.

It is doubtful if the true *A. cœrulea* (Sm.) has been found in this county. Most, if not all, those gathered were blue-flowered *A. arvensis.*

1. Linton; R. B. S.—3. Fox-hole's-down Farm, Barrington. N.—4. By Madingley road!; C. Ingram.—5. Exning; E. S.—6. By Roswell Pits, Ely.

2. A. tenélla Linn. *Bog Pimpernel. Purple Money-wort.*

Nummularia minor flore purpurascente, R. C. 104. M. M. 72. *Lysimachia tenella,* M. Pl. 5. *A. tenella,* Relh. 88.

Spongy bogs. P. July, August.

1. On the Common at one mile on London road, Cambridge!; H. Sawston Moor; S. W. W. Shelford Common; Fulbourn Moor; W. H. C. Ditch by Hinton road; J. W. *Trumpington, Hinton, and Teversham Moors;* Relh.— 2. Peat-holes, Triplow; N. Sawston and Ickleton; G. S. G. —3. Gamlingay. Croydon; N.—5. Bottisham Fen.— 7. Doddington Turf-fen.

Centúnculus Linn.

1. **C. mínimus** Linn. *Bastard Pimpernel.*

Relh. ed. 3, 64.
Boggy places. A. June, July.
3. *Gamlingay Bogs;* Relh.

Glaux Linn.

1. **G. marítima** Linn. *Black Salt-wort.*

Lyons, 25. M. Pl. 5. Relh. 102.
Salt marshes.
8. Foul Anchour, on both sides of the river.

Sámolus Linn.

1. **S. Valerándi** Linn. *Brook-weed.*

Anagallis aquatica rotundifolia, R. C. 11. *Samolus Valerandi,* M. M. 72. M. Pl. 5. Relh. 97.

Damp, watery places. P. July, August.
1. Coldham's Common, and on the Common at one mile on London road, Cambridge. Hinton; S. W. W. *Teversham;* J. M.—2. Peat-holes, Triplow. Sawston Moor; G. S. G.—3. Gamlingay. Comberton; Kingston!; Bourn; Gravel-pit between Caxton and Eltisley; N. Sheeps' Green, Cambridge; W. H. C. *Trumpington;* J. M.—4. Pits near the Observatory. Waterbeach Fen. *Histon road;* J. M.—5. Burwell Fen; N. Bottisham; Swaffham Prior!; H.

Swaffham Bulbeck. Wicken Fen. By water in Chippenham Park.—6. Stretham!; Rev. H. Baber. Near Sandy's Cut; N.—7. Doddington Fen.—8. Wisbech. Four Gouts.

PLUMBAGINACEÆ.

Státice Linn.

1. S. Limónium Linn. *Sea Lavender.*

Relh. ed. 2, 126; ed. 3, 131.
First found by Mr Skrimshire.
Muddy salt-marshes. P. July to September.
8. Wisbech!; H. and A. P.

2. S. cáspia Willd.

S. reticulata, Eng. Bot. fol. 328. Relh. 131.
Muddy sea-shores. P. July, August.
8. *Tydd Marsh;* Mr Skrimshire (Bot. Guide, 50). *Below Wisbech;* Rev. J. Hemsted, who sent the specimen from thence, which is figured in Eng. Botany. By the riverside, recently; A. P.

Arméria Willd.

1. A. marítima Willd.

Statice Armeria, Lyons, 29. M. Pl. 7. Relh. 131.
First found by Prof. J. Martin.
Muddy sea-shores. P. April to September.
8. River-side below Wisbech!; H. and A. P.

PLANTAGINACEÆ.

Plantágo Linn.

1. P. Corónopus Linn. *Buckshorn Plantain.*

Coronopus, R. C. 39. *P. foliis laciniatis Coronopus dicta*, M. M. 19. *P. coronopifolia*, Relh. ed. 1, 62. *P. Coronopus*, M. Pl. 3. Relh. 63.

Dry, gravelly places, and by the sea. A. June, July.

3. Gamlingay.—4. *Hill of Health;* J. M.—8. River-side, Wisbech. Foul Anchour.

2. P. marítima Linn.

Relh. ed. 1, Suppl. ii. 9; ed. 3, 63.
By the sea. P. June to September.

8. On both sides of the river at Foul Anchour.

3. P. lanceoláta Linn. *Ribwort.*

P. quinquenervia, R. C. 121. *P. angustifolia, officinarum,* M. M. 19. *P. lanceolata,* M. Pl. 5. Relh. 62. *P. succisa,* Lyons, 20.

Pastures and waste ground. P. June, July.
Common throughout the county.

P. succisa of Lyons is a state in which the spike is very short, and there is much woolly hair at the crown of the root. He found it on the Gogmagog Hills and Hill of Health.

4. P. média Linn. *Lamb's-tongue.*

P. incana, R. C. 120. M. M. 19. M. Pl. 3. *P. media,* Relh. 62.

Dry meadows and pastures. P. June to September.
Common in the (1) Cambridge, (2) Royston, (3) Wimpole, and (4) Cottenham Districts.—5. Horningsey. Upware. Chippenham. Exning.—6. Haddenham. Witchford. Sutton. Ely.—8. Wisbech; A. P.

5. P. major Linn. *Way-bread.*

P. latifolia vulgaris, R. C. 120. M. M. 19. *P. major,* M. Pl. 3. Relh. 62. *P. major panicula sparsa,* R. C. 120. M. M. 19.

Way-sides and waste places. P. June to August.

Not unfrequent throughout the county. The panicled state is accidental.

LITTORÉLLA Linn.

1. L. lacústris Linn.

Holosteum minimum palustre capitulis longissimis filamentis donatis, R. Cat. Angl. ed. 1, 169. *Plantago uniflora*, Lyons, 21. *L. lacustris*, Relh. 390.

Margins of ponds. P. June, July.

1. *Hinton Moor, plentifully;* Ray.—3. *Gamlingay bogs;* Relh.

AMARANTHACEÆ.

AMARÁNTHUS Linn.

[1. A. Blitum Linn.

Blitum rubrum minus, R. C. 23. M. M. 18. *A. Blitum*, Relh. ed. 1, 360; ed. 3, 392.

Waste places near towns. A. August.

1. *In the outskirts of Cambridge;* Relh. An escape from gardens.]

CHENOPODIACEÆ.

SUÆDA Forsk.

1. S. marítima Dun.

Chenopodium maritimum, Lyons, 27. M. Pl. 6. Relh. 106.

First found by Prof. J. Martyn.

Sea-shore. A. July to September.

8. River-side below Wisbech.

CHENOPODIACEÆ.

CHENOPÓDIUM Linn.

1. C. ólidum Curt. *Stinking Orache.*

Atriplex olida, R. C. 17. *C. fœtidum*, M. M. 17. *C. Vulvaria*, M. Pl. 6. *C. olidum*, Relh. 106.

Dry waste places near houses. A. August, September.

1. Under the old Abbey wall, Barnwell; By Jesus College, Under wall by Cow Fen, In Russell Street, Cambridge. Parker's Piece, Cambridge; H.—3. Mount Pleasant, Cambridge.

2. C. polyspérmum Linn.

Blitum album minus, R. C. 22. *C. polyspermum*, Lyons, 26. M. Pl. 6. Relh. 106.

Damp waste places and recently cut woods. A. August, September.

3. Comberton!; Toft; In a field opposite to the south lodge at Croxton; N. *Kingston Wood after it had been newly felled; Gamlingay, by road-side;* Ray.—4. Near the river, Milton; Near the spring, Comberton; Cottenham; Rampton; Relh.—5. Upware.—6. Stretham Ferry; At the junction of Witchford and Cambridge road at Ely; H. By Ely bridge. Stuntney; N.—7. March!; H.

3. C. úrbicum Linn.

Lyons, 25. Relh. 104. *C. u. β. intermedium*, Bab. Man. ed. 4, 276.

Rich waste ground and dunghills. A. August.

1. *Barnwell;* Relh. *In Hinton;* Lyons.—3. *Coton;* Relh.—4. *Cottenham;* Relh.

4. C. álbum Linn. *Fat-hen.*

Atriplex sylvestris, R. C. 17. *C. folio sinuato candicante*, M. M. 17. *C. album*, M. Pl. 6. Relh. 105.

Waste and cultivated land. A. July, August.
Common throughout the county.
β. *C. viride* Linn.
4. Landbeach.—8. Wisbech.—And probably elsewhere.

5. C. ficifolium Sm.

C. serotinum, Lyons, 26. M. Pl. 6. Relh. ed. 1, 104. *G. ficifolium,* Relh. ed. 3, 105..

First noticed by Prof. J. Martyn.

Arable and waste ground. A. August, September.

1. Near Hinxton; N.—2. Near Royston; N.—3. Cambridge. Toft; Eversden!; Gamlingay; N.—4. Near Upware Ferry. By the river below Waterbeach, plentifully. Chesterton. By Arbury Banks.—5. Wicken. Clayhythe. Reche. Swaffham Bulbeck. Upware.—6. West Fen, and other places about Ely. Barraway. Witchford. Stuntney; By Sandy's Cut; N.—7. Between Chatteris and Carter's Bridge. Doddington.—8. Wisbech.

6. C. murale Linn. *Sowbane.*

Atriplex sylvestris latifolia, R. C. 18. *C. pes anserinus* I, M. M. 17. *C. murale,* Relh. 104.

Rich waste ground and dunghills. A. August.

4. By Chesterton road!; Mr A. G. More.—Also, probably, elsewhere. It appears to have been more common formerly, for Ray and Relhan do not mention any exact localities for it.

7. C. hybridum Linn.

Chenopodium Stramonii folio, Dill. in R. Syn. ed. 3, 154. *C. hybridum,* M. Pl. 6. Relh. 105.

Damp places. A. August.

1. *Near Maids' Causeway, Cambridge, in* 1817, Relh.— 6. Common about Ely. Haddenham; N.

8. C. rubrum Linn.

Atr. sylvestris latifolia altera, R. C. 18. *C. pes anserinus* II, M. M. 17. *C. rubrum*, M. Pl. 6. Relh. 104.

Rich waste ground and dunghills. A. August, September.

1. Sawston; G. S. G.—3. Lane between the Madingley road and St John's Farm, in 1852. By the brook, Barton road, in 1853. Clayhythe. Coton.—4. Waterbeach Fen.—5. Horningsey. Wicken. Swaffham Bulbeck.—6. Ely!; H.—7. Chatteris.—8. By the river at Wisbech.

†9. C. Bonus-Henricus Linn. *Allgood.*

Bonus-Henricus, R. C. 23. *C. folio triangulo*, M. M. 17. *C. Bonus-Henricus*, M. Pl. 6. Relh. 103.

Waste places near houses. P. May to August.

1. Churchyard, Fulbourn; N. *Teversham;* J. M.—2. Guilden Morden; N. Royston; H. F. Odsey; A. M. B.—3. Harston. Toft; Orwell; Harlton; Caxton; Barrington; Eltisley; Gamlingay; N.—4. Between Impington church and Histon. By the river below Waterbeach. Swavesey. Long Stanton; N. *Madingley;* J. M.—5. Newmarket. Fen Ditton. Upware.—6. Sutton.

BÉTA Linn.

1. B. maritima Linn. *Sea Beet.*

Relh. ed. 2, 103; ed. 3, 107.

Near the sea. P. July to September.

8. *Below Wisbech;* Mr Skrimshire.

SALICÓRNIA Linn.

1. S. herbácea Linn. *Glasswort.*

Relh. ed. 1, Suppl. ii. 1; ed. 3, 2.

Muddy sea-shores. A. August, September.

8. By the river below Wisbech. On both sides of the river at Foul Anchour.

A′TRIPLEX Linn.

1. A. littoralis Linn.

A. patula β, Relh. ed. 1, 379. *A. littoralis*, Relh. 416.
Salt marshes. A. July to September.
8. By the river below Wisbech. Tydd Marsh.

2. A. angustifolia Sm.

A. sylvestris angustifolia, R. C. 17. M. M. 17.
A. patula, M. Pl. 23. Relh. ed. 1, 379. *A. patula β*, Relh. ed. 2, 396. *A. angustifolia*, Relh. ed. 3, 416.

Cultivated and waste ground. A. July to October.
1. Cambridge. Stetchworth. Balsham. Hildersham.—2. Near Ashwell; N.—3. Cambridge. Barton. Grantchester. Kingston. Long Stow; Gamlingay; Croxton; Caxton; Toft; N.—4. Cambridge. Dry Drayton. Madingley. Waterbeach Fen. Elsworth; N.—5. Chippenham. Wicken.—6. Barraway.—7. Doddington. Chatteris.—8. Newton. Wisbech. Foul Anchour.

3. A. erécta Huds.

Lyons, 56. M. Pl. 23.
First found by Prof. J. Martyn.

Waste gravelly places and arable land. A. July to October.
1. Fields between Stetchworth and Wood Ditton. Hildersham.—2. Steeple Morden; Near Ashwell; N.—3. Near Hayley Wood; Harlton; Great Eversden; Bourn; Croxton; Toft; Barrington; N.—4. Waterbeach Fen. Elsworth; N.—6. Stuntney; N.—7. Between Chatteris and

CHENOPODIACEÆ. 199

Carter's Bridge.—8. In fields and by the river-side at Wisbech.

4. A. deltoïdea Bab.

First noticed in this county in 1853 by C. C. B.
Waste and arable land. A. June to October.
1. Near Hinxton; N.—3. Haslingfield!; Toft!; N.—4. Cottenham. Mare Way. Fen Drayton.—5. Horningsey. Upware.—6. Witchford.—7. Common about Chatteris and Doddington.—8. In fields and by the river-side at Wisbech.

5. A. hastata Linn.

A. sylvestris altera, R. C. 18. *A. s. folio hastato seu deltoïde*, M. M. 17. *A. hastata*, M. Pl. 23. Relh. ed. 1, 379. *A. patula*, Relh. ed. 3, 415.

Waste and cultivated land. A. June to October.
1. Cambridge. Wilbraham.—3. Cambridge. Wimpole. Croydon. Bourn; N.—4. Mare Way. Cottenham. Waterbeach. Elsworth; N.—5. Upware. Reche. Swaffham Bulbeck.—6. Stretham!; H.—7. Doddington. Chatteris.

6. A. Babingtónii Woods.

A. hastata β, M. Pl. 23 and 43. Relh. ed. 1, 379. (*A. maritima nostras*, R. Cat. Angl. ed. 1, 35.)

Sea-shore. A. July to October.
First noticed by Prof. J. Martyn.
8. By the river-side both above and below Wisbech. On both sides of the river at Foul Anchour.
It is probable that Ray's plant was the same as ours.

OBIÓNE Gaert.

1. O. pedunculáta Moq.

Atriplex pedunculata, Relh. ed. 2, 327; ed. 3, 416.
Salt marshes. A. August, September.

8. *By the river-side a little below Wisbech, and in the salt marshes;* Mr Skrimshire.

2. O. portulacoïdes Moq.

A. portulacoïdes, Lyons, 56. M. Pl. 23. Relh. 415.
First found by Prof. J. Martyn.
Salt marshes. P. August to October.

8. By the river below Wisbech. On both sides of the river at Foul Anchour.

POLYGONACEÆ.

Rúmex Linn.

1. R. marítimus Linn. *Golden Dock.*

Lapathum folio acuto flore aureo, R. C. App. ii. 11. M. M. 14. *R. maritimus*, M. Pl. 9. Relh. ed. 3, 149. *R. aureus*, Relh. ed. 1, 147.
Wet places. B. July, August.

1. Jesus Piece and other places about Cambridge; H. *Hinton Moor;* Relh.—3. Eltisley; N. *Behind Grantchester Mill;* Relh.—5. Cowbridge, Swaffham; H.—6. Ely; Stretham!; H. Barraway. *North Fen, Ely;* Dent.—8. Wisbech; J. M.

2. R. palústris Sm.

Lapathum aquaticum sive Hydrolapathum minus, R. C. 85. M. M. 14. *R. palustris*, Relh. 149.
Wet places. B. July to September.

1. Near St Peter's College garden, Cow Fen, Cambridge. *Hinton;* Relh.—4. By the river above Upware. Clayhythe. Bottisham!; H.—6. Ely. Barraway. Aldreth. By Sandy's Cut; N.—7. By Vermuden's Drain, near Chatteris.—8. Four Gouts. Horse-shoe Corner, and above the town by the river, Wisbech.

3. R. conglomerátus Murr.

Lapathum acutum, R. C. 84. *L. a. officinarum*, M. M. 13. *R. acutum*, M. Pl. 8. Relh. 148.

Waste ground, banks. P. June to August.

Common throughout the county, but least so in the Fens.

4. R. sanguíneus Linn.

Relh. ed. 1, Suppl. 1, 12; ed. 3, 147.

Woods and banks. P. June to August.

Common in the (1) Cambridge, (2) Royston, (3) Wimpole, and (4) Cottenham Districts.—5. Horningsey. Chippenham. Snailwell. Swaffham Bulbeck.—6. Stuntney; N.

It is the *R. viridis* (Sibth.) that is common. Mr Newbould finds the typical *R. sanguineus* in Eversden and Kingston Woods.

5. R. púlcher Linn. *Fiddle Dock.*

Lapathum pulchrum Bononiense sinuatum, R. C. App. ii. 11. M. M. 14. *R. pulcher*, M. Pl. 8. Relh. 148.

Dry waste ground, especially near villages. P. July to September.

1. Fulbourn. Hildersham. Trumpington. *In the courts of Trinity and Caius Colleges;* Dent.—2. Triplow. Hinxton; N.—3. Cambridge. Gamlingay. Arrington. Kingston. Haslingfield. Harlton; Barton; Comberton; Toft; Croydon; N. — 4. Long Stanton. Swavesey. Landbeach. Mount Pleasant and Castle Hill, Cambridge. Oakington; N.— 5. Fen Ditton. Chippenham. Snailwell. Swaffham Bulbeck. Eastwards from Newmarket.—6. Witchford. Haddenham. —8. Wisbech; A. P.

6. R. obtusifólius Linn.

Lapathum sylvestre folio minus acuto, R. C. 85. M. M. 14. *R. obtusifolius*, M. Pl. 8. Relh. 148.

Waste ground. P. July to September.

Probably common throughout the county, although but few localities are recorded in the Fens.

7. R. praténsis M. and K.

First noticed by Mr Newbould in 1848.

1. Near Hinxton; N.—2. Royston; N.—3. Comberton. Eversden; Toft; Kingston!; Crane's Lane; Bourn; Eltisley; Gamlingay; N.—4. Near the Observatory. By the Railway bridge, Chesterton. By the river above Upware. Wash of the old Ouse river below Aldreth Bridge, plentifully. Childerley; Elsworth; N.—5. By the river at Upware.—6. Near Roswell Pits, Ely.

8. R. crispus Linn.

Lapathum folio acuto crispo, R. C. App. ii. 11; Cat. Angl. ed. 1, 187. *R. crispus*, M. Pl. 8. Relh. 147.

Waste and cultivated ground. P. June to August.
Common throughout the county.

9. R. Hydrolápathum Huds. *Great Water Dock.*

Lapathum aquaticum I, *sive aquaticum folio cubitali*, R. C. 85. M. M. 14. *R. Britannica*, M. Pl. 9. *R. aquaticus*, Relh. ed. 3, 149. *R. Hydrolapathum*, Relh. ed. 1, 491.

Banks of streams and ditches. P. July, August.

1. Cow Fen, Cambridge.—2. Peat-holes, Triplow. Riverside, Ickleton. Between Whittlesford and Shelford.—3. Sheep's-Green, Cambridge. Harston; N.—4. Chesterton. Waterbeach. Swavesey. Fen Drayton. Aldreth Bridge.—5. Fen Ditton. Wicken Fen.—6. Ely. Aldreth. Witchford. By Sandy's Cut; N.—7. Doddington Turf-fen. Chatteris.—8. Newton. Wisbech.

10. R. Acetósa Linn. *Common Sorrel.*

Acetosa vulgaris, R. C. 3. *Lapathum acetosum vulgare*, M. M. 14. *R. Acetosa*, M. Pl. 9. Relh. 150.

Meadows and pastures. P. May, June.

1. Cambridge. — 2. Ickleton. Whittlesford. Foxton; Gatwell End, Steeple Morden; N. Royston; H. F.— 3. Cambridge. Gamlingay. Croydon. Toft; Eversden; Comberton; Harlton; Bourn; Caxton; N.—4. Histon. Girton. Long Stanton. Elsworth; N.—5. Horningsey. Snailwell. Newmarket; N.—8. Newton. Wisbech.

11. R. Acetosélla Linn. *Sheep's Sorrel.*

Acetosa arvensis lanceolata, R. C. 3. *A. a. officinarum*, M. M. 12. *R. Acetosella*, M. Pl. 9. Relh. 150.

Dry gravelly ground. P. May to July.

1. Cambridge. Stetchworth. Furze-hills, Hildersham. Shuckburgh Castle, Newmarket Heath. Allington Hill. Six-mile-Bottom. — 2. Odsey; A. M. B.— 3. Cambridge. Gamlingay. Eltisley. Toft; N.—4. Swavesey. *Hill of Health;* J. M.—5. Chippenham. Snailwell. Wicken. Stetchworth; N.

Polýgonum Linn.

1. P. Bistórta Linn. *Snakeweed.*

Bistorta major, R. C. 22. *B. officinarum*, M. M. 15. *P. Bistorta*, M. Pl. 9. Relh. 163.

Moist meadows. P. June and October.

1. Fulbourn!; H. *Shelford;* J. M. *Hinton;* Relh.— 2. Copse near Railway-station at Harston; N.—3. Whitwell; S. W. W.—4. *Closes near Howe's House;* Relh. *Hill of Health;* J. F.—5. Swaffham Prior; H.

2. P. amphibium Linn.

Potamogeton angustifolium, R. C. 124. *Persicaria Salicis folio Potamogeton angustifolium dictum*, M. M. 15. *P. amphibium*, M. Pl. 9. Relh. 161.

Rivers, ditches, and bogs. P. July, August.

1. Paper Mills, Cambridge.—2. Near Duxford; N. Peat-holes, Triplow.—3. Gamlingay. Sheep's-Green, Cambridge. Coton. Between Caxton and Eltisley; N.—4. By Madingley road. Waterbeach Fen. Aldreth Bridge. Fen Drayton. Oakington; N.—5. Bottisham Fen. Upware. Reche.—6. Ely. Witcham.—7. Doddington Turf-fen.—8. Wisbech; J. B. Near Peterborough.

3. P. lapathifólium Linn.

P. pensylvanicum, Lyons, 33. Relh. ed. 1, 33. *P. lapathifolium*, Relh. ed. 3, 162.

On rubbish and damp cultivated land. A. July, August.

Tolerably common throughout the county in both its forms, viz. *P. pallidum* (With.) and *P. nodosum* (Pers.), but more especially the former.

4. P. láxum Reichenb.

Suppl. to Eng. Bot. fol. 2822.

First found in 1836 by C. C. B.

Damp gravelly places. A. July to September.

1. Stourbridge Fair Green.—4. River-side below Waterbeach, plentifully.—5. Wicken.—6. River-side below Barraway.—8. Newton. Wisbech.

5. P. Persicária Linn.

Persicaria mitis maculosa, R. C. 116. *Per. maculata officinarum*, M. M. 15. *P. Persicaria*, M. Pl. 9. Relh. 161.

On rubbish and damp places. A. June to September.

Tolerably common throughout the county.

6. **P. míte** Schrank.

Suppl. to Eng. Bot. fol. 2867.
First noticed in 1836 by C. C. B.
Wet places. A. August, September.

1. River-side at Barnwell.—4. River-side below Waterbeach, plentiful.—5. Upware.—7. Doddington.—8. Wisbech.

7. **P. Hydropíper** Linn. *Water Pepper.*

Persicaria urens, R. C. 117. *P. non maculata officinarum*, M. M. 15. *P. Hydropiper*, M. Pl. 9. Relh. 162.
Wet places. A. August, September.

1. Cambridge.—3. Cambridge. Old Mill-dam, Grantchester. Gamlingay. Toft; N.—4. Gravel Hill near the Observatory.—5. Snailwell. Wicken. Reche.

9. **P. minus** Huds.

Persicaria pusilla repens, R. C. 116. M. M. 15. *P. minus*, M. Pl. 9. Relh. 163.
Wet gravelly places. A. August, September.

4. At the end of Waterbeach Lode, next to the town; Ray.

10. **P. aviculáre** Linn. *Knot-grass.*

P. mas vulgare, R. C. 122. *P. officinarum*, M. M. 15. *P. aviculare*, M. Pl. 9. Relh. 163.
Waste and cultivated gravelly ground. A. May to October.
Common throughout the county.

Mr Newbould believes that there are two quite distinct species confounded under this name, which are about equally abundant in this county. Their times of flowering differ slightly, as do the forms of their nuts and colour of their perianths.

11. **P. Convólvulus** Linn. *Black Bindweed.*

Convolvulus niger, R. C. 38. *Fegopyrum scandens sylvestre*, M. M. 14. *P. Convolvulus*, M. Pl. 9. Relh. 164.

Cultivated and waste land. A. July to September. Common throughout the county.

Fagopýrum Gaert.

[1. **F. esculéntum** Moench. *Buckwheat. Brank.*

Fegopyrum, R. C. 53. M. M. 14. *Polygonum Fagopyrum*, M. Pl. 9. Relh. ed. 1, Suppl. iii. 3; ed. 3, 164.

Cultivated, and remaining for a few years afterwards. A. July, August.

No certain station can be mentioned, and it is now rarely found.]

THYMELACEÆ.

Dáphne Linn.

1. **D. Lauréola** Linn. *Spurge Laurel. Lowry.*

Laureola, R. C. 85. *L. officinarum*, M. M. 121. *D. Laureola*, M. Pl. 9. Relh. 160.

Woods and thickets. Sh. February to April.

1. Hinton. Gramham's Camp, Shelford. Linton. Fulbourn. *Teversham;* J. M. *Hildersham;* Relh.—3. Hardwick. Gamlingay Wood. Wimpole. Coton. Harlton; Long Stow. Croxton; Kingston Wood; N.—4. Madingley. Between Moor Barns and Huntingdon road; W. H. C.— 6. Near the south end of Long Drove, Witchford.—8. Wisbech; J. B.

SANTALACEÆ.

Thésium Linn.

1. **T. humifúsum** Cand. *Bastard Toadflax.*

Linaria adulterina, R. C. 88. *Knawel montanum calycè specioso lacteo*, M. M. 19. *T. linophyllum*, M. Pl. 5. Relh. 102.

Chalky places. P. June, July.

1. Gogmagog Hills. Devil's Ditch. Allington Hill. Furze Hills, Hildersham. Sawston!; N. *In a pit between Hinton and Fulbourn;* J. M. *Teversham;* Relh.—2. Newton. To the east of Chrishall Grange. Ickleton; G. S. G. Royston; D. B. Odsey; A. M. B. *Triplow;* J. M.— 5. Devil's Ditch. Snailwell Heath. Bottisham; N. Exning; E. S. *Chippenham;* Relh.

ARISTOLOCHIACEÆ.

Aristolóchia Linn.

*1. **A. Clematítis** Linn. *Birthwort.*

R. C. App. ii. 2. M. Pl. 20. Relh. 368. *A. C. officinarum,* M. M. 74.

Naturalized near ruins. P. July, August.

1. *In several hedges at Whittlesford;* Dent.—4. *Milton;* Relh.

EUPHORBIACEÆ.

Euphórbia Linn.

1. **E. Helioscópia** Linn. *Sun Spurge.*

Tithymalus helioscopius, R. C. 163. M. M. 70. *E. Helioscopia,* M. Pl. 11. Relh. 191.

Waste and cultivated ground. A. June to September. Thinly spread throughout the county.

2. **E. platyphýlla** Koch.

Tithymalus segetum longifolius, R. C. App. ii. 17. M. M. 70. *E. segetalis,* M. Pl. 11. *E. verrucosa,* Relh. ed. 1, 186. *E. platyphylla,* Relh. ed. 3, 101.

Arable land. A. June to August.

3. Barton; Toft; Caldecot!; Between Harlton and Comberton!; Long Stow!; Near Hayley Wood; Eltisley; N. Coton and Whitwell!; H. Near Kingston Wood. *Near Gransden Lodge; To the north of Eversden Wood;* Relh.—4. Near the Observatory; Childerley; Elsworth; N.

3. E. amygdaloides Linn. *Wood Spurge.*

Tith. Characias amygdaloïdes, R. C. 153. M. M. 70. *E. amygdaloides,* M. Pl. 11. Relh. 192.

Woods and thickets. P. March, April.

1. Wood Ditton Park Wood. Devil's Ditch, near Camois Hall. *Cheveley Park;* Ray. *Catlidge;* Relh.—3. Gamlingay.

4. E. Péplus Linn. *Petty Spurge.*

Peplus sive Esula rotunda, R. C. App. i. 7. *Tith. parvus annuus foliis subrotundis non crenatis Peplus dictus,* M. M. 70. *E. Peplus,* M. Pl. 11. Relh. 190.

Cultivated land. A. July, August.

1. Cambridge.—2. Steeple Morden; Foxton; N. Odsey; A. M. B.—3. Comberton. Cambridge. Gamlingay. Haslingfield; Toft; Eversden; Barton; Croxton; Barrington; N.—4. Histon. Oakington; Elsworth; N.—5. Swaffham Bulbeck. Chippenham.—6. Ely.—8. Wisbech.

5. E. exígua Linn.

Esula exigua, R. C. 50. *Tithymalus leptophyllos,* M. M. 70. *E. exigua,* M. Pl. 11. Relh. 190.

Cultivated land. A. June to August.

1. Stetchworth. Newmarket. Hildersham. Balsham. Shudy Camps. Fulbourn. Hinton. Brinkley.—2. Near Royston; Ashwell; Foxton; N. Odsey; A. M. B.—3. Coton. Grantchester. Barton. Arrington. Haslingfield. Kingston. Eversden; Harlton; Comberton; Caldecot;

Caxton; Toft; Eltisley; Croxton; N.—4. Madingley. Dry Drayton. Both the Papworths; Graveley; Elsworth; N.—5. Chippenham. Horningsey. Upware. Newmarket. Bottisham; H.

MERCURIÁLIS Linn.

1. **M. perénnis** Linn. *Dog's Mercury.*

Cynocrambe mas et fœmina, R. C. 43. *M. perennis repens Cynocrambe dicta,* M. M. 11. *M. perennis,* M. Pl. 22. Relh. 410.

Woods and thickets. P. April, May.

1. Back lane, Hinton. Wood Ditton. Dullingham. Deersley's Wood, Newmarket. Burrough-Green. Brinkley. Shudy Camps.—3. Cambridge. Kingston and Hardwick Woods. Croydon. Arrington. Gamlingay. Hayley Wood; Caxton; Eversden; Wimpole; Eltisley; N.—4. Madingley Wood.

2. **M. ánnua** Linn. *French Mercury.*

M. mas et fœmina, R. C. App. ii. 12. M. M. 11. *M. annua,* M. Pl. 22. Relh. 410.

Cultivated ground. A. August, September.

6. Ely.—Dent and J. Martyn say that this plant was frequent as a weed. Now it is very rare.

CERATOPHYLLACEÆ.

CERATOPHÝLLUM Linn.

1. **C. démersum** Linn. *Hornwort.*

Equisetum palustre ramosum aquis immersum, R. C. 49. *Dichotophyllum,* M. M. 12. *C. demersum,* M. Pl. 21. Relh. 392.

Ponds and ditches. P. June, July.

1. Fulbourn; N.—2. Sawston Moor; G. S. G.—3. Barrington; N.—4. In the river, Waterbeach. Swavesey. Fen Drayton. Willingham; N.—5. Snailwell and Wicken Fens. Fen Ditton. Horningsey.—8. By Roman bank on the way from Four Gouts to Newton.—Possibly the plants which grow at Barrington and Newton may be *C. submersum;* but it is improbable.

CALLITRICHACEÆ.

Callítriche Linn.

1. **C. vérna** Linn. *Water Starwort.*

Stellaria aquatica, R. C. 160. M. M. 77. *C. verna* and *C. autumnalis,* M. Pl. 1. Relh. ed. 1, 1. *C. autumnalis,* Lyons, 1. *C. aquatica,* Relh. ed. 3, 4.

Stagnant and slowly running water. A. April to September.

Common throughout the county.

2. **C. platycárpa** Kütz.

Suppl. to Eng. Bot. fol. 2864.

First noticed in 1840 by C. C. B.

Stagnant water, and on mud. A. May to September.

1. Dullingham.—2. Hauxton.—3. Gamlingay. Eltisley. Grantchester.—4. Cuckoo Lane, Histon. Mare Way. Near Aldreth Bridge. Fen Drayton. Swavesey.—5. Bottisham Fen. East Brook at Chippenham.—6. Roswell Pits, Ely. Aldreth Causeway. Witchford.

URTICACEÆ.

Parietária Linn.

1. **P. erécta** Koch. *Wall Pellitory.*

First noticed in 1856 by C. C. B.

Old walls. P. July to September.

1. Burrough-Green.—4. Churchyard wall, Swavesey.—6. Churchyard wall, Sutton.

2. **P. diffúsa** Koch. *Wall Pellitory.*

Parietaria, R. C. 113. *P. officinarum,* M. M. 18. *P. officinalis,* M. Pl. 23. Relh. 65.—Probably all these include *P. erecta.*

Old walls. P. July to September.

1. Great Wilbraham; N. Churchyard, Trumpington; W. H. C. — 2. Ickleton church. Litlington church. — 3. Back gate of Clare College.—4. Chesterton churchyard; W. H. C. Willingham church.—5. Chippenham. Exning. Fen Ditton churchyard wall.—6. Ely. Sutton churchyard wall. Haddenham.

Urtíca Linn.

†1. **U. pilulifera** Linn. *Roman Nettle.*

Ann. Nat. Hist. i. 196.

About towns. A. June to August.

8. Wisbech!; Rev. Dr Jermyn. Upwell (possibly not in this county); Rev. L. Jenyns. Both of these were the *U. Dodartii* Linn.

2. **U. úrens** Linn.

U. minor, R. C. 180. M. M. 11. *U. urens,* M. Pl. 21. Relh. 391.

Cultivated and waste land. A. June to September.

Found throughout the county, but not very common.

3. **U. dioïca** Linn. *Common Nettle.*

U. vulgaris urens, R. C. 180. M. M. 11. *U. dioïca,* M. Pl. 21. Relh. 391.

Waste ground, hedge-banks, &c. P. June to September. Common throughout the county.

Húmulus Linn.

1. H. Lúpulus Linn. *Hop.*

Lupulus, R. C. 91. *L. officinarum*, M. M. 10. *H. Lupulus*, M. Pl. 22. Relh. 407.

Hedges and thickets. P. July.

Common in the (1) Cambridge, (2) Royston. (3) Wimpole, and (4) Cottenham, Districts.—5. Horningsey. Fen Ditton. Chippenham.—8. Wisbech; J. B.

ULMACEÆ.

U'lmus Linn.

1. U. suberósa Ehrh. *Common Elm.*

U. vulgatissimus folio lato scabro, R. C. 178. M. M. 124. *U. campestris*, M. Pl. 6. Relh. 107.

Woods and hedges. T. March to May.

1. Cambridge. Stetchworth. Linton. Shudy Camps. West Wratting.—3. Cambridge. Wimpole. Eversden. Toft. Kingston. Croydon.—4. King's Hedges.—5. Chippenham. Bottisham!; H.—7. Doddington.

β. *U. glábra* Sm.

U. folio glabro, R. C. 178. M. M. 124. *U. glabra*, Relh. ed. 1, 107. *U. montana* β, *glabra*, Relh. ed. 3, 108.

5. Bottisham!; H.

2. U. montana With. *Wych Elm.*

U. folio latissimo scabro, R. C. 178. M. M. 124. *U. glabra* β, M. Pl. 6. *U. montana*, Rclh. 108.

Woods and hedges. T. March, April.

1. *By the road-side, within half a mile of Bartlow, and elsewhere;* Ray.—5. Bottisham; H.—It is doubtful if the true plant has been found in the county.

AMENTIFERÆ.

Sálix Linn.

1. S. frágilis Linn. *Crack Willow.*

S. folio lato splendente fragilis, R. C. 143. M. M. 116.
S. fragilis, M. Pl. 22. Relh. 402.
 Damp meadows. T. April, May.
 2. Whittlesford.—3. Gamlingay.
β. *S. Russelliána* Sm.
 1. *Near the first mile-stone on the London road;* Relh.
—3. *Gamlingay;* Relh. By Hardwick brook above Coton.
—5. Chippenham. Bottisham Fen.
 Smith states that *S. decipiens* (a variety of *S. fragilis*) is cultivated in this county.

2. S. álba Linn. *White Willow.*

S. folio utrinque glauco viminibus albidioribus, R. C. 142.
S. officinarum, M. M. 115. *S. alba*, M. Pl. 22. Relh. 406.
 Wet places. T. May.
 2. Whittlesford.—3. Grantchester.—4. Fen Drayton. *Madingley Road;* Relh.—5. Chippenham.
β. *S. vitellína* Sm.
S. folio utrinque glauco viminibus rubris, R. C. 142. M. M. 115.
 3. *Queens' Green;* Relh.—8. Newton.

‡3. S. unduláta Ehrh.

S. lanceolata, Relh. 403.
 Sides of ditches. T. April, May.
 4. *Chesterton;* Relh.—6. *Aldreth Causeway;* Relh.

4. S. triándra Linn.

Relh. ed. 1, Suppl. iii. 7; ed. 3, 401.
Wet woods and banks of ditches. T. April, May.

1. *Cow Fen!*; H.—2. *Melbourn Common.*—3. *Behind Grantchester Mill;* Relh.—7. *Doddington.*

β. *S. Hoffmanniána* Sm.

1. *Cow Fen!*; H. *By the footway to Ditton*, Rev. J. Holme (Sm. Eng. Fl. iv. 168).—2. *Whittlesford.*—5. *Chippenham.*

γ. *S. amygdalína* Sm.

S. folio splendente auriculato flexilis, R. C. 144. *S. amygdalina*, Relh. 401.

1. *Near Cambridge;* Rev. J. Holme (l. c.).—4. *Mare Way.*—6. *Ely.*

5. S. purpúrea Linn.

S. folio longo et folio et vimine subluteo non auriculato, and *S. f. l. non auriculato vimine rubro*, R. C. 146. M. M. 116. (Ray combined these in his Synopsis; T. Martyn placed them under *S. purpurea*, but Smith believed them to be *S. vitellina*). *S. purpurea*, M. Pl. 22. Relh. 402.

Marshes and river-banks. T. March, April.

6. *In osier-holts, and by the river Cam's side;* Ray.

β. *S. Hélix* Linn. *Rose Willow.*

S. humilior foliis angustis, subcœruleis, ut plurimum sibi invicem oppositis, R. C. 144. *S. Helix*, M. Pl. 22. Relh. 402.

1. *Cambridge!*; H. *By the horse way-side to Hinton, in the close just by the water which you pass over to go thither* [that is, Coldham's Lane]; Ray.

6. S. rúbra Huds.

Relh. ed. 1, Suppl. iii. 8; ed. 3, 403.
Low meadows. T. April, May.

1. Coldham's Lane, Cambridge. *By the river-side at Jesus-Green, Cambridge;* Relh.—5. *Bullen Close, Fen Ditton;* Relh.—6. Ely.

β. *S. Forbiána* Sm.

S. fissa, Relh. ed. 2, 385. *S. Forbyana,* Relh. ed. 3, 403.

3. Toft!; Bourn; Caldecot; N.—6. *Prickwillow, near Ely;* Rev. J. Hemsted.

7. S. viminális Linn. *Osier.*

S. folio longissimo, R. C. 146. M. M. 116. *S. viminalis,* M. Pl. 22. Relh. 405.

Wet places. T. April, May.

2. Melbourn Common.—3. Kingston. Comberton; Toft!; N.—4. Pit to the south-west of Madingley Park. Mare Way.

8. S. acumináta Sm.

Relh. 405.

Wet places. Sh. or T. April.

1. *Hinton;* Relh.—3. *Hauxton;* Relh.—4. *Chesterton;* Relh.—5. Wicken Fen.

9. S. cinérea Linn. *Sallow.*

Relh. ed. 2, 387.

Wet places. T. or Sh. March, April.

1. Wilbraham Fen.—3. Toft; Comberton!; N.—5. Bottisham and Wicken Fens. Chippenham.—8. Newton.

β. *S. aquática* Sm.

S. folio ex rotundo acuminato auriculata, R. C. 145. *S. aquatica,* Relh. 404.

1. West Wratting. Fulbourn.—3. Kingston.—7. Doddington.

γ. *S. oleifólia* Sm.

1. Shudy Camps. Balsham.—8. Newton.

10. S. aurita Linn.

S. caprea β, Relh. ed. 2, 387; ed. 3, 404.
Damp woods. Sh. April, May.

1. Coldham's Lane, Cambridge. Fulbourn. *Wood Ditton; Teversham; Babraham; West Wratting; Weston Colville Wood;* Relh.—2. Whittlesford.—3. Gamlingay.—*Grantchester Lane, Cambridge; Eversden;* Relh.—4. *Chesterton;* Relh.—5. Gravel-pit, Chippenham.

11. S. capréa Linn.

S. folio subrotundo auriculata, R. C. 145. M. M. 116. *S. caprea*, M. Pl. 22. Relh. 405.

Woods and hedges. T. April, May.

1. Hinton. Fulbourn. Stetchworth. West Wratting.—2. Triplow. Melbourn Common. Near Little Shelford.—3. Gamlingay. Coton. Eversden. Croydon. Kingston Wood. Toft; Croxton; Whitwell; Hayley and Hardwick Woods; N. *Hauxton;* Relh.—4. Madingley. Cuckoo Lane, Histon. Pit to the south-west of Madingley Park. Fen Drayton. Childerley; N. *Chesterton;* Relh.—5. Bottisham and Wicken Fens.—7. Doddington.—8. Newton.

12. S. répens Linn.

S. humilis repens, and *S. h. r. rotundifolia*, R. C. App. ii. 16. M. M. 116. *S. repens*, M. Pl. 22. Relh. 403.

Boggy heaths. Sh. March, April.

2. Peat-holes, Triplow. — 3. Gamlingay Heath. — 5. Wicken Fen.

Pópulus Linn.

1. P. álba Linn. *White Poplar.* *Abele.*

R. C. 122. M. M. 115. Relh. 408.
Damp woods and hedges. T. April.

1. By the footpath to Hinton near the brook. *Many great trees at Teversham;* Ray.—3. Gamlingay.—4. *Many trees at Chesterton;* Ray.—8. Wisbech; J. B.

†2. **P. canéscens** Sm. *Gray Poplar.*

Relh. 409.

Damp woods and hedges. T. April.

1. Cambridge. *Fulbourn;* Relh. — 2. *Shelford;* Relh. —3. Harlton; N.—4. Huntingdon road.

3. **P. trémula** Linn. *Aspen.*

P. Libyca, R. C. 123. *P. tremula*, M. M. 115. M. Pl. 22. Relh. 409.

Woods. T. March, April.

1. *Teversham;* Ray.—2. Triplow.—3. Gamlingay. By Grantchester footpath. Eversden Wood. Harlton; N.—, 4. Histon road. *Chesterton;* Ray.—7. Doddington Wood.

4. **P. nígra** Linn. *Black Poplar.*

R. C. 123. M. Pl. 22. Relh. 409. *P. nigra officinarum*, M. M. 115.

Wet land and by water. T. March.

1. Fulbourn; S. W. W.—4. Histon. Waterbeach Fen. King's Hedges.—5. Fen Ditton. In the Fens near the river Cam. Upware. Horningsey.—6. *In the meadows about Fordham and Soham, plentifully;* Ray.—8. Wisbech; J. B.

MYRÍCA Linn.

1. **M. Gále** Linn. *Sweet Gale.*

Elæagnus Cordi, R. C. 47. *Gale frutex odoratus Septentrionalium*, M. M. 115. *M. Gale*, M. Pl. 22. Relh. 407.

Wet fens. Sh. May.

6. North Fen, Ely, until recently; W. M. *Abundant in many parts of the Isle of Ely;* Ray. This plant is now destroyed by drainage and cultivation.

Bétula Linn.

1. **B. glutinósa** Fries. *Common Birch.*

Betula, R. C. 21. *B. officinarum*, M. M. 113. *B. alba*, M. Pl. 21. Relh. 390.

Heathy woods. T. April, May.

1. Newmarket Heath.—3. Gamlingay.—5. Chippenham.

[*B. alba* (Linn.) is plentiful, but apparently planted, on Newmarket Heath.]

Álnus Tourn.

1. **A. glutinósa** Gaert. *Alder.*

Alnus, R. C. 7. M. M. 114. *Betula Alnus*, M. Pl. 21. Relh. 390.

Wet places and by water. T. March.

1. Fulbourn. *Wilbraham;* Ray. *Linton Wood;* Relh.—2. Whittlesford.—3. Gamlingay. Cambridge. Wimpole; N. —5. Chippenham. Snailwell.

Fágus Linn.

‡1. **F. sylvática** Linn. *Beech.*

Relh. 395.

Woods, especially on chalk. T. March, April.

1. Planted on the Gogmagog Hills; Newmarket Heath; and elsewhere.—3. May be a native in White Wood, Gamlingay.—4. Planted by the Madingley road.—5. Planted about Chippenham.

AMENTIFERÆ.

Quércus Linn.

1. Q. Róbur Linn. *Oak.*

Q. latifolia, R. C. 129. *Q. officinarum*, M. M. 115. *Q. Robur*, M. Pl. 22. Relh. 395.

Woods. T. April, May.

1. Wood Ditton. Devil's Ditch, near Camois Hall.—2. Whittlesford.—3. Eversden and Kingston Woods. Eltisley. Caxton; N.—4. King's Hedges. Madingley Wood.—Planted in many places.

Neither the *Q. intermedia* nor *Q. sessiliflora* have been found in this county.

Córylus Linn.

1. C. Avellána Linn. *Hazel.*

Corylus sylvestris, R. C. 39. *C. officinarum*, M. M. 114. *C. Avellana*, M. Pl. 22. Relh. 396.

Woods and hedges. Sh. March, April.

1. Devil's Ditch, near Stetchworth. Shudy Camps. Balsham. Long Pasture, Hildersham.—2. Triplow. Odsey; A. M. B.—3. Cambridge. Gamlingay. Plentiful in the western part of the District.—4. Madingley. King's Hedges.—5. Near Newmarket; N.—6. Witcham.—7. Doddington Wood.

Cárpinus Linn.

1. C. Bétulus Linn. *Hornbeam.*

Betulus, R. C. 22. *Carpinus*, M. M. 114. *C. Betulus*, M. Pl. 22. Relh. 396.

Damp woods and hedges on clay. T. May.

1. Balsham Wood. Linton Wood; S. W. W. *In hedges on the Cambridge side of Abington;* Ray.—3. *Grantchester;* Relh.—4. *Moor Barns Thicket;* Relh.

CONIFERÆ.

Táxus Linn.

* 1. T. baccáta Linn. *Yew.*

Taxus, R. C. 161. M. M. 117. *T. baccata*, M. Pl. 23. Relh. 412.

Hedges and churchyards, probably planted. T. March, April.

1. Barnwell Abbey.—3. Eversden. Caldecot. Comberton.—4. Madingley.

Juníperus Linn.

1. J. commúnis Linn. *Juniper.*

Juniperus, R. C. 82. *J. officinarum*, M. M. 117. *J. communis*, M. Pl. 23. Relh. 411.

On chalky hills. Sh. May.

1. Roman road, near Hildersham. The Rivey, Linton. Westhoe, near Linton. In the park, Gogmagogs; W. H. C. *Juniper Hill, Hildersham* (which was so named from the abundance of the plant); Ray.

MONOCOTYLÉDONES OR ENDÓGENÆ.

TRILLIACEÆ.

Páris Linn.

1. **P. quadrifólia** Linn. *Herb Paris. Herb Truelove. One-berry.*

Herba Paris, R. C. 74.   M. M. 62.   *P. quadrifolia*, M. Pl. 9.   Relh. 165.

Damp woods. P. May.

1. Wood Ditton Park Wood. Linton Wood. Hall Wood, West Wratting. Balsham Wood; R. B. S.— 3. Eversden, Gamlingay, and Kingston Woods. Whitwell Wood, near Coton. Hayley Wood; N.

DIOSCORIACEÆ.

Támus Linn.

1. **T. commúnis** Linn. *Black Bryony.*

Bryonia nigra, R. C. 23.   *Tamarum vulgo*, M. M. 61. *T. communis*, M. Pl. 22.   Relh. 408.

Hedges and thickets. P. May, June.

Common in the (1) Cambridge, (2) Royston, (3) Wimpole, and (4) Cottenham Districts.—5. Landwade. Wicken. Horningsey. Quy road.—6. Ely. Sutton. Witchford. Witcham.—8. Wisbech; J. B.

HYDROCHARIDACEÆ.

Hydrócharis Linn.

1. H. Morsus-ránæ Linn. *Frogbit.*

Morsus ranæ, R. C. 100. *Stratiotes foliis Asari semine rotundo*, M. M. 77. *H. Morsus-ranæ*, M. Pl. 23. Relh. 411.

Ponds and ditches. P. July, August.

1. Cow Fen.—3. *Grantchester;* J. M.—4. Chesterton. Milton. Waterbeach. Histon road. Aldreth Bridge.—5. Bottisham and Wicken Fens. Horningsey. Fen Ditton.—6. Ely.—8. Wisbech; J. B. Whittlesey; N.

Stratiótes Linn.

1. S. aloïdes Linn. *Water Soldier.*

Militaris aizoïdes, R. C. 98. *Stratiotes sive Militaris aizoïdes*, M. M. 77. *S. aloïdes*, M. Pl. 12. Relh. 218.

Fen ditches. P. July.

4. *In the ditch on the left-hand side as you go to Stretham Ferry;* Ray. Over; N.—5. In the fen near the village of Wicken.—6. Roswell Pits, Ely. By the railway, Thetford. One mile above Ely, between the river and the road; H. Between Delph drove and the Old Ouse, near Earith; N. By Aldreth Causeway. *Littleport;* Relh. *Mepal;* J. M.—7. Welney; A. P.—8. Whittlesey; J. B.

Anácharis Rich.

* 1. **A. Alsinástrum** Bab. *Water Thyme.*

Bot. Gaz. iii. 135.

Rivers and ditches. P. July to September.

1, 3, 4, 5 and 6. In the river Cam, from Sheep's Green down to the boundary of the county below Littleport. It was introduced into the Botanic Garden in 1847, and escaped in 1848 into the Vicar's Brook and the river. Hinton Brook.—4. By the Ouse, near Swavesey. Has come from the Ouse down the Old river by Aldreth Bridge. Pits at Impington.—6. and 7. In the Bedford rivers.—8. In the Nene at and below Peterborough.

See an interesting account of the introduction and spread of this plant, by Mr W. Marshall, in the Phytologist, iv. 705—715, which is also published separately by Pamplin, under the title of *The New Water Weed*.

ORCHIDACEÆ.

O'RCHIS Linn.

1. O. Mório Linn. *Green-winged Orchis.*

O. Morio fœmina, R. C. 106. M. M. 101. *O. Morio*, M. Pl. 20. Relh. 358.

Pastures. P. May, June.

1. Devil's Ditch. Linton. Shelford. Fulbourn. Furzehills, Hildersham; W. H. C. West Wratting. Weston Colville. Dullingham.—2. Odsey; A. M. B.— 3. Coton. Gamlingay. Croydon. Hardwick. Comberton; Toft; Bourn Park; Orwell; N.—4. Madingley. Moor Barns. King's Hedges. *Chesterton and Girton;* J. M.—5. Closes at Horningsey.—6. Ely; W. M.—8. Wisbech; J. B.

2. O. máscula Linn.

O. Morio mas foliis maculatis, R. C. 106. M. M. 101. *O. mascula*, M. Pl. 20. Relh. 359.

Woods and pastures. P. May.

1. Stetchworth. Devil's Ditch. Yenhall Wood, and Hall Wood, West Wratting. Westley Wood. *Hildersham;* J. M.—3. Hardwick. Whitwell, near Coton. Eversden,

Gamlingay, and Kingston Woods. Comberton. Haslingfield. Trumpington Spinney; W. H. C. Caldecot; East Hatley; N.—4. Madingley Wood.—8. Wisbech; J. B.

3. O. ustuláta Linn.

O. sive Cynosorchis minor Pannonica, R. C. 108. *O. Pannonica*, M. M. 102. *O. ustulata*, M. Pl. 20. Relh. 359.

Open chalky pastures. P. June.

1. Gogmagog Hills. Old chalk-pits at Hinton. Devil's Ditch. Fleam Dyke. Between Stapleford and Babraham; S. W. W. Near Hildersham; Linton; G. S. G.—2. Ickleton; G. S. G.—5. Devil's Ditch; E. S. *Chippenham;* Relh.

4. O. maculáta Linn.

O. palmata fœmina, sive Palma Christi fœmina maculato folio, R. C. 107. *O. p. speciosiore thyrso folio maculato*, M. M. 101. *O. maculata*, M. Pl. 20. Relh. 360.

Damp woods and pastures. P. May, June.

1. Old chalk-pits, Hinton. Fulbourn. Wood Ditton. Shelford chalk-pit; Trumpington; W. H. C. *Hildersham;* J. F.—2. Plantation by Royston Heath; D. B. Odsey; A. M. B.—3. Common in the woods and pastures in the western part.—4. Madingley. Moor Barns. King's Hedges. Dry Drayton; N. Elsworth; T. Y. *Chesterton;* J. M.—5. Anglesey Abbey; Bottisham; H.—6. Ely; W. M.—8. Wisbech; J. B.

5. O. latifólia Linn. *Marsh Orchis.*

O. palmata major mas, sive Palma Christi mas, R. C. 106. *O. p. non maculata*, M. M. 101. *O. latifolia*, M. Pl. 20. Relh. 360. (These synonyms include *O. incarnata.*)

Marshes and damp meadows. P. June.

2. Peat-holes, Triplow.—3. Eltisley. Orwell. Wimpole.

6. O. incarnáta Linn.

First noticed by me in the year 1833, but not then distinguished from *O. latifolia*.

Fens and marshes. P. June.

1. Linton. Dernford Fen; S. W. W.—2. Peat-holes, Triplow. Near the railway, Sawston Moor!; N.—3. Old pits near the Observatory; N.—5. Bottisham and Wicken Fens.

7. O. pyramidális Linn.

O. sive Cynosorchis purpurea spica congesta pyramidali, R. C. 108. *O. purpurea spica congesta pyramidali*, M. M. 102. *O. pyramidalis*, M. Pl. 20. Relh. 358.

Chalky pastures and banks. P. July.

1. Old chalk-pit, Hinton. Chalk-pit, Babraham; Shelford chalk-pit; S. W. W. Devil's Ditch. Abington. Balsham; R. B. S. Fulbourn; N. *Linton;* Relh.—2. Ickleton; G. S. G. Odsey and Steeple Morden plantations; H. F. —3. Between Barrington and Orwell; Bourn; Longstow; Caldecot; Mare Way; N. By Eversden Wood; W. H. C. —4. By the wood and outside the park, Madingley. Dry Drayton; Childerley; N. *Chesterton;* J. M.—5. Chippenham. Bottisham and Swaffham Bulbeck; H.—8. Wisbech; J. B.

GYMNADÉNIA R. Br.

1. G. conopséa R. Br. *Red-handed Orchis.*

O. palmata rubella cum longis calcaribus rubellis, R. C. 107. M. M. 102. *O. conopsea*, M. Pl. 20. Relh. 361.

Chalky pastures. P. July.

1. Old chalk-pits, Hinton. Devil's Ditch. Fulbourn. Chalk-pit near Shelford; W. H. C. Balsham Heath and Fleam Dyke; R. B. S.—2. Triplow!; Near the railway, Sawston!; 'N.—3. Comberton; Mr A. G. More. Long Stow; Bourn; N.—4. King's Hedges; S. W. W. *Chesterton;* J. M. —5. Anglesey Abbey; H. Devil's Ditch!; Rev. J. Downes.

ORCHIDACEÆ.

A′ceras R. Br.

1. A. anthropóphora R. Br.

Ophrys anthropophora, R. ed. 1, 338; ed. 3, 364.
Dry chalky places, P. June.
1. Linton. Barrington Hill (until recently). Furzehills, Hildersham. Devil's Ditch; Westley; H. Babraham; Mr Josh. Clark.—3. Old Quarry, Haslingfield. Wimpole; N.

Habenária R. Br.

1. H. víridis R. Br. *Frog Orchis.*

Orchis palmata flore viridi, R. C. 107. *O. p. f. luteoviridi,* M. M. 102. *Satyrium viride,* M. Pl. 20. Relh. 361.
Damp pastures. P. June, July.
1. Linton. Burrough-Green. Fulbourn. Abington Park; S. W. W. Devil's Ditch!; H. *Pit near Chalk-pit-close, Hinton;* J. M. *Wood Ditton; Catlidge;* Relh.—3. Between Bourn and Long Stow; N.—4. King's Hedges. Chesterton!; H.—5. Baitsbite. Horningsey Fen. Anglesey Abbey!; H.

2. H. chloránthsa Bab. *Butterfly Orchis.*

·*Orchis Serapias bifolia vel trifolia minor,* R. C. 109. M. M. 102. *O. bifolia,* M. Pl. 20. Relh. 357.
Clayey woods. P. May, June.
1. Linton, Wood Ditton Park and Westley Woods. Hall Wood, West Wratting. *Barrington Hill;* Relh.—3. Whitwell, near Coton. Gamlingay, Eversden, and Kingston Woods. Gayne's Copse, Comberton; S. W. W. Barton!; H. Caldecot; Harlton; N.—4. Plantations by road a little beyond the Observatory. In the Wood and round the Park, Madingley. Dry Drayton; N.

O'PHRYS Linn.

1. O. apífera Huds. *Humble-bee Orchis.*

Orchis sphegodes sive fucum referens, R. C. 109. *Or. fuciflora galea et alis purpurascentibus*, M. M. 102. *Op. apifera*, M. Pl. 20. Relh. 365.

Chalky pastures and pits. P. June, July.

1. Between Little Wilbraham and Fulbourn. Lately in the Avenue and fields at Brooklands, Cambridge; S. W. W. Old chalk-pits, Hinton. In the plantations by the road crossing the railway, Trumpington. Stetchworth. Thickets near the brook at Fulbourn. Westhoe Park near Linton; W. H. C. Furze-hills, Hildersham; Rev. Dr Cookson. *Burrough-Green;* Ray. *Abington;* J. F. *Teversham;* Relh. — 2. Triplow!; Near the railway, Sawston; N. Odsey; H. F. Shepreth; N. W. *Shelford;* J. F.—3. Comberton; Harlton; Kingston!; Barrington; Haslingfield; Great Eversden; Orwell; N. Trumpington Spinney; W. H. C. Bourn; Rev. E. A. Powell. Long Stow; Rev. J. Rushton. —4. Copse at further end of Moor Barns Thicket; Honey Hill, Childerley; Oakington; S. W. W. Madingley Field!; Rev. F. France. Dry Drayton; N.—5. Park and an adjoining close, Bottisham; S. W. W. Anglesey Abbey; H. *Chippenham;* Relh.

2. O. aranífera Huds. *Spider Orchis.*

Orchis sive Testiculus Sphegodes hirsuto flore, R. C. App. i. 7, M. M. 102. *Op. apifera β*, M. Pl. 20. *Op. aranifera*, Relh. 366.

Chalky and gravelly places. P. April, May.

1. On a balk towards the Roman road, Abington. Dry open chalky banks near Hildersham, sparingly; G. S. G. but it has not been recently found. *In a bushy close by the foot-way between the two Abingtons;* T. M. *Near Bartlow;*

Relh.—2. *In an old gravel-pit near Shelford by the foot-way from Trumpington to the Church, we found hundreds of them,* A.D. 1660—63; Ray. This pit is now obliterated by the plough.

3. O. muscífera Huds. *Fly Orchis.*

O. myodes, R. C. 106. M. M. 101. *O. muscifera*, M. Pl. 20. Relh. 365.

Damp chalky thickets and pastures. P. May, June.

1. In a field by the southern end of the Devil's Ditch at Stetchworth. Hall Wood, West Wratting. Linton. Thickets near the brook at Fulbourn. *Hinton; Chippenham; J. F. Teversham;* Relh.—2. Plantation by railway station, Harston; S. W. W. Shelford; H. Steeple Morden, D. B.—3. Harlton.—5. Swaffham Prior. Plantation by road near Biggin Abbey, which was grubbed up in 1859. Bottisham!; H. *Chippenham:* Relh.

HERMÍNIUM R. Br.

1. H. Monórchis R. Br. *Musk Orchis.*

Orchis pusilla odorata, R. C. App. i. 7. *O. odorata moschata sive Monorchis*, M. M. 102. *Ophrys Monorchis*, M. Pl. 20. Relh. 364.

Old grassy chalk-pits. P. June, July.

1. In a close near Linton!; H. *Chalk-pit-close, Hinton; Old pits on the Gogmagog Hills;* Ray. *Old pits above Hinton; Old gravel-pit near Mr Keene's house at Westhoe;* Relh.

SPIRÁNTHES Rich.

1. S. autumnalis Rich. *Lady's Tresses.*

Orchis spiralis alba odorata, R. C. 109. M. M. 102. *Ophrys spiralis*, M. Pl. 20. Relh. 363.

Dry, chalky, and gravelly places. P. August, September.

1. Coldham's Common; S. W. W. Shelford Common!; H. Linton; Sawston; G. S. G. *By Hinton Moor;* J. M. *By Teversham Moor towards the Gogmagogs;* Ray.—2. Sawston Moor; N. *Five miles from Cambridge on the road to Barkway;* Relh.—4. *Hill of Health; In a field to the left of the road from Histon to the Ely road;* Relh.—5. *Moor near Snailwell; Newmarket Heath;* Relh.—8. Wisbech!; H. and J. B. Roman banks near Newton; W. M.

Lístera R. Br.

1. **L. ováta** R. Br. *Tway-blade.*

Ophrys, R. C. 105. M. M. 103. *Ophrys ovata*, M. Pl. 20. Relh. 362.

Woods, copses, and shady places. P. May, June.

Common in the (1) Cambridge, (2) Royston, (3) Wimpole, and (4) Cottenham Districts.—5. Bottisham!; H. Chippenham. Fen Ditton.—8. Wisbech; J. B.

Neóttia Linn.

1. **N. Nidus-ávis** Linn. *Bird's-nest.*

Ophrys Nidus-avis, Relh. ed. 1, 336; ed. 3, 362.

Dense woods. P. June.

1. Linton Wood!; S. W. W. Wood Ditton Park Wood. Balsham Wood!; R. B. S.—4. *Madingley Wood;* Relh.

Epipáctis Rich.

1. **E. média** Fries.

Helleborine latifolia montana, R. C. 73. M. M. 103. *Serapias latifolia*, M. Pl. 20. Relh. 367.

Woods. P. August.

1. Abington and Linton Woods; S. W. W.—3. Eversden Wood; W. H. C. Kingston Wood (This is rather *E. purpurata* than *E. media* when transplanted into the rectory garden; N.).—4. *Madingley Wood;* Relh.—5. Round Wood by Chippenham Avenue.

I have not seen the *E. latifolia* in this county, and suspect that all the stations recorded for it belong to *E. media*. Ray's station is *Kingston Wood*, which belongs to *E. media*. Relhan's are also *Linton and Madingley Woods*, from whence I have not seen specimens. Nor have I seen the plant of Messrs. Wanton and Coleman's stations, which they named *E. latifolia* at a time when *E. media* was not distinguished from it.

2. E. palústris Sw.

Helleborine angustifolia, R. C. 72. *H. palustris nostras*, M. M. 103. *Serapias latifolia* γ, M. Pl. 20. *S. palustris*, Relh. 367.

Wet springy ground. P. July, August.

1. Dernford Fen; S. W. W. *Hinton and Teversham Moors;* Ray.—2. Peat-holes, Triplow; S. W. W. Near the railway in Sawston Moor!; N. Foulmire Moor; N. W.—3. *Gamlingay bogs;* Relh.—5. Bottisham Fen!; H.—6. *In many places in the Isle of Ely in great plenty;* Ray.

Cephalanthéra Rich.

1. C. grandiflóra Bab.

Helleborine minor flore albo, R. C. App. ii. 8. *H. flore albo*, M. M. 103. *Serapias longifolia*, M. Pl. 20. *S. grandiflora*, Relh. 368.

Dry chalky woods. P. June.

1. Babraham in 1859!; Mr Josh. Clarke.—2. Steeple Morden Plantations, once found; H. F.—5. In a copse between Bottisham Park and the road, in 1850; S. W. W.

Dent in his Appendix to Ray's *Catalogus* gives, "on Teversham Moor, nigh Quy Water," as a station for this plant; and Relhan quotes Ray as finding it in the Isle of Ely, in neither of which places it is at all likely to have been found. Apparently both of those localities belonged to *E. palustris*, and therefore Mr Wanton was the discoverer of *C. grandiflora* in this county.

MALÁXIS Sw.

1. M. paludósa Sw.

Bifolium palustre, R. Hist. Pl. ii. 1233; Fasc. Stirp. Brit. 3. *Ophrys minor palustris*, M. M. 103. *O. palustris*, M. Pl. 20. *O. paludosa*, Relh. ed. 1, 337; ed. 2, 351. *M. paludosa*, Relh. ed. 3, 366.

First found by Messrs F. Dale and Peter Dent in 1684.
Spongy bogs. P. July to September.

1. *Hinton Moor;* Dr Manningham (in Bot. Guide i. 64). —3. Bogs, Gamlingay, formerly very abundant, nearly extirpated in 1855.

STÚRMIA Reichenb.

1. S. Loesélii Reichenb.

Orchis lilifolius minor sabuletorum Zelandiæ et Bataviæ, R. C. 105. *Pseudo-Orchis bifolia palustris*, M. M. 103. *Ophrys lilifolia*, M. Pl. 20. Relh. ed. 1, 337. *O. Loeselii*, Relh. ed. 1, Suppl. ii. 15; ed. 3, 363.

Spongy bogs. P. June.

1. *In the watery places of Hinton and Teversham Moors;* Ray. *Fulbourn Moor; Sawston Moor;* Relh.—5. Burwell Fen, not far from Reche in 1835 and 1836, for the last time. Bottisham Fen; H.

There is a specimen from Hinton Moor in the late Mr Edw. Forster's collection, gathered probably about 1800, and Prof. Martyn found it on Teversham Moor in 1793; N.

IRIDACEÆ.

Iris Linn.

1. I. Pseud-ácorus Linn. *Yellow Flag.*

Acorus palustris, R. C. 4. *Iris lutea officinarum*, M. M. 100. *I. Pseud-acorus*, M. Pl. 2. Relh. 19.

Wet places. P. June, July.

1. Quy Bridge. Hinton. Little Linton.—2. By the way from Chesterford to Ickleton. Peat-holes, Triplow. Newton. Meldreth; D. B. Shepreth; N. W.—3. Sheep's-Green, Cambridge. By Arrington Bridge. Barrington. Kingston. The Moats, Caxton; Great Eversden; N.—4. Mare Way. Histon and Histon road. Waterbeach. Swavesey.—5. Wicken, Bottisham, and Snailwell Fens. Horningsey. Fen Ditton.—6. Ely. Aldreth. Witchford. —8. By the canal, Wisbech. Near Peterborough.

2. I. fœtidíssima Linn. *Gladdon.*

Xyris, R. C. 182. *I. sylvestris quam Xyrim vocant*, M. M. 100. *I. fœtidissima*, M. Pl. 2. Relh. 19.

Woods and thickets. P. May to July.

1. Hall Wood, West Wratting. Copse by Hogram's house at Fulbourn; S. W. W. *Teversham;* Ray. *In a little grove to the right hand of the road from Coldham's to Hinton;* J. M.—2. *Triplow;* Relh.—3. Eversden Wood!; S. W. W.—4. Near Oakington Church, and by the road-side towards Histon; W. H. C.—5. Burwell; H.

[*Crocus sativus* (Linn.). *Crocus*, R. C. 41. *C. officinarum*, M. M. 99. *C. sativus*, M. Pl. 2. Relh. 18. *Saffron.*

It was formerly cultivated, especially about Hinton. It is not a native, and is now never found even naturalized.]

AMARYLLIDACEÆ.

NARCÍSSUS Linn.

1. N. Pseudo-narcíssus Linn. *Daffodil.*

N. pallido-luteus longo calyce, sive sylvestris Anglicus, R. C. 103. *N. medio-luteus vulgaris,* M. M. 99. *N. Pseudo-narcissus,* M. Pl. 8. Relh. 137.

Thickets and pastures. P. March, April.

2. Whittlesford! H., seen there in 1858 by Mrs Thrupp; N. Copse at Abington Pigotts; Plantation one mile from Royston on the Newmarket road; D. B.—3. Pasture in front of the house at Whitwell, near Coton, and in bushy ground adjoining.—8. Abundant in a field at Leverington; A. P.

[*N. medio-luteus,* R. C. App. ii. 13. *N. sylvestris,* M. M. 99. *N. poeticus,* M. Pl. 8. Dent states that this was found "In Barnwell Abbey, and other places in this county." Relhan excluded it from his Flora. I know nothing of it.]

ASPARAGACEÆ.

ASPÍRAGUS Linn.

[1. A. officinális Linn.

Lyons, 29.

Hedge-banks. P. August.

1. Established in the hedge by the road-side, near to Mr Okes's house at Hinton.—2. *Great Shelford churchyard;* Lyons.—4. In the hedge on the right-hand side of the road from the Huntingdon road to Dry Drayton.

This plant has no claim to be considered as a native of Cambridgeshire.]

ASPARAGACEÆ.

Convallária Linn.

†1. C. majális Linn. *Lily of the valley.*

Lyons, 30, M. Pl. 8. Relh. 140.
First found by Prof. J. Martyn.
Woods. P. May.
3. White Wood, Gamlingay.

Rúscus Linn.

‡1. R. aculeátus Linn. *Butcher's Broom. Kneeholm.*

Ruscus, R. C. App. i. 8. *R. officinarum*, M. M. 62. *R. aculeatus*, M. Pl. 23. Relh. 412.

Thickets. Sh. March, April.

4. Near the church, Madingley; W. H. C.—5. About Anglesey Abbey!; H.

LILIACEÆ.

Fritillária Linn.

1. F. Meleágris Linn. *Fritillary. Snake's-head.*

Relh. ed. 1, 137; ed. 3, 139.
Pastures. P. May.
1. *In some closes at Westhoe near Linton;* Relh.

Ornithógalum Linn.

‡1. O. umbellátum Linn. *Star of Bethlehem.*

Relh. ed. 1, Suppl. ii. 11; ed. 3, 139.
Orchards and pastures. P. May.

1. In Mr Hicks's grounds at Wilbraham; S. W. W. Shuckburgh Castle, Newmarket Heath. *Fulbourn;* Relh. —3. Copse to the east of the church at Little Eversden; Field between the church and brook at Toft; N.—4. Chesterton. Swaffham Bulbeck!; H.—8. Wisbech; J. B.

[**O. nútans** (Linn.)

By the north-east corner of Jesus College Grove, Cambridge; In a field opposite to Mr Townley's house at Fulbourn; S. W. W. Abundant for one season (1821) between Cambridge and Trumpington, but did not occur again!; H. Not a native of this county.]

2. **O. pyrenáicum** Linn. *Spiked Star of Bethlehem.*

O. angustifolium majus floribus ex albo virescentibus, R. C. App. ii. 14. M. M. 100. *O. pyrenaicum*, M. Pl. 8. Relh. 139.

Woods and thickets. P. June.

3. *In a bushy close in Little Eversden near the church,* Dent. Relhan says that the close is on the eastern side of the church, and also (MSS.) that Messrs Davies and Green saw it there in 1774. Mr Newbould cannot now (1859) find it. [It grows at Bassmead near St Neots in Bedfordshire.]

A′LLIUM Linn.

1. **A. vineále** Linn. *Crow Garlic.*

A. sylvestre, R. C. 6. *Cepa juncifolia minor purpurascens*, M. M. 101. *A. vineale*, M. Pl. 8. Relh. 138.

Waste ground and dry fields. P. July.

1. Between Hinton and Fulbourn; Mr A. G. More. Furze-Hills, Hildersham!; Rev. Dr Cookson. *Wall of Jesus College; Foot-way to Hinton, about a gravel-pit and in many other places;* Ray.—2. In a field between Royston and Bassingbourn; D. B.—3. Hill-side above the road from Coton to Barton toll-gate. Caldecot. Barrington. Toft. Kingston. Comberton; Eversden; Croydon!; N.—4. Moor Barns. In the lane near Arbury. By path to Chesterton church; W. H. C. Dry Drayton; N. *Girton;* J. M.— 5. Chippenham.

2. **A. oleráceum** Linn.

Relh. 137.

Fields. P. July, August.

1. *In a meadow at Hinton;* Relh.—3. Hill-side above the road from Coton to Barton toll-gate; S. W. W. but may not this have been *A. vineale?*

3. **A. ursínum** Linn. *Ramsons.*

A. sylvestre latifolium, R. C. 6. M. M. 101. *A. ursinum*, M. Pl. 8. Relh. 138.

Damp woods and hedges. P. May, June.

1. Wood Ditton Park Wood. *Hall Wood, Ditton;* Relh. *Christ's Piece;* T. M. *Hinton;* J. M.—3. Eversden Wood; N.—5. *At Ditton, on a little woody hillock near the riverside;* Ray.

ENDÝMION Dumort.

1. **E. nútans** Dum. *English Hyacinth. English Harebells.*

Hyacinthus anglicus, R. C. 78. *H. officinarum*, M. M. 99. *H. non-scriptus*, M. Pl. 8. Relh. ed. 1, 136. *Scilla nutans*, Relh. ed. 3, 140.

Woods and thickets. P. May.

1. Hinton. Dullingham. Wood Ditton. Linton. Bartlow. West Wratting.—2. Foxton; N. Steeple Morden; D. B.—3. Coton. Whitwell. Kingston, Eversden and Hardwick Woods. Wimpole. Comberton. Long Stow; Hayley Wood; N.—4. Madingley Wood. Moor Barns. Graveley; N.—5. Plantation (just destroyed) by road near Biggin Abbey. Dullingham.

MUSCÁRI Tourn.

‡1. **M. racemósum** Mill. *Grape Hyacinth.*

Hensl. Cat. ed. 2, 55.

Hedge-banks. P. April, May.

1. Under hedges between Hinton and the Gogmagog Hills, in several places; also between Red Cross Turnpike and the Gogmagogs, abundantly. Near Sawston (Miss Plowden); G. S. G.

This plant is abundant about Cavenham and other places in Norfolk bordering upon Cambridgeshire. It may be an escape in this county; but the only *Muscari* that I can find in cultivation is *M. botryoïdes*. I gathered *M. racemosum* near Hinton on April 11, 1828, but do not know who first found it, in that or the preceding year.

COLCHICACEÆ.

Cólchicum Linn.

1. **C. autumnále** Linn. *Meadow Saffron.*

Relh. ed. 1, Suppl. i. 12; ed. 3, 152.

Meadows. P. September, October.

1. *In a close on the south side of Mr Eaton's house at Wood Ditton;* Relh.

JUNCACEÆ.

Narthécium Huds.

1. **N. ossífragum** Huds. *Bog Asphodel.*

Asphodelus Lancastriæ, R. C. App. ii. 2. *Phalangium anglicum palustre Iridis folio*, M. M. 100. *Anthericum ossifragum*, Relh. ed. 1, 138. *N. ossifragum*, M. Pl. 8. Relh. ed. 3, 140.

Turfy bogs. P. June, July.

3. Bogs, Gamlingay.

JUNCACEÆ.

JÚNCUS Linn.

1. J. effúsus Linn. *Common Soft Rush.*

J. lævis vulgatior panicula sparsa, R. C. 81. M. M. 111. *J. effusus.* M. Pl. 8. Relh. 142.

Wet ground. P. July.

1. Stetchworth. Little Linton.—3. Gamlingay. Eversden; Croxton; N.—4. Waterbeach and Cottenham Fens. Mare Way.—5. Wicken Fen. By Swaffham Lode!; H.—7. Doddington Wood.—8. Near Whittlesey; N.

2. J. conglomerátus Linn. *Round-headed Soft Rush.*

J. lævis panicula non sparsa, R. C. 81. M. M. 111. *J. conglomeratus*, M. Pl. 8. Relh. 142.

Wet ground. P. July.

1. Balsham. Brinkley.—3. Gamlingay. Croydon. Eversden. Kingston; Near Hayley Wood; Eltisley; N.—4. Waterbeach.—5. Anglesey Abbey!; H.—6. Haddenham. —7. Doddington. Chatteris.—8. Near Whittlesey; N.

3. J. glaúcus Sibth. *Hard Rush.*

J. acutus, R. C. 81. M. M. 111. *J. inflexus*, M. Pl. 8. Relh. ed. 1, 141. *J. glaucus*, Relh. ed. 3, 142.

Damp places, by road-sides. P. July.

Common throughout the county.

4. J. obtusiflórus Ehrh.

Relh. 143.

Wet places. P. July to September.

1. Teversham; N.—2. Melbourn Common. Sawston; G. S. G. Gatwell End, Steeple Morden; N.—3. Comberton; N.—4. Pits near the Observatory. Sand-pits by the railway, Waterbeach. Waterbeach Fen.—5. Chippenham.

Snailwell. Bottisham and Wicken Fens. Reche. Swaffham Bulbeck.—6. West Fen, Ely. Sandy's Cut; Black Bank; N. — 7. Doddington Turf-fen. March; N. — 8. Near March; N.

5. J. acutiflórus Ehrh.

Gramen junceum aquaticum magis sparsa panicula, R. C. 68. *J. nemorosus folio articulato*, M. M. 112. *J. articulatus β*. M. Pl. 8. *J. acutiflorus*, Relh. 143.

Wet places. P. June to August.

1. Wood Ditton.—3. Gamlingay. Eltisley. Wimpole. Kingston. Croxton; N.—4. Graveley; N.—5. Wicken and Bottisham Fens.

6. J. lamprocárpus Ehrh.

Gramen junceum aquaticum Bauhini, R. C. 68. *J. foliis articulatis floribus umbellatis*, M. M. 112. *J. articulatus a.* M. Pl. 8. *J. lampocarpus*, Relh. 143.

Swampy ground. P. July, August.

1. Balsham.—3. By the Barton road. Comberton; Bourn; Caldecot; Eversden; Kingston; Croxton; Eltisley; Near Hayley Wood; N—5. Wicken Fen. Swaffham Bulbeck. Chippenham.—7. Doddington Turf-fen.

Probably more common than is shown by these localities.

7. J. supínus Moench.

Gramen junceum capsulis triangulis minimum, R. C. App. ii. 8. *J. minor angustifolius panicula sparsa*, M. M. 112. *J. viviparus*, Relh. ed. 1, 143. *J. uliginosus*, Relh. ed. 3, 144.

Boggy places. P. June to August.

3. Gamlingay. Croxton; N. — 4. Cottenham Fen. Graveley; N.—5. Wicken and Bottisham Fens. Chippenham Park.—6. Ely!; H.—7. Doddington Turf-fen.— 8. Newton.

8. J. squarrósus Linn. *Moss Rush. Goose-corn.*

J. acutus Cambro-Britannicus, R. C. App. ii. 11; R. Cat. Angl. ed. 2, 172. *J. montanus palustris*, M. M. 112. *J. squarrosus*, M. Pl. 8. Relh. 143.

Wet heaths and moors. P. June, July.

1. *Hinton and Teversham Moors;* Relh.—3. Gamlingay Heath.

9. J. compréssus Jacq.

Gramen junceum maritimum, R. C. 68. *J. parvus cum pericarpiis rotundis*, M. M. 111. *J. bulbosus*, M. Pl. 8. Relh. 144.

Damp places. P. June to August.

1. *Shelford Moor;* Relh.—3. Road-side between Barton and Comberton. *Gamlingay;* Relh.—5. Anglesey Abbey!; H.—8. Road-side between Wisbech and Tydd. Tydd Marsh. On the eastern side of the river at Foul Anchour.

10. J. bufónius Linn. *Toad-grass.*

Gramen junceum, R. C. 68. *J. palustris humilior erectus*, M. M. 111. *J. bufonius*, M. Pl. 8. Relh. 144.

Spots where water has stagnated. A. July, August.

1. Gravel-pits by Grantchester foot-path.—3. Gamlingay. Barton; Eversden; Comberton; Caldecot; Kingston; Bourn; Caxton; Eltisley; Hayley Wood; N.—4. Bird's Pastures, Childerley; Graveley; N.—5. Snailwell Fen. Upware. Chippenham.—7. Doddington Wood.—8. Newton. Wisbech. Tydd Marsh.

Lúzula Cand.

1. L. sylvática Bich. *Wood-rush.*

Hensl. Cat. ed. 2, 56.

First found by the Rev. J. Downes in 1832.

Shady places. P. April to June.
1. Wood Ditton !; Rev. J. Downes; H.

2. L. pilósa Willd.

Juncus pilosus, Lyons, 31. M. Pl. 8. Relh. 145.
Woods. P. May.
First found by Mr Lyons.
1. Wood Ditton Park Wood. Linton; S. W. W.—
3. Eversden, Kingston, and Gamlingay Woods.—4. *Madingley;* Relh.

3. L. campéstris Willd.

Gramen exile hirsutum, R. C. 67. *Juncus villosus capitulis Psyllii*, M. M. 111. *J. campestris*, M. Pl. 8. Relh. 145.

Dry pastures. P. April, May.
1. Hinton. Gogmagog Hills. Fulbourn. Six-mile-Bottom. Allington Hill. Newmarket Heath. Shudy Camps. West Wratting. Weston Colville. Westley Wood.—2. Heydon Ditch. Litlington. Odsey; H. F.—3. Cambridge. Comberton. Gamlingay. Toft; Harlton; N. — 4. *Hill of Health;* J. M.—8. Wisbech; A. P.

4. L. multiflóra Lej.

Gramen hirsutum majus panicula juncea compacta, R. C. 68. *Juncus campestris β*, M. Pl. 8. Relh. 145.

Moorish ground. P. June.
1. By the brook, Fulbourn.—3. Gamlingay.—5. Bottisham and Wicken Fens.

ALISMACEÆ.

ALÍSMA Linn.

1. A. Plantágo Linn. *Great Water Plantain.*

Plantago aquatica, R. C. 120. *Alisma*, M. M. 56. *A. Plantago*, M. Pl. 9. Relh. 152.

Ponds and wet ditches. P. July, August.
Common throughout the county.

2. A. ranunculoïdes Linn.

Plantago aquatica minor, R. C. 120. *A. angustifolium capitulis rotundis echinatis*, M. M. 56. *A. ranunculoïdes*, M. Pl. 9. Relh. 153.

Turfy bogs and wet ditches. P. July, August.

1. Shelford Moor. Wilbraham Fen. Near the brook, Fulbourn. Dullingham. Hinton Moor; W. H. C.— 2. Near the railway in Sawston Moor!; S. W. W.— 3. Sheep's Green, Cambridge.—4. Waterbeach Fen. Near Stretham Ferry; S. W. W.—5. Wicken, Bottisham, and Swaffham Fens. Quarry at Upware.—6. Near Ely; S.W.W. —7. Doddington Turf-fen.

SAGITTÁRIA Linn.

1. S. sagittifólia Linn. *Arrowhead.*

Sagitta minor and *S. major*, R. C. 141. M. M. 56. *S. sagittifolia*, M. Pl. 22. Relh. 394.

Wet ditches and slow streams. P. August.

1. Conduit stream by Trumpington road, also in Cow Fen, Cambridge. Hinton. Wilbraham Fen.—2. Sawston Moor. Duxford. Shepreth; N. W.—3. Sheep's Green, Cambridge. Bourn Brook, from near its source to its mouth. Barrington. Malton.—4. Madingley chalk-pit. Waterbeach. Cottenham.—5, 6, 7, and 8. Common throughout the Fens.

BÚTOMUS Linn.

1. B. umbellátus Linn. *Flowering Rush.*

Gladiolus palustris, R. C. 62. *Butomus*, M. M. 65. *B. umbellatus*, M. Pl. 9. Relh. 168.

Wet ditches and streams. P. July.

1. *Teversham Moor;* J. M. *In the Stour, by the Paper Mills, Barnwell;* J. F.—3. In the Bourn Brook at intervals, from Bourn to its mouth. In the Cam about Grantchester, both upwards and downwards; W. H. C. Sheep's Green, Cambridge.—4. Near the railway-station, Waterbeach, and in the Fen. Milton Fen. Chesterton. Fen Drayton. Westwick.—5. Wicken Fen. By the foot-way from Cambridge to Fen Ditton. Clayhythe; W. H.C.—6. Wilburton!; H. Aldreth. Witchford.— 8. Ditches at North Brink, Wisbech; A. P. Near Whittlesey.

Triglóchin Linn.

1. T. marítimum Linn.

Relh. ed. 2, 145; ed. 3, 151.
First noticed by Mr Skrimshire.
Salt marshes. P. July, August.

8. River-side, below Wisbech!; H. *Tydd Marsh;* Mr Skrimshire.

2. T. palústre Linn.

Gramen marinum spicatum, R. C. 69 (not of his other works). *G. junceum spicatum seu triglochin*, R. Cat. Angl. ed. 1, 151. *Juncago palustris et vulgaris*, M. M. 112. *T. palustre*, M. Pl. 8. Relh. 151.

Marshy and fenny ground. P. June, July.

1. On the Common at one mile on London road, Cambridge. Shelford and Teversham Moors; W. H. C. *Hinton Moor;* Ray. — 2. Sawston Moor; G. S. G. and S. W. W. Newton; N.—3. Grantchester Meadows.—4. Milton Fen; W. H. C.—5. Ely; S. W. W.—7. Doddington Turf-fen. —8. Newton. Wisbech.

TYPHACEÆ.

Týpha Linn.

1. T. latifólia Linn. *Great Reed-mace.*

Typha, R. C. 173. M. M. 113. *T. latifolia*, M. Pl. 21. Relh. 375.

Ponds. P. June, July.

1. By Cow Fen, Cambridge; W. H. C. Hildersham; J. W. *In a pond at the hithermost end of Teversham Moor;* Ray.—2. Duxford; G. S. G.—3. Sheep's Green, Cambridge; W. H. C. Near the old Mansion, Haslingfield. Eltisley. Brick-pits, Gamlingay. Great Eversden; N. Malton; D. B. —4. Brick-pits near the Observatory, and by the Chesterton road. Waterbeach. Pit at Dry Drayton. Histon road. Sand-pits by railway in Waterbeach Fen. *Between Histon and Girton;* Relh.—5. Bank of river Cam by Swaffham Fen!; Bottisham Park!; H. Wicken.—7. Pit by the roadside near Doddington. Between March and Sixteen-foot river near the railway; N.—8. Between March and Wisbech. North Brink, Wisbech; A. P. Near Peterborough; N.

2. T. angustifólia Linn.

Relh. ed. 1, 348; ed. 3, 375.

Ponds. P. June, July.

2. Peat-holes, Triplow.—3. Bourn!; H.—4. Brick-pits near the Observatory, and by the Ely road. Waterbeach. Elsworth; N.—5. Bank of Cam below Fen Ditton. In a pit by the way from Chippenham to Herringswell. Ditch by Spinny Bank, Upware.—6. Roswell Pits, Ely.—8. Newton. North Brink, Wisbech; A. P.

TYPHACEÆ.

Sparganium Linn.

1. S. ramósum Huds. *Bur-reed.*

R. C. 158. M. M. 113. Relh. 376. *S. erectum* α, M. Pl. 21.

Wet ditches and ponds. P. July.
Common throughout the county.

2. S. simplex Huds.

S. non ramosum, R. C. 159. M. M. 113. *S. erectum* β, M. Pl. 21. *S. simplex*, Relh. 376.

Wet ditches and ponds. P. July.

1. Wilbraham Fen.—3. Brook by the Barton road. Near Toft bridge. Ditches by Grantchester foot-path. Kingston; Long Stow; Barrington; N.—4. Waterbeach Fen. By the river above Clayhythe.—5. Snailwell Fen.—6. Ely. Stretham Fen!; H.—7. Vermuden's Drain, Chatteris.

3. S. mínimum Fries.

S. natans, Relh. ed. 1, Suppl. ii. 16; ed. 3, 376.

Figured in English Botany (tab. 273) from a specimen sent from Burwell Fen by Mr Hemsted.

Wet ditches. P. July, August.

1. Wilbraham Fen.—2. Peat-holes, Triplow. *Sawston Moor;* Relh.—5. Burwell, Bottisham, and Wicken Fens.—6. Ely!; W. M. Aldreth.

ARACEÆ.

A'corus Linn.

‡ 1. A. Calamus Linn. *Sweet Flag.*

A. verus sive Calamus officinarum, R. Syn. ed. 1, 206. *A. Calamus*, Lyons, 30. M. Pl. 8, and 28.

Very wet places. P. June.

3. *In a ditch by Great Founder's Closes near House-in-the-fields, Cambridge;* Relh.—4. On the wash of the Cam, half a mile below Waterbeach. Marsh by the river at Chesterton. By the river nearly opposite Fen Ditton; W. H. C. On the river-bank in Chesterton parish, opposite to the Barnwell Gas Works; H. *Planted in a ditch by Dove-House Close near Jesus College by Mr Dent* (Lyons) *or Dr Heberden* (T. M.).—5. Wash of river above Upware.—6. River-bank below Barraway.

A′RUM Linn.

1. **A. maculátum** Linn. *Wake-Robin. Cuckoo-pint.*

Arum, R. C. 16. *A. officinarum*, M. M. 63. *A. maculatum*, M. Pl. 20. Relh. 394.

Hedge-banks and copses. P. April, May.

Common in the (1) Cambridge, (2) Royston, (3) Wimpole, and (4) Cottenham Districts.—5. Horningsey. Upware.—6. Witchford.—7. Doddington.—8. Wisbech; J. B.

LÉMNA Linn.

1. **L. trisúlca** Linn. *Water Ivy.*

Hederula aquatica, R. C. 72. *Lenticula aquatica trisulca*, M. M. 10. *L. trisulca*, M. Pl. 21. Relh. 374.

In stagnant water. A. June.

1. Shudy Camps.—2. Triplow.—3. Cow Fen, Cambridge. Comberton. Eltisley. Croydon.—4. Pits near the Observatory. Baitsbite. Dry Drayton. Waterbeach Fen. — 5. Horningsey. Wicken Fen.—6. Ely. Aldreth. Witchford. —7. Doddington.—8. Newton.

2. **L. minor** Linn. *Duck-weed.*

Lens palustris, R. C. 86. *Lenticula palustris vulgaris*, M. M. 10. *L. minor*, M. Pl. 21. Relh. 374.

On stagnant water. A. June, July.
Common throughout the county.

3. L. polyrrhíza Linn.

Lyons, 49. M. Pl. 21. Relh. 374.
First noticed by Mr Lyons.
On water. A. Flowers have not been seen.

1, 4, 5, 6. In still spots by the side of the river Cam and ditches adjoining, abundantly.—3. Eltisley.—5. Reche Lode!; H. Wicken Fen.—7. Doddington.

4. L. gíbba Linn.

Relh. ed. 1, 346; ed. 3, 374.
On stagnant water. A. June to August.

1. Hinton.—2. Sawston; G. S. G.—3. Sheep's Green, Cambridge. Eltisley. Eversden; N.—4. Dry Drayton. Waterbeach Fen.—5. Horningsey. Wicken Fen.—6. Ely. —7. Doddington.—8. Newton. Wisbech.

POTAMOGETONACEÆ.

Potamogéton Linn.

1. P. nátans Linn. *Pond-weed.*

P. foliis latis splendentibus, R. C. 123. *P. rotundifolium,* M. M. 16. *P. natans,* M. Pl. 4. Relh. 67.
Ponds and still waters. P. June, July.

1. Wood Ditton. Stetchworth. Willingham. Fulbourn. Dullingham. *Hinton Moor;* J. M.—2. Octagon Pond in Wimpole Avenue!; Haslingfield!; N.—3. Gamlingay. Sheep's Green, Cambridge. Pit near the spring-head at Hardwick. Eversden. Kingston. Haslingfield. Croydon. Grantchester!; H. Harston; Barrington; Bourn; Caxton; Eltisley; Croxton; N.—4. Pit to the south-west of Madingley Park. Dry Drayton.—5. Wicken and Bottisham Fens.—6. Ely!; W. M.

[**P. polygonifólius** (Pourr.)

I have a note of finding this plant at Gamlingay in May, 1852, but do not possess a specimen. It is probable that it was the true plant, but uncertain.]

2. **P. plantagíneus** Ducr.

Ann. Nat. Hist. ii. 350. Suppl. to Eng. Bot. fol. 2848.
First found by the Rev. Leonard Jenyns in 1827.
Peaty ditches. P. June, July.
2. Melbourn Common. Sawston Moor; N. In the Ruddry Brook near Ashwell; Fl. Hertf.—3. Great Eversden; N.—4. Brick-pits by the Ely road.—5. Bottisham and Wicken Fens.—6. West Fen, Ely. By Sandy's Cut; N.—7. Doddington Turf-fen.

3. **P. ruféscens** Schrad.

Hensl. Cat. ed. 2, 58.
Ditches. P. July.
5. Bottisham Fen. Burwell; G. S. G.

3. **P. heterophýllus** Schreb.

P. palustre, Relh. ed. 2, 64. *P. heterophyllum*, Relh. ed. 3, 67.
Ponds and ditches. P. June, July.
1. Wilbraham Fen!; H. *Shelford Common;* Relh. West Wratting ?.—3. Gamlingay; H.—5. Wicken, Swaffham, and Bottisham Fens.

4. **P. lúcens** Linn.

P. longis acutis foliis, R. C. 124. *P. aquis immersum folio pellucido lato oblongo acuto*, M. M. 16. *P. lucens*, M. Pl. 4. Relh. 68.
Deepish water. P. June.

1, 3, 4, 5, 6. In the river Cam throughout the county.
—1. Wilbraham Fen.—4. Waterbeach Fen. River Ouse, near Swavesey. Pits near the Observatory.—5. Wicken Fen.
—6. Aldreth. By Quaveney Drove, Stuntney; N.—7. Vermuden's Drain, Chatteris.—8. Wisbech; J. B.

5. P. prælóngus Wulf.

Bot. Gaz. i. 276.
First noticed by C. C. B. in 1849.
Streams and ditches. P. June.

1. Often found floating down the river.—3. In the river just above Sheep's Green, Cambridge.—4. In the ballast-pits near the railway junction at Chesterton; Rev. Dr Cookson.—5. Baitsbite.—6. Roswell Pits, Ely. By Sandy's Cut; N.

6. P. perfoliátus Linn.

R. C. 124. M. M. 16. M. Pl. 4. Relh. 67.
Streams and ditches. P. July.

1, 3, 4, 5, 6. In the river Cam throughout the county.—
4. Old Ouse, near Aldreth Bridge. Ouse, near Swavesey.
—5. Swaffham!; H.—6. Roswell Pits, Ely. Witcham. By Quaveney Drove, Stuntney; N.—7. Vermuden's Drain, Chatteris.

7. P. críspus Linn. *Water Caltrops. Frog's Lettuce.*

Tribulus aquaticus minor quercus floribus Clusii, R. C. 165. *P. seu Fontalis crispa*, M. M. 16. *P. crispum*, M. Pl. 4. Relh. 68.
Ditches and pits. P. June.

1. Coldham's Common, Cambridge. Hinton. Fulbourn. Brinkley. West Wratting. Little Linton.—2. Peat-holes, Triplow.—3. Sheep's Green, Cambridge. Arrington. Great Eversden. Croydon. Barton. Gamlingay. Toft. Tad-

low; N.—4. Histon road. Waterbeach Fen. Westwick Field. Swavesey.—6. Ely. Aldreth. Witchford. Stuntney; By Sandy's Cut; N.—7. Pit in a field between Doddington and the wood.

8. P. zosterifólius Schum.

P. compressum β, Relh. ed. 2, 66. *P. folio quam in P. compresso latiora, longiora* (sic), Relh. ed. 3, 69. *P. zosterifolius*, New Bot. Guide, 153.

Rivers and streams. P. June.

3. Queens' Ditch, and River Cam at Sheep's Green, Cambridge.—5. Baitsbite.—6. Ely.

[*P. gramíneus* (Linn.) is said by Relhan to have been found in Cow Fen. Was not this *P. zosterifolius* which grows close to that place?]

9. P. compréssus Linn.

P. ramosum caule compresso, folio graminis canini, nondum descriptum, R. C. 124. M. M. 16. *P. compressum*, M. Pl. 4. Relh. 69.

Ditches. P. June, July.

1. *Cam, near Cambridge;* Ray.—4. On the way to the ferry, and by Middle Hill Drove, Waterbeach.—5. Reche Lode.

10. P. pusíllus Linn.

P. pusillum gramineo folio, caule rotundo, nondum descriptum, R. C. 127. M. M. 16. *P. pusillum*, M. Pl. 4. Relh. 69.

Ponds and ditches. P. June.

1. *Hinton Moor;* Relh.—2. Shepreth. Sawston; G. S. G. —3. Gamlingay. Malton; N.—4. Waterbeach.—5. Bottisham, Snailwell, and Wicken Fens. *Near Paper Mills, Cambridge;* J. F.—6. Ely!; H.

11. P. pectinátus Linn.

Millefolium tenuifolium, R. C. 99. *P. millefolium seu foliis gramineis ramosum*, M. M. 16. *P. pectinatum*, M. Pl. 4. Relh. 69. *P. marinum*, Relh. ed. 1, Suppl. ii. 9.

Ponds and streams. P. June, July.

3. Harston; N.—4. In the river near Chesterton. Waterbeach. Old Ouse, near Aldreth Bridge. River Ouse, near Swavesey.—5. Wicken Fen. Clayhythe. In Bottisham Lode, near its mouth.—6. Ely. *Stretham Ferry;* J. M.—7. Vermuden's Drain, Chatteris.—8. Newton. *In some pits below Wisbech in* 1787 !; Relh.

[In all probability *P. flabellatus* inhabits this county. It grows in Wick Fen, near Nordelph, Norfolk, which is only about four miles from the boundary of Cambridgeshire.]

12. P. dénsus Linn. *Small Water Caltrops.*

Tribulus aquaticus minor muscadellæ floribus, R. C. 165. *P. seu Fontalis media lucens*, M. M. 16. *P. densum*, Relh. 68.

Ditches. P. June, July.

Common, except (?) in the Fens. There are no recorded stations for it in the (7) Chatteris, nor (8) Wisbech Districts.

Rúppia Linn.

1. R. rostelláta Koch. *Tassel Pond-weed.*

R. maritima, Relh. ed. 2, 66; ed. 3, 70.

Salt marshes. July, August.

8. *Ditches below Wisbech;* Mr Skrimshire. It is probable that this was *R. rostellata*.

ZANNICHÉLLIA Linn.

1. Z. palústris Linn. *Horned Pond-weed.*

Potamogeito affine gramen aquaticum, R. C. 125. *Aponogeton aquaticum graminifolium staminibus singularibus,* M. M. 13. *Z. palustris,* M. Pl. 21. Relh. 373.

Stagnant water. P. May to August.

Generally distributed throughout the county, but not very abundant, although perhaps often overlooked.

CYPERACEÆ.

SCHŒNUS Linn.

1. S. nígricans Linn. *Bog Rush.*

Juncus palustris panicula glomerata ex rubro nigricante, R. C. 81. *Scirpus palustris capitulo nigricante semine splendente,* M. M. 111. *S. nigricans,* M. Pl. 2. Relh. 20.

Peaty moors and fens. P. June.

1. By the brook, Fulbourn. Teversham and Shelford Moors; Side of road to Hinton; W. H. C.—2. Melbourn Common. Peat-holes, Triplow. Sawston Moor.—3. Gamlingay.—4. Pits near the Observatory.—5. Quy, Bottisham, Wicken, and Horningsey Fens.

CLÁDIUM Pat. Br.

1. C. Maríscus R. Br. *Common Sedge.*

Cyperus longus inodorus sylvestris, R. C. 43. *Pseudo-Cyperus palustris foliis et carina serratis,* M. M. 110. *Schœnus Mariscus,* M. Pl. 2. Relh. 20.

Peaty moors and fens. P. July.

1. Wilbraham Fen. By the Quy Water. Fulbourn and Teversham Moors; S. W. W. *Hinton Moor;* Ray.—

2. Peat-holes, Triplow. Melbourn Common. Sawston Moor.—4. Pits near the Observatory.—5. Reche and Burwell Fens. Anglesey Abbey; H. Still exceedingly abundant in Wicken Fen, where it is cut as a crop. *Chippenham Moor;* Relh.—6. West Fen, Ely.

This plant was formerly far more abundant than at the present time throughout the Fens. It was considered as a valuable natural crop in the undrained districts, and used largely for lighting fires at Cambridge and other places.

RHYNCHÓSPORA Vahl.

1. **R. álba** Vahl.

Gramen junceum leucanthemum, R. C. App. ii. 8. *Cyperus minor palustris hirsutus paniculis albis paleaceis,* M. M. 109. *Schœnus albus,* M. Pl. 2. Relh. 21.

Bogs. P. July.

3. Bogs on Gamlingay Heath until very recently, but, perhaps, now lost.

ELEÓCHARIS R. Br.

1. **E. palústris** R. Br.

Juncus aquaticus minor capitulis equiseti, R. C. 81. *Scirpus Equiseti capitulo majori,* M. M. 111. *S. palustris,* M. Pl. 2. Relh. 21.

Wet and marshy places. P. June.

Not uncommon throughout the county, although I have no recorded station for it in the (2) Royston District.

2. **E. multicaúlis** Sm.

Scirpus multicaulis, Relh. 22.

Marshy places. P. July.

2. Ickleton; G. S. G.—3. Bogs, Gamlingay.

3. E. acicularis Sm.

Sc. acicularis, Relh. 23.

Watery places. P. July, August.

5. Cowbridge, Swaffham!; H. *In the quarry at Upware;* Rev. John Holme (Relh.).—6. Roswell Pits, Ely.

Scírpus Linn.

1. S. maritimus Linn.

Gramen cyperoïdes palustre panicula sparsa, R. C. 66. *Cyperus longus inodorus latifolius spicis tumidioribus minus sparsis*, M. M. 108. *S. maritimus*, M. Pl. 2. Relh. 24.

Marshes. P. July.

6. Ely; W. M. *Stretham Ferry; Littleport;* Relh.—8. By the Railway between March and Whittlesey. Newton. Wisbech. By Burnt-House Drove; N.

2. S. lacústris Linn. *Bullrush.*

Juncus aquaticus maximus, R. C. 81. *S. palustris altissimus*, M. M. 110. *S. lacustris*, M. Pl. 2. Relh. 23.

In water. P. June, July.

1. Wilbraham Fen. Quy Water. Little Linton. *Between Stourbridge and Hinton;* J. F.—3. Impington; Grantchester; Hardwick; Long Stow; N.—4. Old Ouse, near Aldreth Bridge. Ouse near Swavesey.—5. Upware. Horningsey.—7. Near Carter's Bridge.

3. S. Tabernæmontáni Gm.

S. lacustris β, M. Pl. 2. Relh. 23.

In water. P. June, July.

3. Pit by Edge Hill, near Kingston Wood.—8. By the river opposite Guyhirne, above Wisbech. Tydd Common!; A. P.

4. S. cæspitósus Linn.

Lyons, 3. M. Pl. 2. Relh. 22.
Peaty moors and fens. P. June to August.

1. *Near Stourbridge;* Relh. and J. F. *Hinton Moor;* Relh.—5. Wicken Fen.

5. S. pauciflórus Lightf.

Relh. 22.
Peaty moors and fens. P. June to August.

1. Shelford Common. *Hinton Moor;* Relh.—2. Peatholes, Triplow. Near the Railway in Sawston Moor!; N.—3. Gamlingay.

6. S. fluítans Linn.

Isolepis fluitans, Hensl. Cat. ed. 2, 59.
Ponds and ditches. P. June, July.

3. Gamlingay.—6. Grunty Fen!; W. M.—7. By Blackbush Drove, Whittlesey!; W. M.

6. S. setáceus Linn.

Juncellus omnium minimus, R. C. App. ii. 11; Cat. Angl. ed. 2, 174. *S. foliaceus humilis,* M. M. 111. *S. setaceus,* M. Pl. 23. Relh. 23.

Wet, sandy, and gravelly places. P. July.

1. Shelford Common; H. *Coldham's Common;* Relh.—3. Gamlingay.—5. Cowbridge, Swaffham; H.

BLÝSMUS Panz.

1. B. compréssus Panz.

Schœnus compressus, Lyons, 2. M. Pl. 2. Relh. 20.
Boggy pastures. P. June, July.

1. The Common one mile from Cambridge towards London. *Coldham's Common; Hinton Moor;* Relh.—3. *Between*

Little Shelford and Whittlesford; Lyons.—5. *Near Bottisham Lode;* Relh.

Erióphorum Linn.

1. E. angustifólium Roth. *Cotton-grass.*

Gramen tomentosum et Linogrostis, R. C. 70. *Linagrostis panicula ampliore,* M. M. 112. *E. angustifolium,* M. Pl. 2. Relh. 24.

Bogs and fens. P. May, June.

1. Shelford Common. The Common one mile from Cambridge towards London. *Hinton and Teversham Moors;* J. M.—2. Ickleton; Sawston; G. S. G.—3. Sheep's Green, Cambridge. Gamlingay.—5. Quy Fen; S. W. W. Bottisham Fen.—8. Near March; A. P.

2. E. latifólium Hoppe.

E. pubescens, Sm. Engl. Flora, i. 68. Hensl. Cat. ed. 1, 25.

First found by the Rev. J. Holme.

Bogs. P. May, June.

1. Hinton Moor; H. Dale Moor, Shelford!; Mr J. Ball.

Cárex Linn.

It is probable that the plants of this genus have not been properly noticed in the Fens, unless they are far less numerous there than is supposed to be the case.

1. C. dioica Linn.

Lyons, 49. M. Pl. 21. Relh. 377.

Spongy bogs. P. May, June.

1. *Hinton Moor;* Relh.—3. Gamlingay bogs, now extinct there.—5. *Moor between Snailwell and Exning;* Relh.

2. C. pulicáris Linn.

Gramen cyperoïdes pulicare, R. Cat. Angl. ed. 1, 148. Lyons, 50. Relh. 377.

Bogs and fens. P. June.

1. *Hinton Moor; Shelford Common;* Relh. *Quy Water;* Ray.—5. Bottisham Fen.

[*C. divisa* (Huds.), M. Pl. 21. *Gr. cyperoïdes ex monte Ballon, spica divulsa*, R. C. 67. *Scirpoïdes spica ex pluribus spicis composita*, M. M. 110, is not in Relhan's Flora nor likely to inhabit this county.]

3. C. dísticha Huds.

Lyons, 50. M. Pl. 21. *C. intermedia*, Relh. 379.

Marshy places. P. May, June.

1. The Common one mile from Cambridge on the London road. Between Fulbourn and Little Wilbraham. *Hinton;* Relh.—2. Octagon pond in Wimpole Avenue!; N.— 3. Brick-pits, Gamlingay. Barrington; N. *Grantchester;* Relh.—4. Chalk-pit, Madingley. Pits near the Observatory. —5. Wicken Fen. Quarry, Upware.—6. Ely.

4. C. vulpína Linn.

Gramen cyperoïdes palustre majus spica compacta, R. C. 67. *Scirpoïdes palustre majus spica compacta*, M. M. 110. *C. vulpina*, M. Pl. 21. Relh. 380.

Wet places. P. June.

Common in the (1) Cambridge, (3) Wimpole, and (4) Cottenham Districts.—5. Horningsey.—6. Ely; Aldreth. Witchford.—7. Doddington.—8. Newton. Wisbech.

5. C. muricáta Linn.

Gr. cyperoïdes palustre minus, R. C. 67. *Scirpoïdes palustre minus spicis minoribus minusque compactis*, M. M.

110. *C. spicata*, M. Pl. 21. *C. muricata*, Lyons, 51. Relh. 380.

Gravelly pastures. P. June.

1. Hinton!; H. Hildersham. Westley Wood. — 2. Hauxton; Hinxton; N. — 3. Comberton; S. W. W. Croydon. Haslingfield. Malton. Gamlingay. Caldecot; Toft; Eversden; Croxton; N. *In the closes near Grantchester Meadows;* Lyons.—4. Madingley Wood; W. H. C. Girton. Fen Drayton. Dry Drayton; Oakington; Elsworth; N.—5. Bottisham Fen. Chippenham. Snailwell.—6. Witchford.—8. Newton.

6. **C. divúlsa** Good.

C. canescens, Lyons, 51. M. Pl. 21. *C. divulsa*, Relh. 380.

Moist shady places. P. June.

1. *Hall Wood, Wood Ditton; Weston Colville;* Relh.— 2. Hauxton; N.—3. Comberton. Caldecot. Toft. Kingston. Bourn; Eversden; Hardwick; Near the turnpike by Wimpole Avenue; N. *Gamlingay;* J. M.—4. Papworth St Everard; N. *King's Hedges; Madingley Wood;* Relh.

7. **C. teretiúscula** Good.

Relh. ed. 2, 365; ed. 3, 381.

Boggy places. P. June.

1. *Near the footpath to the old mill, Fulbourn;* Relh. —2. Peat-holes, Triplow.—3. *Amongst some low bushes in Grantchester Meadows;* Relh.

8. **C. paniculáta** Linn.

Lyons, 50. M. Pl. 21. Relh. 381.
First found by Prof. J. Martyn.
Bogs. P. June.

1. By the brook between Fulbourn and Wilbraham.—2. Peat-holes, Triplow.—3. Gamlingay.—5. Bottisham and Snailwell Fens.

9. C. axilláris Good.

Relh. ed. 2, 364; ed. 3, 379.
Wet sides of ditches. P. June.
1. *Hall Wood, Wood Ditton;* Relh.

10. C. remóta Linn.

Gr. cyperoïdes angustifolium panicula multiplici, R. C. App. i. 5. *Scirpoïdes angustifolium spicis sessilibus in foliorum alis,* M. M. 110. *C. remota,* M. Pl. 21. Relh. 397.

Damp hedge-banks. P. June.
1. Wood Ditton. Balsham.—2. Between Foulmire and Shepreth; W. H. C.—3. Eversden and Kingston Woods. Bourn; Long Stow; Caxton; Eltisley; Hayley Wood; N. —4. *Madingley Wood;* Relh.—7. Doddington Wood.

11. C. stellulàta Good.

C. muricata, Relh. ed. 1, 350. *C. stellulata,* Relh. ed. 3, 378.

Damp places. P. May, June.
2. Triplow Heath.—3. Gamlingay. *In the closes near Grantchester Meadows;* Relh.—4. *Madingley Wood;* Relh. —5. Chippenham.

12. C. cúrta Good.

C. canescens, Relh. ed. 1, Suppl. iii. 16. *C. curta,* Relh. ed. 3, 378.

Bogs. P. June.
3. Gamlingay.

13. C. ovális Good.

C. leporina, R. ed. 1, Suppl. i. 15. *C. ovalis*, Relh. ed. 3, 378.

Meadows. P. July.

2. Sawston; G. S. G.—3. Gamlingay.—4. *Between Histon and Rampton;* Relh.

14. C. strícta Good.

Gr. cyperoïdes angustifolium spica spadiceo-viridi minus et majus, R. C. 66. *Cyperoïdes nigro-luteum vernum majus*, M. M. 109. *C. cæspitosa*, M. Pl. 21. Relh. ed. 1, 354. *C. stricta*, Relh. ed. 3, 386.

Wet places. P. June.

1. Hinton; H. Wilbraham Fen. Fulbourn. Temple, Wilbraham. Spring Head, Bottisham.—2. Peat-holes, Triplow. Melbourn Common.—3. *Grantchester Meadows;* Relh. —4. Swavesey. *Impington road;* Relh.—5. Reche, Burwell, and Bottisham Fens. Quy; H.—6. Ely.

15. C. acúta Linn.

Gr. cyperoïdes majus angustifolium, R. Cat. Angl. ed. 1, 143. *C. gracilis*, Relh. ed. 1, 357. *C. acuta*, Relh. ed. 3, 387.

Wet places. P. June.

1. Wilbraham Fen.—5. Bottisham Fen. Quy!; H.— 6. Ely.

16. C. vulgáris Fries.

C. cæspitosa, Relh. ed. 2, 370; ed. 3, 386.

Wet places. P. May, June.

1. Wood Ditton. Cow Fen, Cambridge!; H. *Hinton, Sawston, and Teversham Moors;* Relh.—3. St Neots Road, Cambridge. Gamlingay.

17. C. palléscens Linn.

Relh. ed. 1, Suppl. iii. 7; ed. 3, 385.
Wet places. P. June.
1. Wood Ditton Park Wood. Westley Wood. *Linton Wood;* Relh.—3. Eversden and Gamlingay Woods. *Kingston Wood;* Relh.

18. C. pánicea Linn.

Lyons, 53. M. Pl. 21. Relh. 384.
First noticed by Prof. J. Martyn.
Boggy and fenny places. P. June.
1. Wilbraham Fen. By the brook, Fulbourn.—2. Peatholes, Triplow. Ickleton; G. S. G.—3. Gamlingay. Croydon. Eltisley. Comberton.—5. Bottisham, Reche, and Wicken Fens.

19. C. strigósa Huds.

Relh. ed. 2, 366; ed. 3, 382.
Woods. P. May, June.
1. *Hall Wood, Wood Ditton;* Relh.

20. C. péndula Huds.

Gr. cyperoïdes spica pendula longiore, R. C. App. i. 5. *Cyperoïdes spica pendula longiore et angustiore*, M. M. 109. *C. pendula*, M. Pl. 21. Relh. 381.
Borders of woods. P. May.
1. Wood Ditton Park Wood. *Catlidge;* Relh. *Fulbourn; Teversham;* J. M.

21. C. præcox Jacq.

Gr. nigro-luteum vernum, R. C. 66. *Cyperoïdes nigro-luteum vernum minus*, M. M. 109. *C. globularis*, Lyons, 52. *C. saxatilis*, M. Pl. 21. *C. montana*, Relh. ed. 1, 353. *C. præcox*, Relh. ed. 3, 382.

Dry places. P. April, May.

1. Gogmagog Hills. Newmarket Heath. Hinton. Furze-hills, Hildersham; W. H. C. Devil's Ditch.—2. Litlington. Odsey; H. F.—3. Gamlingay. Great Eversden towards Toft!; Comberton; Long Stow; N.—5. Devil's Ditch. Kennet Heath.

22. C. pilulífera Linn.

Lyons, 53. Relh. 386.
Damp heathy places. P. May.

1. Devil's Ditch. Gogmagog Hills. Fulbourn.—3. Gamlingay.—5. Devil's Ditch; H.

23. C. glaúca Scop.

C. recurva, Relh. ed. 1, 356; ed. 3, 384.
Damp places. P. June.

Probably common throughout the county. Few stations are recorded in the Fens, and none in the (8) Wisbech District.

24. C. flává Linn. *Hedge-hog-grass.*

Gr. palustre echinatum, R. C. 70. *Cyperoïdes palustre aculeatum panicula breviore,* M. M. 109. *C. flava,* M. Pl. 21. Relh. 382.
Wet places. P. May, June.

1. The Common at one mile on London road from Cambridge. Stourbridge Fair Green. Fulbourn. Shelford Common!; H.—2. Triplow. Sawston!; N.—3. Gamlingay.—5. Wicken and Bottisham Fens.

25. C. Œderi Ehrh.

C. extensa, Relh. ed. 2, 367. *C. Œderi,* Relh. ed. 3, 383.
Boggy places. P. June, July.

1. Near the Nine Wells, Shelford. Wilbraham Fen.—2. Sawston; G. S. G.—3. Gamlingay.—4. Between Water-

beach and the Lock. *Ditch near King's Hedges;* Relh.—
5. Bottisham and Wicken Fens. By Bottisham Lode near North Hill Farm. Quarry, Upware. Pit by Chippenham Avenue. Cowbridge, Swaffham!; H.—7. Doddington Turf-fen.

26. C. fúlva Good.

C. flava β, Relh. ed. 2, 367. *C. fulva*, Relh. ed. 3, 383.
Peaty ground and fens. P. June.
1. Shelford Common. By Quy Water. By brook near Fulbourn.—2. Peat-holes, Triplow. Sawston Moor!; N.—3. Comberton. By Kingston Wood!; N.—4. *King's Hedges;* Relh.—5. Bottisham and Wicken Fens.

Most of the specimens of this plant belong to the state called *C. Hornschuchiana* (Hoppe).

27. C. dístans Linn.

Lyons, 54. Relh. 383.
Peaty ground. P. May.
1. Shelford Common.—3. Gamlingay. Comberton.—4. Madingley; H.

28. C. binérvis Sm.

Relh. 384.
Peaty heaths. P. June, July.
1. Spring Head, Bottisham. Long Pasture, Hildersham. *Shelford;* Relh.—2. Peat-holes, Triplow.—3. Gamlingay. Little Eversden. Between Cl pton and Croydon Farms, Croydon. Wimpole.—4. *Chesterton;* Relh.

29. C. sylvática Huds.

Gr. cyperoïdes sylvarum tenuis spicatum, R. C. 67. *Cyperoïdes sylvarum tenuis spicatum,* M. M. 109. *C. sylvatica,* M. Pl. 21. Relh. 384.

Damp woods. P. May.

1. Hinton. Balsham. Wood Ditton Park Wood. Lenhall Wood, West Wratting. Westley Wood.—3. Whitwell Wood, near Coton. Comberton. Eversden, Kingston, Hardwick, and Gamlingay Woods. Orwell; Caldecot; Harlton; Croxton; N.—4. Madingley Wood. Moor Barns Thicket.

30. C. Pseudo-cypérus Linn.

Gr. cyperoïdes spica pendula breviore, R. Cat. Angl. ed. 1, 147. *C. Pseudo-cyperus*, M. Pl. 21. Relh. 385.

Wet places. P. June.

1. Wood Ditton. By the foot-path to the old mill at Fulbourn; W. H. C. *Catlidge; Pampisford;* Relh.—3. In a swamp on the north side of Kingston Wood; W. H. C. Eltisley; N—5. *Anglesey Abbey; Burwell;* Relh.—8. Wisbech; A. P.

31. C. filifórmis Linn.

Relh. 389.

Peaty places and fens. P. May.

1. By the brook near Fulbourn and Wilbraham.—3. *Gamlingay;* Relh.—4. Pits near the Observatory; W. H. C.—5. Wicken and Bottisham Fens. Cowbridge, Swaffham!; H.

32. C. hírta Linn. *Hammersedge.*

Lyons, 55. M. Pl. 21. Relh. 389.

Wet places. P. April.

1. The Common at one mile from Cambridge, towards London. Wilbraham. West Wratting. Westley Wood. Long Pasture, Hildersham. Little Linton; S. W. W. Deersley's Wood, Newmarket.—3. Gamlingay. Eltisley. Croydon. Haslingfield. Comberton; Harston; Barton; Kingston; Tadlow; Toft; Eversden; Caxton; Croxton;

N.—4. Madingley. Swavesey.—5. Chippenham. Bottisham and 'Wicken Fens. Quarry at Upware.—6. Ely. Witchford.

33. C. ampullácea Good.

C. rostrata, Relh. ed. 1, Suppl. iii. 41. *C. ampullacea* Relh. ed. 3, 388.

Figured in Eng. Bot. (t. 780) from a Cambridge specimen sent by the Rev. J. Holme.

Wet bogs and in water. P. June.

1. The Common at one mile on the London road from Cambridge. By Quy Water. *By Coldham's Lane;* Relh. —2. *Foulmire Common;* Relh.—3. Gamlingay.—5. Snailwell and Bottisham Fens. Quy!; H. By Burwell Lode, at Upware. *Chippenham Moor;* Relh.—6. Ely; W. M.

34. C. vesicária Linn.

Relh. ed. 1, Suppl. i. 15; Suppl. iii. 41; ed. 3, 388.
First found by the Rev. Mr Newton of Jesus College.
Wet bogs and in water. P. May.

3. *Paradise, Cambridge;* Relh.—5. Bottisham Fen. —6. Ely.—8. *In a field near the New Common bridge, Wisbech;* Mr Skrimshire.

35. C. paludósa Good.

C. acuta, M. Pl. 21. Relh. ed. 1, 357. *C. paludosa*, Relh. ed. 3, 387.

By water. P. May.

1. Hinton.—3. Gamlingay.—5. Fen Ditton. Bottisham and Wicken Fens.—7. Chatteris.—8. Wisbech!; A. P.

36. C. ripária Curt.

Gr. cyperoïdes cum paniculis nigris, R. C. 66. *Cyperoïdes latifolium spica rufa sive caule triangulo*, M. M. 109. *C. riparia*, Relh. 387.

By water. P. May.

1. Hinton. Linton.—2. Odsey; H. F.—3.' Gamlingay. Croydon. Harston. Harlton; Great Eversden; N.— 4. Pit south-west from Madingley Park. Waterbeach Fen. Swavesey.— 5. Fen Ditton. Snailwell, Bottisham, and Wicken Fens.—6. Witchford.—8. Wisbech!; A. P.

GRAMINEÆ.

SETÁRIA Pal. de Beauv.

‡ 1. **S. víridis** Beauv.

Panicum viride, Relh. ed. 1, Suppl. i. 7; ed. 3, 26.

Fields. A. July, August.

5. *In the gravel-pits at Chippenham, and corn-fields adjoining;* Relh.

PHÁLARIS Linn.

1. **P. arundinácea** Linn. *Great Reed-grass.*

Gramen arundinacea, acerosa gluma nostras, R. C. 65. *Calamagrostis aquatica vulgaris,* M. M. 106. *P. arundinacea,* M. Pl. 2. Relh. 25.

By water. P. June, July.

1. Fulbourn. Little Linton.—3. Comberton. Gamlingay. Wimpole. Bourn Brook. Eversden; Barrington; Kingston; Caxton; Bourn; Barton; Near Hayley Wood; N.—4. Middle-hill Drove, Waterbeach. Mare Way. Westwick; Papworth St Everard; N.—5. Fen Ditton. Horningsey. Upware. Chippenham. Swaffham Prior!; H.— 6. Ely. Aldreth. Witchford.—8. Leverington; A. P.

The stripe-leaved state has been found (1) *on the left-hand side of the footway leading to the mill at Fulbourn;* Relh., and (2) at Sawston!; N.

[*P. canariensis* (Linn.) is sometimes cultivated, and may therefore occasionally be found in a half-wild state.]

GRAMINEÆ.

ANTHOXÁNTHUM Linn.

1. A. odorátum Linn. *Sweet Vernal-grass.*

Lyons, 1. M. Pl. 1. Relh. 13.
Pastures. P. May, June.
Common in the (1) Cambridge, (2) Royston, (3) Wimpole, and (4) Cottenham Districts.—5. Fen Ditton. Chippenham.—8. Wisbech; A. P. Woodhouse, Elm; J. B.

PHLÉUM Linn.

[1. P. ásperum Jacq.

P. paniculatum, Relh. ed. 1, Suppl. i. 8; ed. 3, 27.
Dry fields. A. July.
1. *Gogmagog Hills;* Mr Woodward (Wither, Bot. Arr. ii. 117). *Bourn Bridge;* Mr Crowe (Relh.). *Newmarket Heath,* Mr Miller (Huds. Fl. Angl. ed. 2, 26). Relhan never found this grass, nor has any person since his time.]

2. P. Boéhmeri Wibel.

Phalaris phleoïdes, Relh. ed. 1, 22; ed. 3, 26.
Gathered in this county by Mr Lyons before 1780 (Eng. Bot. fol. 459).
Dry sandy fields. P. July.
1. Furze-hills and by the road-side near to them, Hildersham. Barrington Hill, near Linton; H., but now lost there. *Newmarket Heath;* Relh. *Devil's Ditch;* J. F.— 5. *Chippenham Park wall;* Rev. J. Hemsted, who supplied the specimen figured in Eng. Bot. tab. 459.

3. P. arenárium Linn.

Phalaris arenaria, Eng. Bot. 222. Relh. 25.
Sandy places. A. June.

5. *Newmarket Heath;* Rev. J. Hemsted, who sent it from thence to be figured in Eng. Bot. tab. 222. Probably his specimens grew on the Chippenham side of the Heath.

4. P. praténse Linn. *Cat's-tail grass.*

Gr. typhinum maximum, R. C. 70. *Typhoïdes maximum spica longissima*, M. M. 106. *P. pratense*, M. Pl. 2. Relh. 26. *Gr. typhoïdes medium sive vulgatissimum*, R. C. 71. *Typhoïdes minus spica breviore*, M. M. 106. *P. nodosum*, Relh. ed. 1, 23.

Pastures. P. June.

Common throughout the county, except on the peat soil of the Fens. •

It is possible that several species may be confounded under this name.

ALOPECÚRUS Linn.

1. A. praténsis Linn. *Fox-tail grass.*

M. Pl. 2. Relh. 27.

Meadows and pastures. P. April to June.

1. Cambridge. Linton. West Wratting. Brinkley.— 2. Odsey; H. F.—3. Cambridge. Comberton. Orwell. Wimpole. Croydon. Gamlingay. Haslingfield. Kingston. Toft. Caldecot. Bourn; Harlton; Caxton; N. — 4. Madingley. King's Hedges. Mare Way. Swavesey. Childerley; Oakington; N. — 5. Horningsey. Wicken. — 6. Witchford. Stuntney; N.

2. A. geniculátus Linn.

Gr. aquaticum spicatum, R. C. 65. *Alopecuroïdes aquaticum procumbens spica gracili breviore*, M. M. 106. *A. geniculatus*, M. Pl. 2. Relh. 28.

Damp places and dried-up ponds. P. June, July.

Common throughout the county.

3. A. fúlvus Sm.

Relh. 28.

Wet borders of ponds and ditches. P. June to September.

3. Orwell; Comberton; N.—4. Behind the Hall at Long Stanton.—5. Wash of the river below Clayhithe.—6. Aldreth Causeway. Below Alderforth, Witchford. Hill Row, Haddenham; N.

4. A. agréstis Linn. *Black grass.*

Gr. alopecurinum minus, R. C. 64. *Alopecuroïdes minus spica longa gracili aristis rectis*, M. M. 105. *A. myosuroïdes*, M. Pl. 2. *A. agrestis*, Relh. 28.

Arable land. A. April to September.

A troublesome weed throughout the county, except on the fen land, where it is rare. No locality is recorded in the (7) Chatteris District.

Nárdus Linn.

1. N. strícta Linn. *Mat-grass.*

Spartum parvum Lobelio, R. C. 159. M. M. 106. *N. stricta*, M. Pl. 2. Relh. 24.

Heaths. P. July.

1. *Hills near Hildersham;* Relh.—3. Gamlingay.

Mílium Linn.

1. M. effúsum Linn. *Millet-grass.*

Relh. 28.

Damp shady woods. P. June.

1. Wood Ditton Park Wood. *Linton;* Relh.—3. Eversden Wood. *Kingston Wood;* Relh.—4. *Madingley;* Relh. —7. Doddington Wood.

PHRAGMÍTES Trin.

1. **P. commúnis** Trin. *Common Reed.*

Arundo vallatoria R. C. 16. *A. officinarum*, M. M. 107. *A. Phragmites*, M. Pl. 3. Relh. 47.

Marshes and in water. P. August.
Common throughout the county.

CALAMAGRÓSTIS Adans.

1. **C. lanceoláta** Roth.

Arundo Calamagrostis, Relh. 48.
Wet places. P. July.

1. *In the corner of Wood Ditton Park Wood, next to the common;* Relh. — 2. Melbourn Common. — 4. *Madingley Wood;* J. M. *In the bushy closes at the north side of Hardwick Wood;* Relh.—5. Wicken Fen. Swaffham Bulbeck. Anglesey Abbey!; H.—6. *Prickwillow;* Rev. Dr Goodenough (Bot. Guide, 44).

2. **C. Epigéjos** Roth.

Ar. Calamagrostis, Lyons, 19. M. Pl. 3. Relh. ed. 1, 52. *Ar. Epigejos*, Relh. ed. 1, 51; ed. 3, 47.

Damp shady places. P. July.

1. Teversham Moor; Barrington Hill, Linton; W. H. C. Fulbourn; Borley Wood; N. *Wood Ditton;* Relh. — 2. Sawston Moor!; N. — 3. Eversden, Hardwick, and Kingston Woods. Comberton. Toft; Hayley Wood; Eltisley; Brickfield, Great Eversden; N. — 4. *Madingley Wood;* Relh. and J. F.—6. *Isle of Ely;* Relh.

APÉRA Adans.

1. **A. Spica-vénti** Beauv. *Wind-grass.*

Gr. segetum panicula speciosa, R. Cat. Angl. ed. 1, 154. *Agrostis Spica-venti*, M. Pl. 2. Relh. 29.

Sandy fields. A. June, July.
1. Shudy Camps !; H.—3. Gamlingay; N.

1. A. interrúpta Beauv.

First found in this county by C. C. B. in 1852.
Sandy fields. A. June, July.
1. In the ancient ditch near Pampisford Hall !; Mr Jas. Stratton, 1855.—5. Between Chippenham and Badlingham in 1852 and 1859.

AGRÓSTIS Linn.

1. A. canína Linn.

Relh. ed. 1, 26; ed. 3, 29.
Wet, peaty land. P. July, August.
3. Malton. Cow Fen, Cambridge.—4. Mare Way. Cottenham Fen.—5. Abundant in Wicken Fen.

2. A. vulgáris Wither. *Bent-grass.*

Gr. pratense vulgare spica fere arundinacea, R. Cat. Angl. ed. 1, 154. *A. capillaris,* Relh. ed 1, 27. *A. vulgaris,* Relh. ed. 3, 29.
Rather dry places. P. July.
1. Devil's Ditch !; H—2. Royston; H. F.—3. Gamlingay.

β. *A. pumila,* Lightf.
Relh. ed. 1, 28.
3. Gamlingay.
This is rather a diseased plant than a true variety.

3. A. álba Linn. *Fiorin-grass.*

A. palustris, Lyons, 4. *A. stolonifera,* Lyons, 4. M. Pl. 2. Relh. 30. *A. alba,* Relh. 30.
Fields. P. July.
Common throughout the county.

Hólcus Linn.

1. H. lanátus Linn. *Soft grass.*

Gr. pratense paniculatum molle, R. C. 69. *Avena pratensis tomentosa*, M. M. 107. *H. lanatus*, M. Pl. 23. Relh. 33.

Pastures. P. July.
Common throughout the county.

2. H. móllis Linn.

Relh. ed. 1, 376; ed. 3, 33.
Copses and open woods. P. July.
3. Gamlingay!; H. Mr Newbould thinks that he has seen it in Hayley and Eversden Woods. *Kingston Wood;* Relh.—4. *Madingley Wood;* Relh.

Aíra Linn.

1. A. cæspitósa Linn. *Corn-grass.*

Gr. paniculatum arvense III. *sive Gr. segetum altissimum panicula sparsa*, R. C. 69. *Milium sylvestre panicula speciosa*, M. M. 107. *A. cæspitosa*, M. Pl. 2. Relh. 31.

Pastures, thickets, road-sides. P. July.
Common in the (1) Cambridge, (3) Wimpole, and (4) Cottenham Districts.— 2. Whittlesford. Shelford.— 5. Chippenham. Horningsey. Wicken.— 6. Aldreth.— 7. Chatteris.

2. A. flexuósa Linn.

Lyons, 6. M. Pl. 2. Relh. 32. *A. montana*, M. Pl. 2. Relh. ed. 1, 30.
Heaths. P. July.
3. Gamlingay Heath!; H.

3. A. caryophyllea Linn.

Gr. montanum panicula spadicea delicatiore, R. C. 69.
Milium montanum panicula spadicea delicatiore, M. M. 107.
A. caryophyllea, M. Pl. 2. Relh. 33.

Dry, gravelly places. A. June.

1. Furze-hills, Hildersham.—3. Gamlingay.—4. *Hill of Health;* Relh.

4. A. præcox Linn.

Lyons, 7. Relh. 32.
Dry, gravelly places. A. April, May.

1. Furze-hills, Hildersham.—2. Odsey; A. M. B.—3. Gamlingay.—4. *Hill of Health;* Relh.—5. Kennet Heath.

Trisétum Pers.

1. T. flavéscens Beauv. *Yellow Oat-grass.*

Gr. avenaceum panicula flavescente locustis parvis, R. Cat. Angl. ed. 1, 141. Lyons, 17. M. Pl. 3. Relh. 47.

Dry pastures and way-sides. P. July.

Common in the (1) Cambridge, (2) Royston, (3) Wimpole, and (4) Cottenham Districts.—5. Horningsey. Swaffham Bulbeck!; H. Upware. Chippenham.—6. Haddenham. Wentworth. Witcham. Sutton. Stuntney; N.—8. Wisbech.

Avéna Linn.

1. A. fá'tua Linn. *Wild Oats.*

Ægilops bromoïdes, R. C. 4. *A. spuria elatior panicula speciosa*, M. M. 107. *A. fatua*, M. Pl. 3. Relh. 46.

Arable land. A. July.

1. *Hinton;* Relh.—2. Royston and Steeple Morden; N.—3. Footway to Coton, and in many other places.—4. Com-

mon.—5. Quy?; H.—6. Haddenham. Stuntney; N.—8. Newton. Wisbech. Whittlesey; N.

2. A. praténsis Linn.

Gr. avenaceum montanum, spica simplici, aristis recurvis, R. Hist. Pl. ii. 1290. Lyons, 17. M. Pl. 3. Relh. 46.

First found by Mr Dale.

Dry chalky banks. P. June.

1. Gogmagog Hills. Newmarket Heath. Hinton. Hildersham. Allington Hill.—2. Pit at Litlington. Triplow. Royston!; N. Odsey; A. M. B.—3. Eversden. Croydon. Haslingfield. Barrington; Toft; Barton; N.—4. Dry Drayton; N. *Hill of Health;* Relh.—5. Chippenham. Snailwell Heath. High Ditch lane, Fen Ditton. Devil's Ditch.

3. A. pubéscens Linn.

Lyons, 18. Relh. 46.

Gravelly and chalky banks. P. July.

1. Hinton. Devil's Ditch. Allington Hill. Fulbourn. West Wratting. Burrough-Green. Brinkley.—2. Triplow. Royston; H. F. Odsey; A. M. B.—3. Eltisley. Croydon. Eversden; N.—4. Gravel Hill, near the Observatory. Madingley.—5. North-hill Farm, Horningsey. Chippenham. Anglesey Abbey!; Newmarket Heath!; H.

ARRHENATHÉRUM Beauv.

1. A. avenáceum Beauv. *Oat-grass.*

Avena elatior, Lyons, 16. M. Pl. 3. Relh. 45. *A. nodosa*, Lyons, 16. *A. elatior* β, M. Pl. 3. *Holcus avenaceus*, Relh. ed. 2, 30.

Fields and hedges, P. June.

Common throughout the county.

GRAMINEÆ. 275

Triódia R. Br.

1. T. decúmbens Beauv. *Heath-grass.*

Festuca decumbens, Lyons, 13.   Relh. ed. 1, 43.   *Poa decumbens*, Relh. ed. 3, 37.

Heaths and moors.   P. July.

1. Shelford Moor!; H. *Hinton Moor; Gogmagog Hills;* Relh.— 2. Triplow Heath; W. H. C.   Sawston!; N.— 3. Gamlingay.   Toft; N.—4. *King's Hedges;* Lyons.— 5. Kennet Heath.   Bottisham Fen.

Koeléria Pers.

1. K. cristáta Pers. *Crested Hair-grass.*

Poa cristata, Relh. ed. 1, 37, and Suppl. ii. 8.   *Aira cristata*, Lyons, 4.   Relh. 31.

Chalky and gravelly places.   P. June, July.

1. Devil's Ditch. Newmarket. Gogmagog Hills. Wool Street. Linton Wood; S. W. W. Shudy Camps. Furze-hills, Hildersham. Old chalk-pits Hinton. Fleam Dyke. Sawston; N.—2. Triplow. Newton. Hauxton. Royston! and Sawston; N. Odsey; H. F.—3. Great Eversden. Croydon. Haslingfield. Harston; Hardwick; Caldecot; N.— 4. Impington. *Hill of Health;* Lyons.—8. Newton.

Mélica Linn.

1. M. uniflóra Retz.

Relh. ed. 1, Suppl. i. 8; ed. 3, 34.

Woods.   P. June.

1. Wood Ditton.   By the footpath from West Wratting to Weston Colville.   *Catlidge;* Relh.

Molínia Schrank.

1. M. cærulea Moench.

Aira cærulea, Lyons, 5. M. Pl. 2. *Melica cærulea*, Relh. 34.

Wet heaths. P. July, August.

1. Fulbourn; S. W. W. *Hinton and Teversham Moors;* Relh.—2. Peat-holes, Triplow. Ickleton; G. S. G. Sawston Fen; N.—3. Toft; N—5. Bottisham and Wicken Fens. Swaffham Bulbeck.

Póa Linn.

1. P. ánnua Linn.

Gr. pratense paniculatum minus, R. C. 69. *Eragrostis pratensis minor vulgaris*, M. M. 108. *P. annua*, M. Pl. 2. Relh. 37.

Waste grounds and walls. A. March to October.
Common throughout the county.

2. P. nemorális Linn. *Wood Meadow-grass.*

Relh. ed. 1, 36; ed. 3, 37.
Shady places. P. June, July.

1. Devil's Ditch!; H.—3. Eversden, Kingston, and Hardwick Woods. Cambridge. Gamlingay.—4. Madingley Wood.

3. P. triviális Linn. *Rough Meadow-grass.*

Lyons, 7. M. Pl. 2. Relh. 36.
Moist and shady places. P. June.

Common throughout the county, although less so in the Fens. I have no recorded station for it in the (7) Chatteris District.

GRAMINEÆ.

4. P. praténsis Linn. *Common Meadow-grass.*

Gr. pratense, R. C. 69. *Eragrostis pratensis major vulgaris*, M. M. 108. *P. pratensis*, M. Pl. 2. Relh. 37. *P. angustifolia*, Lyons, 8. M. Pl. 2. Relh. ed. 1, 34.

Dry pastures. P. June, July.

1. Cambridge. Shudy Camps. Allington Hill. Brinkley. Burrough-Green. Hildersham. West Wratting.—2. Royston.—3. Cambridge. Gamlingay. Eltisley. Wimpole. Croydon. Haslingfield. Caldecot. Comberton; Toft; Caxton; N.—4. Swavesey. Fen and Dry Drayton. Chesterton. Papworth; Graveley; N.—5. Swaffham!; Newmarket Heath!; H. Fen Ditton.—6. Witchford.

5. P. compréssa Linn.

Lyons, 8. Relh. 36.

Dry places, tops of walls. P. July.

1. Walls in Tennis-Court road, and at St John's College, Cambridge. Gogmagog Hills. Fulbourn. Stetchworth.—2. Royston; Hinxton; Foxton; Near Ashwell; N.—3. Caldecot. Toft. Coton. Eversden; Bourn; East Hatley; Barton; Harlton; Eltisley; Croxton; N.—4. Dry Drayton; Papworth St Everard; N. Chalk-pit, Madingley. Wall at Impington.—5. Swaffham Prior!; H.—6. Ely.

Most, if not all, of the stations belong to the form called *P. subcompressa* by Parnell.

GLYCÉRIA R. Br.

1. G. aquática Sm. *Great Water Reed-grass.*

Gr. aquatica majus, R. C. 65. *Eragrostis major latifolia aquatica*, M. M. 108. *Poa aquatica*, M. Pl. 2. Relh. 34.

In water. P. July.

1. Little Linton.—3. Haslingfield. Barrington. Malton. Harston. Toft; Great Eversden; Gamlingay; N.—Common in the rest of the county.

2. G. fluitans R. Br. *Float-grass.*

Gr. aquaticum cum longissima panicula, R. C. 65. *Eragrostis aquatica elatior locustis quasi in spicas digestis*, M. M. 108. *Festuca fluitans*, M. Pl. 3. *Poa fluitans.* Relh. 35.

Ponds and ditches. P. June to September.
Common throughout the county.

3. G. plicáta Fries.

First noticed in this county in 1845 by C. C. B.
Ponds and ditches. P. June to August.

Common probably throughout the county, although I have only one station, Hauxton, recorded in the (2) Royston District, and none in (8) Wisbech District.

β. *G. pedicelláta*, Towns.
Annals of Nat. Hist. Ser. 2, v. 105.
First noticed by Mr Townsend in 1846.

1. Ditch at the entrance of Fulbourn from Cambridge. Wood Ditton.—6. Ely.

Sclerόchloa Beauv.

1. S. maritima Lindl.

Poa maritima, Lyons, 9. M. Pl. 3. Relh. 35.
First found by Prof. J. Martyn.
Sea-coast, in damp places. P. June, July.
8. Wisbech!; W. M.

Prof. T. Martyn has a *P. maritima* β (Pl. 3), which is undeterminable.

2. S. distans Bab.

Poa distans, Relh. ed. 1, Suppl. i. 9 ?; ed. 3, 35.

Sea-shore. P. June to August.

1 and 4. *By the sides of dunghills, common near Cambridge, and Cottenham;* Relh. Probably some other plant.—8. Foul Anchour in 1853. Wisbech.

3. S. rígida Link.

Gr. exile duriusculum in muris et aridis proveniens, R. Cat. Angl. ed. 1, 148. *Poa rigida*, Lyons, 9. M. Pl. 3. Relh. 36.

Walls and very dry places. A. June.

Common in the (1) Cambridge, (2) Royston, (3) Wimpole, and (4) Cottenham Districts.—5. Chippenham. Bottisham. Swaffham Prior!; H.—6. Ely. Witcham.

4. S. loliácea Woods.

Poa loliacea, Relh. ed. 1, 37. *Triticum loliaceum*, Relh. ed. 3, 52.

Sandy sea-coast. A. June, July.

8. *Wisbech;* Relh.

Bríza Linn.

1. B. média Linn. *Quaking-grass. Lady's Hair.*

Gramen tremulum, R. C. 70. *Eragrostis pratensis major locustis tremulis*, M. M. 108. *Briza media*, M. Pl. 2. Relh. 38.

Pastures, chiefly on dry ground. P. June.

Common in the (1) Cambridge, (2) Royston, (3) Wimpole, and (4) Cottenham Districts.—5. Horningsey. Upware. Newmarket. Chippenham. Snailwell. Wicken Fen.—6. Haddenham.—8. On the river-bank, Wisbech; A. P.

GRAMINEÆ.

Catabrósa Beauv.

1. C. aquática Presl.

Gr. paniculatum aquaticum minus, R. Cat. Angl. ed. 1, 154. *Aira aquatica*, Lyons, 6. M. Pl. 2. Relh. 31.

Ponds and ditches. P. June, July.

1. The Common at one mile on the London road from Cambridge. Fulbourn; S. W. W. Shelford!; H.—3. Grantchester. By the Barton road, and pit at the end of Burrell's Walk, Cambridge. Coton. Haslingfield; Orwell; N.—4. By Trinity Conduit Head, Cambridge.—5. Bottisham Fen.

Cynosúrus Linn.

1. C. cristátus Linn. *Dog's-tail-grass.*

Gr. alopecuroïdes minus spica aspera brevi, R. C. 65. *Lolium pratense glumis cristatis*, M. M. 105. *C. cristatus*, M. Pl. 3. Relh. 39.

Pastures. P. August.

Common in the (1) Cambridge, (2) Royston, (3) Wimpole, and (4) Cottenham Districts.—5. Newmarket. Chippenham.—8. Newton. Wisbech.

Dáctylis Linn.

1. D. glomeráta Linn. *Cock's-foot-grass.*

Gr. asperum, R. C. 66. *Hippagrostis vulgaris folio aspero*, M. M. 107. *D. glomerata*, M. Pl. 2. Relh. 39.

Meadows and pastures. P. June.
Common throughout the county.

Festúca Linn.

1. F. sciuroïdes Roth.

First found in this county by C. C. B. in 1852.
Gravelly and sandy places and walls. A. June, July.

1. Linton; G. S. G.—2. Near the railway at Royston; N. — 3. Gamlingay. — 5. Kennet Heath. — 6. Sutton. Witcham.

2. F. Myúrus Linn.

Lyons, 11. Relh. 40.

Sandy and gravelly places and walls. A. June, July.

1. *Wall at Barnwell;* Lyons. Linton; G. S. G. Deserted railway, near Chesterford; N. —3. Abundant at Gamlingay!; N.—4. *Hill of Health;* Relh.

3. F. ovína Linn. *Sheep's Fescue-grass.*

Lyons, 10. M. Pl. 3. Relh. 40.

Dry pastures. P. June.

1. Gogmagog Hills. Fulbourn. Newmarket Heath. Devil's Ditch. Allington Hill. Furze-hills, Hildersham. Burrough-Green. Sawston; Near Hinxton; N.—2. Newton. Hauxton. Royston. Sawston!; N. Odsey; A. M. B. —3. Gamlingay. Wimpole. Croydon. Haslingfield. Caldecot. Comberton; Harlton; Bourn; Kingston; Croxton; N.—4. Cottenham. Dry Drayton. Elsworth; N.—5. Snailwell Heath. Newmarket Heath. Chippenham.—6. Stuntney; N.—8. Newton. Foul Anchour.

β. *F. duriúscula* Linn.

Lyons, 10. Relh. 40.

3. Gamlingay.—5. Swaffham. White Fen, near Bottisham Lode.—Probably in many other places.

4. F. rúbra Linn.

First noticed in this county by C. C. B. in 1855.

Dry, gravelly, and sandy places. P. June.

1. Streetway Hill, near Great Wilbraham. West Wratting.—3. Gamlingay. Toft!; N.—4. Outside Madingley Park.—8. Wisbech.

The *F. rubra* of Lyons (11) and Relh. (ed. 1, 41) seems to have been the *F. rubra* of Withering, which is a variety of *F. ovina*.

5. F. gigántea Vill.

Bromus giganteus, M. Pl. 3. Relh. ed. 1, 46. *F. gigantea*, Relh. ed. 3, 40. *Bromus longifolius*, Lyons, 15.

Damp shady places. P. July.

1. Shudy Camps. Balsham Wood.—2. Royston; Steeple Morden; Whittlesford; N.—3. Behind the Colleges, Cambridge. Barton. Grantchester. Caldecot. Kingston. Malton. Harston. Toft; Great Eversden; Bourn; Wimpole; Croydon; Eltisley; Caxton; Croxton; Hatley St George; Hayley Wood; N.—4. Girton. Histon. Moor Barns Thicket. Madingley. Dry Drayton. Papworth St Agnes; Graveley; Elsworth; N.—5. Chippenham. Horningsey. High Ditch, Fen Ditton.—7. Doddington Wood.

6. F. arundinácea Schreb.

F. elatior, Lyons, 13. M. Pl. 3. Relh. 41.

Damp pastures. P. June, July.

1. By the New road from Trumpington to Hinton. Balsham. Fulbourn.—2. Near Hinxton; N.—3. Old chalkpit, Haslingfield. Caldecot. Kingston. Barrington. Harston. Hayley Wood; Great Eversden; Barton; Caxton; N. —4. Madingley. Cuckoo Lane, Histon. Childerley; N.— 5. Chippenham. Bottisham Fen. High-ditch Lane, Ditton. Quy!; H.—6. Haddenham.—8. Wisbech!; A. P.

Our plant is not the typical *F. arundinacea*, but the *F. elatior* of authors.

7. F. praténsis Huds. *Meadow Fescue-grass.*

Lyons, 12. Relh. 40.

Damp meadows. P. June, July.

Common in the (1) Cambridge, (2) Royston, (3) Wimpole, and (4) Cottenham Districts.—5. Fen Ditton. Horningsey!; H. Upware. Wicken Fen. Chippenham.—6. Witchford.—8. Newton.

β. *F. loliácea* Huds.
Relh. ed. 2, 38; ed. 3, 41.
1. Linton. *Hinton;* Relh.—3. *Grantchester Meadows;* Relh. Barrington; N.—5. Wicken Fen.—6. West Fen, Ely.

BRÓMUS Linn.

1. B. eréctus Huds.

Relh. ed. 1, Suppl. ii. 8; ed. 3, 43.
Dry, chalky, and sandy places. P. June, July.
1. Chalk-pit, Gogmagog Hills. Hinton. Between Fulbourn and Little Wilbraham. Between Hinton and Teversham. Allington Hill. *Barnwell Gravel-pit;* Relh.—2. Between Known's Folly and Royston; N.—3. Wimpole Park. Eversden; N.—5. Bottisham Hall. North-hill Farm, Horningsey. High-ditch Lane, Fen Ditton. Gravel-pit, Chippenham.

2. B. ásper Linn.

B. giganteus, Lyons, 14. *B. hirsutus*, Relh. ed. 1, 48. *B. asper*, Relh. ed. 3, 44.
Thickets and damp hedges. P.? July.
Common in the (1) Cambridge, (2) Royston, (3) Wimpole, and (4) Cottenham Districts.—5. Horningsey. Fen Ditton. Chippenham.

3. B. stérilis Linn.

Festuca et Avena græca, R. C. 53. *Fest. avenacea sterilis elatior*, M. M. 108. *B. sterilis*, M. Pl. 3. Relh. 44.

Dry banks and waste places. A. June.

Common in the (1) Cambridge, (2) Royston, (3) Wimpole, and (4) Cottenham Districts. — 5. Fen Ditton. Swaffham Bulbeck. Chippenham.— 6. Sand-pit, Haddenham. Barraway. Ely. Sutton. Stuntney; N. — 7. Doddington.—8. Wisbech.

Serrafálcus Parl.

[1. **S. secalínus** Bab.

Fest. altera, M. M. 108. *Bromus secalinus*, M. Pl. 3. Relh. 42.

Arable land. A. June, July.

1. Hildersham. *Gogmagog Hills;* Relh.—4. *Between Chesterton and Milton;* Relh.—5. Bottisham. Chippenham. Swaffham!; H.—6. *Aldreth;* Relh.

β. *Br. velutínus* Sm.
Br. multiflorus, Relh. ed. 2, 39; ed. 3, 42.
3. *Paradise, Cambridge;* Relh.—6. *Aldreth;* Relh.]

2. **S. commutátus** Bab.

Br. arvensis, Relh. ed. 2, 40. *Br. pratensis,* Relh. ed. 3, 43. Judging from a specimen, it is the *Br. racemosus* of Henslow's Catalogues.

Borders of fields. B. June, July.

1. *Hinton;* Relh.—2. Royston; N. Odsey; A. M. B. —3. Coton. Comberton. Toft. Barrington. Barton. Gamlingay. Harlton!; N.—4. Madingley. Long Stanton. Dry and Fen Drayton. Willingham. Waterbeach Fen. Childerley; N. *By the Impington road;* Relh.—5. Horningsey. Upware.

3. **S. racemósus** Parl. (See Appendix VI.)

Br. racemosus, Relh. 43.

Borders of fields and roads. B. June.

1. Near Cambridge?. *Barnwell;* Relh. Brinkley. Fulbourn.—2. Royston.—3. Toft; Wimpole; Bourn; Barrington; Caldecot; N.—5. Swaffham!; H.

4. S. móllis Parl.

Festuca altera, R. C. 53. *Festuca avenacea hirsuta paniculis minus sparsis,* R. Cat. Angl. ed. 1, 111. *Br. mollis,* Relh. 42.

Meadows, pastures, way-sides. A. May, June.
Common throughout the county.

* 5. S. arvénsis Godr.

First noticed in this county by Mr Newbould in 1856.
Waste ground. A. July, August.

2. In fields and hedges near Royston; N.—3. On some gravelly waste ground by a field between Toft and Caldecot, where it has been noticed for three successive years in abundance. Long Stow; Caxton; N.—4. Bird's Pastures, near Childerley; N. By the Huntingdon road.

Brachypódium Beauv.

1. B. sylváticum R. and S.

Gr. avenaceum dumetorum spica simplici, R. Cat. Angl. ed. 1, 140. *Festuca sylvatica,* M. Pl. 3. Relh. ed. 1, 44. *Bromus sylvaticus,* Lyons, 15. Relh. ed. 3, 44.

Hedges and thickets. P. July.

Common in the (1) Cambridge, (2) Royston, (3) Wimpole, and (4) Cottenham Districts.—5. Horningsey. Fen Ditton. Chippenham.

2. B. pinnátum Beauv.

Festuca pinnata, Relh. ed. 1, 44. *Bromus pinnatus,* M. Pl. 3. Relh. ed. 3, 45.

Dry, chalky ground. P. July.

1. Gogmagog Hills. By Wort's Causeway, Cambridge. Fulbourn; W. H. C. Hildersham; Babraham; G. S. G.—2. Sawston!; Guilden Morden; Royston; N.—3. About Hardwick, Eversden, and Kingston Woods. Between Croydon and Clapton Farms, Croydon. Road-side near Eltisley. Caldecot. Toft; Crane's Lane, Kingston; Near Hayley Wood; Tadlow; Caxton; Wimpole; N.—4. Dry Drayton; Childerley; N.—5. Devil's Ditch.

TRÍTICUM Linn.

1. **T. canínum** Huds.

M. Pl. 3. Relh. 51.

Banks and hedges. P. July.

Common in the (1) Cambridge, (2) Royston, (3) Wimpole, and (4) Cottenham Districts.—5. Fen Ditton. Chippenham.—7. Doddington.

2. **T. répens** Linn. *Couch-grass.*

Gr. caninum, R. C. 66. *T. sylvestre radice repente quod Gramen caninum officinarum*, M. M. 104. *T. repens*, M. Pl. 3. Relh. 51.

Waste and cultivated land. P. July.

Common throughout the county, although I have no station recorded in the (8) Wisbech District.

3. **T. púngens** Pers. (See Appendix VII.)

T. junceum, Lyons, 19. Relh. ed. 1, 55. *T. repens* γ, Relh. ed. 1, Suppl. ii. 9; ed. 3, 51.

First noticed by Prof. J. Martyn.

Near the sea. P. July.

8. On both sides of the river at Foul Anchour. Above and below the town of Wisbech abundantly.

[T. Martyn and Ray include the cultivated forms of *Triticum*, the various kinds of Wheat, in their lists of our plants.]

HÓRDEUM Linn.

1. H. praténse Huds. *Barley-grass.*

H. murinum β, M. Pl. 3. *H. pratense*, Relh. 50.
Damp meadows. P. July, August.
Appears to be common throughout the county.

2. H. murinum Linn. *Wall Barley. Way-bennet.*

H. spurium, R. C. 77. *H. murinum*, M. M. 105. M. Pl. 3. Relh. 50.
Waste sunny places and under walls. P. June, July.
Tolerably common throughout the county.
[T. Martyn and Ray include cultivated Barley in their catalogues.]

3. H. marítimum Wither.

Hensl. Cat. ed. 1, 28.
Near the sea. A. June.
8. Wisbech!; H.

LEPTÚRUS R. Br.

1. L. incurvátus Trin.

Rotbollia incurvata, Relh. ed. 1, Suppl. ii. 8; ed. 3, 49.
Near the sea. A. July.
8. Wisbech; Near Foul Anchour!; W. M.

Lólium Linn.

1. L. perénne Linn. *Red Darnel. Rye-grass.*

L. rubrum, R. C. 90. M. M. 105. *L. perenne,* M. Pl. 3. Relh. 48.

Pastures. P. June.

Common throughout the county.

[2. **L. temuléntum** Linn. *Darnel.*

L. album, R. C. 90. M. M. 105. *L. temulentum,* M. Pl. 3. Relh. 49.

Corn-fields. A. June to August.

1. *Corn-fields near Cambridge; Barnwell Gravel-pits;* Relh.—3. Between Barton and Coton. Between Harlton and Eversden!; Toft; N.

β. *L. arvénse* Wither.

1. *Gogmagog Hills;* Relh.—3. Eversden!; N.]

[3. **L. itálicum** A. Braun. *Italian Rye-grass.*

Cultivated land. B. June.

This has several times been found, but does not continue in any place, being always an escape from cultivation.]

ACOTYLÉDONES OR CRYPTOGÁMEÆ.

EQUISETACEÆ.

Equisétum Linn.

1. E. arvénse Linn. *Corn Horse-tail.*

E. arvense longioribus setis, R. C. 48. M. M. 12. *E. arvense,* M. Pl. 23. Relh. 423.

Damp meadows. P. April.

Common throughout the county.

2. E. Telmatéja Ehrh. *Greater Marsh Horse-tail.*

E. primum, R. C. 49. *E. officinarum,* M. M. 12. *E. fluviátile,* M. Pl. 23. Relh. 424.

Wet places. P. April.

3. By a water-course between Eversden Wood and a clunch-pit in that parish!; N.—6. Ely!; H.

3. E. sylváticum Linn.

M. Pl. 23. Relh. 423.

Wet shady places. P. April, May.

3. *Madingley Wood;* J. M.—4. *Chesterton;* Relh.

4. E. limósum Linn.

M. Pl. 23. Relh. 424.

In stagnant water. P. June, July.

1. In the brook, and in a pond, Fulbourn. Wilbraham Fen. Linton; G. S. G.—2. Peat-holes, Triplow.—3. Gam-

lingay. Eltisley. By the Mare Way above Cobb's Pound. *Grantchester;* J. M. Kingston; Croydon; Barrington; N. —4. Pits near the Observatory.—5. In a pit by the way from Chippenham to Herringswell. Swaffham Bulbeck. Bottisham Fen. Quarry, Upware.

β. *E. fluviátile* Linn.

3. In the Bourn Brook by the road to Haslingfield. Kingston!; N.—6. Aldreth. Roswell Pits, Ely.—7. Turf-fen, Doddington.

5. E. palústre Linn.

M. Pl. 23. Relh. 423.

Spongy bogs. P. June, July.

1. Fulbourn. Brinkley. Cow Fen, Cambridge; W. H. C. —2. Peat-holes, Triplow. Whittlesford. Sawston Moor!; N.—3. Gamlingay. Barrington. Old brick-pits between Bourn and Kingston Wood; Comberton; Toft; N.—4. In small quantity in a clay-pit at Elsworth; N.—5. Clayhithe. Bottisham. Chippenham.—6. Witchford.

6. E. hyemále Linn.

E. nudum, R. C. 48. M. M. 12. *E. hyemale,* M. Pl. 23. Relh. 424.

Wet places. P. June, July.

3. *Gamlingay bogs;* Relh.—6. *Stretham Ferry;* Relh.

FILICES.

Polypódium Linn.

1. P. vulgáre Linn. *Common Polypody.*

Polypodium, R. C. 122. *P. officinarum,* M. M. 6. *P. vulgare,* M. Pl. 24. Relh. 429.

Shady banks, walls, and old trees. P. August to October.

1. Wood Ditton. Shudy Camps!; H. *On Trinity Hall wall in Garrett Hostel Lane, Cambridge;* Ray. *Trumpington church;* Relh.—2. Sand-pit plantation at Odsey; H. F.—3. On a tree in Whitwell Wood, near the brook.—4. Girton church; On a willow by the river opposite Baitsbite; W. H. C. Clunch-pit, Elsworth; T. Y. *Madingley and Chesterton churches;* Relh. — 6. *About the cathedral, Ely;* J. M.—8. *Wisbech;* T. M.

[*P. calcareum* (Sm.), *Dryopteris Tragii* (Ray) is said by Dent (App. ii. 6) to grow on King's College walls, but there must have been some mistake.]

LÁSTREA Presl.

1. L. Thelýpteris Presl.

Bot. Gaz. ii. 305.

First noticed in this county by Mr W. Marshall.

Wet fens. P. July, August.

5. Wicken Fen, abundantly.—6. Recently in the West Fen, Ely, which is now drained; W. M.

2. L. Oreópteris Presl.

Filix palustris seu aquatica, R. C. App. ii. 7; Cat. Angl. ed. 2, 108. *F. minor non ramosa,* M. M. 7. *Acrosticum Thelypteris,* M. Pl. 23. *Polyp. Thelypteris,* Relh. ed. 1, 391; ed. 2, 410. *Aspidium Oreopteris,* Relh. ed. 3, 430.

Heaths. P. July.

3. *Gamlingay* (in company with *Athyrium Filix-fœmina*); Dent.

Relhan states that it grew by the footpath to the mill, on both sides of the way, at Fulbourn; and on Fulbourn and Teversham Moors. It is probable that these plants were the true *L. Thelypteris,* although he latterly placed them under *L. Oreopteris* in his Flora. Dent's original mistake in quoting (although with doubt) the *F. minor non ramosa* of Bauhin (which is *L. Thelypteris*) as synonymous with his *F. palustris seu aquatica* has caused much confusion.

3. L. Filix-más Presl.

Filix-mas, R. C. 53. *F. mas officinarum*, M. M. 7. *Polyp. Filix-mas*, M. Pl. 24. Relh. ed. 1, 392. *Aspid. Filix-mas*, Relh. ed. 3, 431.

Hedge-banks. P. June, July.

1. By road to Kirtling Green from Wood Ditton. Borley and Balsham Woods. Lenhall Wood, West Wratting. Devil's Ditch; Abington Park; S. W. W. *Linton;* Relh. —3. Comberton; N. By Haslingfield church; W. H. C. Gamlingay; T. Y. *On the willow-trees by the old sluice at Grantchester;* Relh.—4. Near the water in Moor Barns Thicket; W. H. C. Near Howe's House; Mr C. Lestourgeon. *King's Hedges;* Relh.—5. By a ditch in the meadows near the brook between Paper Mills and the river!; Mr T. M^{c}K. Hughes. — 8. By ditches near Wisbech, Guyhirne, and Leverington; A. P.

4. L. spinulósa Presl.

Bot. Gaz. ii. 305.

First noticed in this county by Mr Wanton in 1850.

Swampy places. P. August, September.

1. On an Alder-stump near the old mill at Fulbourn!; S. W. W.

5. L. dilatáta Presl.

Filix-mas ramosa pinnulis dentatis, R. Cat. Angl. ed. 2, 108. *Polyp. cristatum*, M. Pl. 24. Relh. ed. 1, 391. *P. dilatatum*, Relh. ed. 3, 432.

Woods and banks. P. August, September.

1. Wood Ditton Park Wood.—3. Gamlingay.

FILICES. 293

POLYSTICHUM Roth.

1. P. aculeátum Roth.

Polyp. aculeatum, M. Pl. 24. Relh. ed. 1, 392. *Aspid. aculeatum*, Relh. ed. 3, 431.

First found by Prof. J. Martyn.

3. *Gamlingay Park;* J. M.

ATHÝRIUM Roth.

1. A. Filix-fœmina Roth. *Lady-fern.*

Filix-mas non ramosa pinnulis angustis raris profunde dentatis, R. C. App. ii. 7. M. M. 7. *Polyp. Filix-fœmina*, M. Pl. 24. Relh. ed. 1, 392. *Asplenium Filix-fœmina*, Relh. ed. 3, 431.

Damp shady places. P. June, July.

3. Gamlingay Park. North side of Haslingfield church; W. H. C.—4. *On the east side of Swavesey church;* Relh., but not there in 1850; S. W. W. Near the water in Moor Barns Thicket; W. H. C.

ASPLÉNIUM Linn.

1. A. Adiántum nigrum Linn. *Black Spleenwort.*

Adiantum nigrum vulgare, R. C. App. ii. 1; Cat. Angl. ed. 1, 95. *Dryopteris nigra*, M. M. 8. *A. Adiantum nigrum*, M. Pl. 24. Relh. 429.

Walls. P. June to September.

1. Hildersham church; S. W. W. *Hinton church;* Relh.—3. *Coton;* J. M.—4. Elsworth church; N.—5. Fen Ditton church.—8. *Wisbech; Tydd St Giles;* J. M.

2. A. Trichómanes Linn. *Common Spleenwort.*

Trichomanes, R. C. 165. M. M. 7. *A. Trichomanes*, M. Pl. 23. Relh. 428.

Walls. P. May to October.

1. *On the south side of Stetchworth church;* T. M.—2. Bassingbourn church. — 4. Swavesey church. *Over church;* Relh.—6. Ely cathedral.—8. *Wisbech; Tydd St Giles;* J. M.

3. **A. Rúta-murária** Linn. *Wall Rue. Tentwort.*

Adiantum album, R. C. 4. *Ruta muraria officinarum*, M. M. 8. *A. Ruta-muraria*, M. Pl. 24. Relh. 428.

Walls. P. May to September.

1. King's College Chapel, Clare Hall Bridge, Senate-House, Cambridge. Hinton church; W. H. C. Westley Waterless church; R. B. S.—2. Litlington church.—3. *Barton church;* J. M.—4. Chesterton church; W. H. C.—6. *Swaffham church;* J. F.—8. *Wisbech;* J. M. Walls at Elm; A. P.

Scolopéndrium Sm.

1. **S. vulgáre** Sym. *Hart's-tongue.*

Phyllitis, R. C. App. i. 7. M. M. 6. *Aspl. Scolopendrium*, M. Pl. 23. Relh. ed. 1, 388. *S. vulgare*, Relh. 427.

Damp shady places. P. July, August.

1. *Hinton church;* Ray. *Hedge by Barnwell pool; In the well at the Duke of Leeds' house, Gogmagogs;* Relh. — 2. *Melbourn churchyard;* Relh.—3. Caxton churchyard; N. Gamlingay; Relh.—4. Over; N. *In a hedge by a grove at Chesterton;* T. M.—6. In a well at Ely; S. W. W.

Bléchnum Linn.

1. **B. boreále** Sw. *Hard-fern.*

Lonchitis aspera, R. C. App. i. 6. M. M. 9. *Osmunda Spicant*, M. Pl. 23. Relh. ed. 1, 387. *B. Spicant*, Relh. ed. 2, 407. *B. boreale*, Relh. ed. 3, 426.

Heathy places. P. July.

3. Gamlingay.

Relhan introduced *Polypodium* [*Asplenium*] *Lonchitis* into his second and third editions, as found by Ray at Gamlingay. He quotes Ray's Cat. App. i. 6, as his authority; but a reference to that place shows that he has mistaken the plant. Ray intended the *Blechnum boreale* or *L. aspera minor* (C. B.) not *L. a. major*.

PTÉRIS Linn.

1. P. aquilína Linn. *Brakes.*

Filix-fœmina, R. C. 54. M. M. 7. *P. aquilina*, M. Pl. 23. Relh. 427.

Woods and heaths. P. July.

1. Wood Ditton Park Wood. Balsham Wood; Mr Jas. Carter. Furze-hills, Hildersham. Linton and Borley Woods; S. W. W. Newmarket Heath. A curious young state was found on a damp wall in a court near St Clement's church, Cambridge, in Oct. 1859, by Mr G. B. P. Fielding of St John's College. — 3. On the Heath and in White Wood, Gamlingay. Triplow; W. H. C.—4. By the foot-way from Fen Ditton to Horningsey; Rev. S. Hiley.—6. On the damp walls of the Cathedral, Ely, for a few years!; W. M.

BOTRÝCHIUM Sw.

1. B. Lunária Sw. *Small Moon-wort.*

Lunaria minor, R. C. App. ii. 12. M. M. 9. *Osmunda Lunaria*, M. Pl. 23. Relh. 426.

Pastures. P. June, July.

1. Little Linton Warren, on the right of the road to Cambridge, at about half a mile from the former town. *Gravel-pit to the left of the road to Balsham, one mile from*

the town; On the first hill next Hildersham, towards Juniper Hill; Dent.—5. *Gravel-pits by Chippenham Park;* Relh.

[*Osmúnda regális* (Linn.) is included in Relhan's Flora, because it is in R. C. App. ii. 7 (*Filix florida seu O. regalis*), where it was inserted by Dent, who stated that it grew at the corner of Gamlingay Park, next to Sandy. No other botanist seems to have found it there, and Ray does not give the locality in his Synopsis nor elsewhere. A mistake is therefore probable.]

OPHIOGLÓSSUM Linn.

1. O. vulgatum Linn. *Adder's-tongue.*

Ophioglossum, R. C. 105. *O. officinarum,* M. M. 9. *O. vulgatum,* M. Pl. 23. Relh. 425.

Pastures. P. May, June.

1. Shelford. Shudy Camps!; H. Castle Camps; Linton; G. S. G. West Wratting. *Hinton Moor;* Relh.—3. Last field by foot-way to Coton. Gamlingay. Near Toft bridge. Harlton; Caldecot; Wimpole!; Orwell; Eversden; Kingston Wood; N. Trumpington Spinney; W. H. C. Whitwell, near Coton; S. W. W. *Grantchester;* J. M.—4. At one mile on the road to Histon; H. In the chalk-pit and behind the Park, Madingley. Between the wind-mill and Madingley Wood; W. H. C. Over!; Mr J. Brown. Dry Drayton; N.—5. Horningsey. Wicken Fen. Quy Hall; Mr R. A. Julian. Fen Ditton; J. M.

LYCOPODIACEÆ.

LYCOPÓDIUM Linn.

1. L. clavátum Linn. *Common Club-moss.*

M. Pl. 24. Relh. 424.

First found by Prof. T. Martyn.

Heaths. P. July, August.
3. Gamlingay Heath.

2. L. inundátum Linn.

Relh. ed. 1, 393; ed. 3, 425.
Boggy heaths. P. August, September.
3. Boggy parts of Gamlingay Heath.

CHARACEÆ.

Chára Linn.

1. C. fléxilis Linn.

Hensl. Cat. ed. 2, 93.
First found in this county by Prof. Henslow in 1831.
Ditches. P. May.
3. By Lord's Bridge, near Barton. — 5. Reche Lode. Bottisham Fen.

2. C. syncárpa Thuil.

Bab. Man. ed. 3, 420.
First found in this county by C. C. B. in 1850.
Stagnant water. A. May.
3. Brick-pits at Gamlingay.

3. C. tenuíssima Desv.

C. gracilis, Hensl. Cat. ed. 1, 28. *C. hyalina*, Hensl. Cat. ed. 2, 93. *C. tenuissima*, Bab. Man. ed. 3, 420.
First found by Prof. Henslow in 1829.
Fen ditches. A. July, August.
5. Bottisham and Burwell Fens.

4. C. polypérma A. Br.

C. nidifica, Relh. ed. 3, 3. *C. polysperma*, Bab. Man. ed. 3, 421.

Clear streams. A. April.

1. In Hobson's Conduit stream, near the Nine Wells. In ponds by the side of the lane leading to the Park Wood, Wood Ditton; H.—3. At one mile on the road from Haslingfield to Barton. *Near a spring in Eversden Wood;* Relh. —4. Knapwell!; Mr J. B. Wilson.

5. C. vulgáris Linn.

Equisetum fœtidum sub aqua repens, R. C. 48. *Hippuris fœtida*, M. M. 13. *C. vulgaris*, M. Pl. 26. Relh. 3.

Ditches and streams. A. June to August.

1. In Hobson's Conduit stream, Cambridge. Brinkley. —2. Sawston Moor; N.—3. Brick-pits, Gamlingay. Hardwick. Toft. Haslingfield. Cambridge. Coton. (The celebrated petrifying spring at Coton was full of this plant. It was situated on the side of the hill above the road from Coton to Barton Toll-gate.)—5. Wicken, Bottisham and Snailwell Fens. Swaffham Bulbeck.—6. Witcham.

6. C. híspida Linn.

C. tomentosa, M. Pl. 26. Relh. ed. 1, 348. *C. hispida*, M. Pl. 26. Relh. 3.

Fen ditches. A. May to August.

1. Near the Old Mill, Fulbourn. Teversham Moor; W. H. C.—2. *Moor near Shelford;* Relh.—3. Sheep's Green, Cambridge. Pit near Kingston Wood.—4. Pits near the Observatory. Waterbeach Fen.—5. Bottisham and Wicken Fens.—6. Witchford. Witcham.—7. Doddington Turf-fen.

APPENDIX.

No. I. On Thalíctrum saxátile.

As our plants were not examined in the flowering state there remains some doubt concerning the species to which they belong. Indeed it is possible that one or more of them may be the *T. flexuosum* (Bernh.) The *T. saxatile* and *T. flexuosum* differ in the following respects.

1. *T. flexuósum* (Bernh.); stem striate; flowers drooping, in a long leafy panicle; carpels narrowly oblong, gibbous.

Stem less zigzag than that of *T. saxatile*, usually much branched. Fruitstalks often patent. Carpels curved, gibbous on one side above on the other below, rather acute at one or both ends.

2. *T. saxátile* (Schleich.); stem scarcely striate; flowers erect, in a nearly leafless panicle; carpels oval.

Stem zigzag, usually unbranched below. Panicle closer. Fruitstalks ascending. Carpels nearly regularly oval, rather blunt at both ends. Leaves smaller and more compact.

The latter seems to be the *T. saxatile* of Reichenbach (Icones, t. 34) and probably of Schleicher, although the latter botanist perhaps confounded *T. Kochii* (Fries) with it. It is much nearer in appearance to *T. minus* than either of the other species, but was placed doubtfully with *T. flexuosum* by me.

If I am correct in believing that its *flowers do not nod* and that its *carpels are nearly exactly oval*, it would seem to be properly separated from all our other species. It is apparently the *T. collinum* of Reichenbach's "Flora exsiccata" No. 691, and has its lower leaves crowded together and nearly sessile when growing amongst short herbage. The *T. collinum* (Wallr.) has *nodding flowers*, according to his description (Sched. 259). The *T. collinum* of Koch (Doutschl. Flora, iv. 130) may be the same as that of Wallroth; but Fries informs us that the latter author denied their identity, and he therefore names the plant *T. Kochii*. I have thought it safest to follow Fries in a case of so much difficulty, and now use the name of *T. Kochii* for the *T. saxatile* of my Manual, ed. 4.

No. II. On Papáver Dúbium.

We have unquestionably two plants confounded under this name in England, both of which occur in Cambridgeshire. It is probable that they have escaped notice owing to the early hour in the day at which the petals fall. When perfectly full-blown flowers are contrasted their differences are conspicuous. Many differences are also to be found in their ripe capsules, the colour of their sap, &c.

Our plants may be characterized and described as follows:

1. *P. Lecóqii* (Lamot.); filaments subulate; stigmatic disk broader than the capsule and folding over its edge, convex-conic, ultimately flat; capsule oblong, club-shaped, suddenly narrowed near the base, which is narrower than the torus; leaves bipinnatifid, with distant, narrow, entire, acute lobes.

P. dubium *Schk. Handb.* t. 140.

Argemone capitulo longiore glabro, *Morris. Oxon.* Pt. 2, p. 279, tab. 14, f. 11.

Sap of the whole plant turning *dark yellow* (ochraceous) in the air. Filaments violet; anthers brownish, just level with the stigmatic disk. Stigmatic disk of the ripe capsule quite flat. Stigmatic rays very nearly, but not quite, reaching to the edge of the disk, which is obscurely but angularly lobed: the lobes are *bluntly triangular* at their end, or even tricrenate (that is, having three crenatures), and deeply divided from each other but rather overlapping below. This disk is broader than the top of the capsule. Seeds reniform, netted. Hairs on the peduncles all adpressed: those on the leaves bubous-based.

It is probable that this is the *P. Lecoqii* (Lamot.), although that plant is stated to have suborbicular petals and stigmatic rays reaching to, or even extending beyond, the edge of the disk. No author except Crepin (Notes sur quelques Plantes rares ou critiques de la Belgique), who is usually very accurate, states anything concerning the colour of the sap of the allied plants. He says that the sap is milky and does not turn yellow in any of them except *P. Lecoqii*. If he is correct in this statement our plant must either be *P. Lecoqii* or be unknown to him. He has favoured me with leaves and stigmatic disks of his *P. Lecoqii* and *P. modestum* (to which latter I was inclined to think that our plant might be referable). Our plant has similar leaves to those of the former; and also similar disks, in all respects except in the *want of a small central conical point*, which he finds to be present upon his plant, and which is very manifest upon the specimens received from him. He is not certain concerning the shape of the petals, but thinks that they are large and suborbicular in his *P. Lecoqii*. He thinks, probably with justice, that little confidence can be placed in their shape. If Jordan is right, as I cannot but suppose, the petals of our plant are exactly like those of *P. modestum;* and if Boreau is correct, as I equally believe, they differ greatly from those of *P. Lecoqii*. When the flower is in perfection the stamens of *P. modestum* "n'atteignent que les $\frac{3}{4}$ de la hauteur de la capsule." The name must therefore remain somewhat doubtful.

This plant is very abundant about Cambridge.

2. *P. Lamóttei* (Bor.); filaments subulate; stigmatic disk broader than the capsule, patent at the edge, its center convex-conic, ultimately nearly flat; capsule clavate, gradually narrowing from near its top to its base, which is as wide as the torus; leaves pinnatifid, with distant, broad entire, bluntish lobes.

P. dubium, *Eng. Bot.* t. 644, *Curt. Fl. Lond.* ii. 104 (fasc. v. 37). *Hayne Gew.* vi. 39.

Sap milky, not turning yellow in the air. Filaments violet; anthers brownish, just raised to the level of the edge of the stigmatic disk. Stigmatic disk of the ripe capsule usually slightly conical in the middle, with turned-up wavy edges, nearly circular. Stigmatic rays not quite reaching to the edge of the disk, which is obscurely lobed; the lobes are separated by very slight notches. In a very young state the stigmatic disk is folded over the ovary, but in the fruit it projects all round like the roof of a house. The ovary is much narrowed in its lower half. The hairs on the peduncles are all adpressed, except for a short distance above its base, where they spread. The petals are of a rather paler red than those of *P. Lecoqii.*

It is found at Chippenham in this county, on a sandy soil. It also occurs in other parts of the kingdom. Mr Newbould found it at Sheffield in 1859.

No. III. Vióla canína Linn.

Ray remarks of his plant, named *V. canina sylvestris*, that it grows "ad sepes, et in dumetis passim. Habetur et in palustribus frequens, nisi forte ea sit distincta species." *Cat. Angl.* ed. 1, 317. The former is therefore our *V. sylvatica*, the latter *V. canina*. Linnæus is supposed by some botanists to have derived his plant from the books of Ray and Gerard; but he quotes neither of them in the *Hortus Cliffortianus* when founding the species. His character, "V. foliis cordatis oblongis, pedunculis fere radicatis," will not apply to our *V. sylvatica*, nor does the cut in Tilland's *Icones Novæ*, 110, represent it, but is a tolerable figure of our *V. canina*. What we have had to determine is not what was the plant of Gerard, which differs in the two editions of his Herbal, but what was really intended by Linnæus. I fully agree with Fries in believing that the type of the Linnæan *V. canina* is the plant which he and I and most of the continental botanists so name. But even if any reasonable doubt remained upon this point we cannot now again alter the names without creating very great confusion.

No. IV. On Arenária serpillifólia.

My attention was first directed to the plant here called *A. leptoclados* by Mr Borrer, whose labours have done so much for the advancement of British botany. He gathered it at Henfield, Sussex, in 1844, and correctly identified it with the *A. serpillifolia β leptoclados* of Reichenbach (*Icones*, v. f. 4941 β). I introduced it into the first and succeeding editions of my *Manual* as the *A. serp.* β *tenuior* of Koch. Recently I have been led to pay more particular attention to it, and incline to the opinion that those botanists are correct who separate it specifically from *A. serpillifolia*. The most palpable difference between them is found in the size of nearly all their organs, although the plants are of about equally vigorous growth. *A. leptoclados* has leaves, flowers, and fruits of about half the size of those of *A. serpillifolia*, which has the effect of giving the plant a much more slender appearance.

Tenore was the first botanist who specifically distinguished the plants; for in his *Relaz. del Viag. di Abruzzo*, p. 66, published apparently in 1830, he described his *A. sphærocarpa* as distinct from what he supposed to be the *A. serpillifolia* of Linnæus. I have not seen that work, but find an account of the two plants in his *Sylloge Pl. Floræ Neapolitanæ*, p. 219. He there tells us that *A. serpillifolia* [our *A. leptoclados*] "a sequente" *A. sphærocarpa* [our *A. serpillifolia*] " dignoscitur laciniis calycinis lanceolato-cuspidatis, nec late ovatis;" and under *A. sphærocarpa* he says, " magnitudine et forma capsularum a simillima *A. serpillifolia* primo intuitu diversa deprehenditur; a qua tamen, sepalis ovatis caule pedunculisque plerumque erectis, aliisque notis facile dignoscitur."

Gussone adopted Tenore's names in his *Fl. Siculæ Synopsis* (i. 495), but in the *Addenda* to that work (p. 824) he alters them, from having learned that the *A. sphærocarpa* is the true *A. serpillifolia* of Linnæus, and gives the name of *A. leptoclados* to his and Tenore's *A. serpillifolia*. This nomenclature and separation of the plants is now followed by continental botanists.

But there remains another plant to be noticed, to which the *A. serpillifolia* which grows near Wisbech (but of which I possess no specimen) may belong. It is the *A. Lloydii* of Jordan, which closely resembles the true *A. serpillifolia*, from which it is very possibly not distinct, but has elevated ribs on its sepals in place of the rather faint nerves of its allies.

These plants may be characterized as follows:

1. *A serpillifolia* (Linn.); leaves ovate, acute, roughish, sessile; petals shorter than the calyx; *sepals ovate-lanceo-*

late acute; *3—5-veined*, hairy on the veins; fruit-stalks erect or patent, straight, longer than the *globose-ovoid capsules, which narrow gradually to their top* and exceed the calyx.

A. serpillifolia, *Linn. Sp. Pl.* 606. *Eng. Bot.* t. 923. *Curt. Fl. Lond.* ii. 87 (fasc. iv. 32). *Reichenb. Icones*, v. t. 216. *Guss. Syn.* ii. 824. *Godr. Fl. Lorr.* ed. 2, i. 122. *Lloyd. Fl. ouest de la Fr.* 77. *Bor. Fl. centr. de la Fr.* ed. 3, 109. *Crep. Pl. rares de la Belg.* 7.

A sphærocarpa, *Ten. Syll.* 219. *Guss. Syn.* i. 495.

Stem much branched, especially in its upper part; flowers from its forks or the axils of the upper leaves; inner sepals with 3, outer with 5, not very strongly marked nor elevated veins; shorter than the capsule. Fruit-stalks once or twice longer than the capsules. Capsule much inflated below, but narrowing gradually to its top, hard and brittle when ripe.

This seems to be the common plant in England.

2 ? *A. Lloydii* (Jord.); leaves ovate, acute, roughish, sessile; petals shorter than the calyx; sepals ovate, acute, *3—5-ribbed*, hairy on the ribs; fruit-stalks erect-patent, straight, short; capsules globose-ovoid, exceeding the calyx.

A. Lloydii, *Jord. Pugil.* 37. *Lloyd. Fl. ouest*, 77. *Bor. Fl. centr.* ed. 3, 109.

A. serpillifolia β macrocarpa, *Lloyd. Fl. Loire*, 42.

This plant is exceedingly like the preceding, and probably not distinct from it. I can detect no tangible difference except that the ribs on the sepals are very manifestly much stronger and more elevated.

It grows in maritime districts on the continent. Mr A. G. More found what may be it in the Isle of Wight.

3. *A. leptocládos* (Guss.); leaves small, ovate, acute, roughish; petals shorter than the calyx; *sepals lanceolate*, acute, *3-veined*, hairy on the veins; fruit-stalks patent, longer than the *ovoid-oblong capsules*.

A. leptoclados, *Guss Syn.* ii. 824. *Godr. Fl. Lorr.* ed. 2, i. 123. *Lloyd Fl. ouest*, 77. *Bor. Fl. centr.* ed. 3, 109. *Crep. Pl. rares*, 7.

A. serpillifolia, *Ten. Syll.* 219. *Guss. Syn.* i. 495. *Lloyd. Fl. Loire*, 42.

A. serpillifolia β leptoclados, *Reichenb. Icon.* v. 32, t. 216.

A. serpillifolia γ tenuior, *Koch. Syn.* ed. 2, 128. *Bab. Man.* ed. 4, 52.

Stems slender, much branched; flowers from the forks or the axils of the leaves. Sepals 3-veined. Fruit-stalks much longer than the capsules, usually (perhaps always) curved at the top when the fruit is not quite ripe, but ultimately straight. Capsules not inflated below, not hard and brittle when ripe, but giving way to pressure; usually (but not always) rather longer than the calyx. All the organs much smaller than those of *R. serpillifolia*, scarcely of half their size.

It remains to be discovered how far this is a common plant in England. My specimens are from Sidmouth, Devon; Clevedon, Somerset; and Henfield, Sussex. Mr Newbould has found it in many places in Cambridgeshire.

No. V. On several Brambles.

There are two species of Bramble named in this *Catalogue*, which do not appear in the fourth edition of the *Manual of British Botany;* I have therefore thought it proper to introduce here some extracts from my manuscript *Monograph of the British Rubi* relating to them.

9. *R. althæifolius* (Host); stem prostrate, slightly angular, with scattered hairs and setæ; prickles many, unequal, slender, patent from an oblong compressed base; leaves quinate or ternate; *leaflets crenately lobed,* pale green, and with hairs on the veins or loose white tomentum beneath; *inferior leaflets of the ternate leaves retrorsely bipartite,* of the quinate leaves sessile, not overlapping the intermediate leaflets; *terminal leaflet rhomboidal-obovate,* subcordate below; panicle leafy, with the axillary branches and top racemose-corymbose, with few very short setæ, the prickles on the middle of the rachis longest and

slender; *sepals* ovate-subacuminate, setose, *loosely adpressed* to the black-blue fruit; petals obovate; styles flesh-coloured at the base.

R. althæifolius, *Host in Trattin Ros.* 37; *Fl. Aust.* ii. 31. *Ser. in DC. Prod.* ii. 562.

R. Wahlbergii β glabratus, *Bell-Salt in Bot. Gaz.* ii. 129; in *Fl. Vect.* 160 (Syn. excl.).

10. *R. tuberculátus;* stem arching very slightly, with scattered short hairs and short setæ; *prickles* many, unequal, slender, patent *from an oblong tubercular base;* leaves ternate or quinate; leaflets rather doubly dentate, hairy on the veins beneath, green on both sides; basal leaflets of the ternate leaves bilobate; terminal leaflet roundish-cordate, subcuspidate; *basal leaflets of the quinate leaves* subsessile, *overlapping the intermediate leaflets;* panicle leafy, with racemose axillary branches and a corymbose top; prickles from the middle to the top of the panicle and peduncles slender and longest; sepals ovate-acuminate, prickly, setose, loosely adpressed to the fruit.

R. nemorosus δ ferox, *Leight. Shropshire Rubi.*

R. dumetorum, *Blox. Fasc. of Rubi.*

Although Mr Leighton named this plant wrong and never published any account of it, it is to him that we are indebted for pointing out its characteristics. This is not the place to enter upon a discussion of the reason for considering that our plant is not the *R. nemorosus* γ *ferox* of Arrhenius, nor the similarly named variety of *R. dumetorum* of Weihe, nor his *R. ferox*. My statement that such is the case must be accepted at as much as it is worth until my projected *Monograph* is published.

The *Rubi cæsii*, as now understood by me, may be shortly distinguished as follows:

1. Sepals reflexed from the fruit .... *R. corylifolius.*
Sepals erect, patent, or loosely adpressed
   to the fruit .... 2.

2. Basal leaflets incumbent .... 3.
Basal leaflets not incumbent .... 4.

3. Prickles from subcompressed bases.
 Styles flesh-coloured.......................*R. Balfourianus.*
 Prickles from tuberculiform bases. Styles
 yellowish-green.............................*R. tuberculatus.*

4. Leaves quinate or ternate. Leaflets crenate-lobate; basal leaflet of ternate leaves retrorsely bipartite. Styles flesh-coloured at the base................*R. althæifolius.*
 Leaves ternate, or rarely quinate-pinnate. Leaflets incised, or coarsely serrate. Styles greenish......................*R cæsius.*

No. VI. On Serrafálcus.

The *S. racemosus* of my *Manual* (ed. 4) is not the true plant, but a state of *S. mollis*. It is only recently that I have learned to know the real *S. racemosus*, of which there is a very fair figure in *English Botany*, t. 1079. It is not represented in Parnell's *British Grasses*, for the plant which is there so named is a glabrous form of *S. mollis*.

There is a very valuable character by which to distinguish some of the species of this difficult genus which has long been pointed out by continental botanists, but which I totally misunderstood until after the issue of the fourth edition of my *Manual*. It is found in the shape of the lower (outer) pale, the sides of which present either a uniform curve from the tip nearly to the base, or have at about a third from the tip an obtuse but very well-marked angle.

Subjoined are the corrected specific characters of *S. mollis* and *S. racemosus*.

1. *S. mollis* (Parl.); panicle close, erect, compound, or rarely simple; spikelets ovate, rather compressed, pubescent; florets closely imbricate, about as long as the straight awn; *sides of lower pale bluntly angular above the middle;* leaves and sheaths hairy or downy.

Bromus mollis, *Eng. Bot.* t. 1078. *Parn. Br. Gr.* t. 116.

One to two feet high. Panicle rather close, and nearly or quite always erect. Spikelets varying greatly in their number of florets. Top of the upper glume reaching halfway to the top of the sixth floret; or a little higher in Parnell's *var. ovalis* (tab. 117), which has short oval spikelets; or about halfway to the top of the eighth floret in Parnell's *var. pratensis* (tab. 118), where the spikelets are longer. Lower pale longer than the upper. Anthers about three times as long as broad. Simple peduncles not longer than their spikelets.

Rarely the spikelets are glabrous when it is *Br. racemosus* of Parnell (tab. 119) and of Babington's *Manual*, but not of Linnæus nor Smith.

This is our common Brome-grass; one of the most frequent of all our grasses.

2. *S. racemósus* (Parl.); panicle long, erect, usually simple; spikelets ovate, rather compressed, glossy; florets imbricate, about as long as the straight awn; *lower pale uniformly rounded on the sides;* leaves and sheaths slightly hairy.

Bromus racemosus, *Eng. Bot.* t. 1079.

B. arvensis, *Eng. Bot.* t. 920.

Often more than two feet high. Panicle long, narrow, loose, nearly or quite erect, close with fruit. Spikelets longer in proportion to their width than those of *S. mollis*. Top of the upper glume reaching halfway to the top of the fourth floret. Lower pale longer than the upper. Anthers four times as long as broad.

In this plant the spikelets are rather rough to the touch, rarely at all hairy, but naked and shining. The herbage is not soft like that of *S. mollis*. It inhabits the borders of fields rather than the field itself.

It only remains to give a table of the characters of these and the allied species as follows:

1. Sides of the lower pale uniformly rounded......2.
   Sides of the lower pale angular...................3.

2. Panicle drooping. Sheaths of the leaves nearly glabrous. Florets loosely imbricate, becoming distinct and cylindrical with fruit..*S. secalinus.*

Panicle erect. Sheaths of the leaves hairy (except sometimes the upper ones). Florets imbricate both with flower and fruit, not cylindrical...*S. racemosus.*

3. Glumes and pales downy........................*S. mollis.*
Glumes and pales nearly or quite glabrous.....4.

4. Pales unequal in length; lower equally ribbed. Branches of panicle bearing one or two spikelets ....*S. commutatus.*

Pales equal in length; lower with two prominent ribs near each margin. Branches of panicle usually with more than two spikelets, very long ....*S. arvensis.*

No. VII. On Trítıcum.

The account of the species of *Triticum* in my *Manual* is incorrect. I am indebted to the description in Boreau's valuable *Flore du centre de la France*, edition 3, for a clear view of them. The following seem correct definitions of our British species.

1. *T. caninum* (Huds.); spike rather close; spikelets 2—5-flowered; 3—5-ribbed glumes and lower pales awned; axis and edges of the rachis hispid; leaves flat, rough on both sides; *root* fibrous.

    T. caninum, *Eng. Bot.* t. 1372. *Parn. Grasses,* t. 62.

Stem erect. Ribs on the upper side of the leaves very slender. Glumes rounded on the back; ribs reaching to the tip and joining to form the short awn. Lower pale shorter than its awn; or in an alpine few-flowered form longer than it.

Hedge-banks and thickets.

2. ***T. répens*** (Linn.); spike rather close; glumes 5—6-ribbed, equalling at least two-thirds of the 4—5-flowered spikelet, rough on the keel; lower pale acuminate; *axis asperous;* rachis with rough angles not brittle; *leaves* mostly flat, *with many slender ribs,* each bearing a row of deciduous hairs above; stoloniferous.

T. repens, *Eng. Bot.* t. 909. *Parn.* t. 62.

Stem erect. Ribs on the upper side of the leaves not much raised, not nearly hiding the intermediate surface of the leaf. Rachis glabrous or downy, with prickles on the angles pointing forwards. Glumes scarcely keeled, acuminate-subulate; ribs reaching to the tip.

β *littoreum;* glaucous, leaves involute, pales mucronate.

Glumes more strongly keeled; pale blunt, although mucronate; otherwise like the type of the species.

Waste and cultivated land, very common; β near the sea.

3. ***T. púngens*** (Pers.); spike close; glumes with 7—9 thick ribs, not exceeding half the length of the 5—12-flowered spikelet, rough on the keel; lower pale acute; axis asperous; rachis nearly or quite smooth, not brittle; *leaves with involute edges, their many thick closely placed ribs* slightly rough, and each bearing a row of acute points above; upper part of leaves wholly involute (subulate and rigid); stoloniferous.

Stem erect. Ribs on the upper sides of the leaves so broad and so elevated as nearly to hide the intermediate part of the leaf. Glumes keeled, ribs reaching to the tip. Lower pale (of our plant) usually awned. Producing erect, barren, leafy, clustered shoots as well as the flowering stems.

Sea-shores, common.

4. ***T. acútum*** (DC.); spike rather close; *glumes with 5—7 slender elevated ribs,* blunt or apiculate, not exceeding two-thirds of the length of the 5—8-flowered spikelet; lower pale blunt, mucronate; axis downy; rachis smooth or slightly rough at the angles, not brittle; leaves flat or with involute edges; their many thick closely placed *ribs rough,*

with minute sharp points (asperous) above; stoloniferous.

T. laxum, *Fries Summa,* 249. *Bab. Man.* ed. 1, 411.

T. acutum, *De Cand. Bot. Gall.* 529, not Fries.

Stem prostrate or ascending. Ribs of the leaves, on each of which there is usually a deciduous row of hairs, not so completely hiding the furrows between them as in *T. pungens.* Glumes keeled, keel reaching the tip or forming a slight mucro there, often bearing bristles pointing forwards. Lower pale rarely slightly awned. Producing decumbent and ascending barren leafy clustered shoots as well as the flowering stems.

Sandy sea-shores, probably common.

5. *T. júnceum* (Linn); spike rather loose; *glumes with 9—11 slender, scarcely elevated ribs,* blunt, equalling at least two-thirds of the 4—8-flowered spikelet, smooth on the keel; lower pale blunt, rarely mucronate; axis smooth or slightly downy; *rachis* smooth, *brittle;* leaves involute, with many thick *ribs with much spreading hair above;* stoloniferous.

T. junceum, *Eng. Bot.* t. 814. *Parn.* t. 63.

Stem prostrate. The short hairs on the ribs of the leaves spread so as to cover the intermediate spaces. Rachis easily separating above each spikelet. Glumes rounded or truncate at the tip; ribs not extending to the tip. Producing decumbent, barren leafy shoots as well as the flowering stem.

Sandy sea-shores, common.

The following table will be useful:

1. Root fibrous, no stoles.................................*T. caninum*.
   Stoles long...2.
2. Lower pale acute or acuminate ....3.
   Lower pale wholly blunt or mucronate ....4.
3. Leaves with slender distant ribs...................*T. repens*.
   Leaves with many thick closely placed ribs.....*T. pungens*.
4. Leaves rough above.....................................*T. acutum*.
   Leaves closely downy above ....*T. junceum*.

No. VIII. On the Vegetation of the Fens.

As the kind of vegetation which formerly occupied the Great Level of the Fens is very little known to botanists, to most of whom the Fens are nearly a "terra incognita," it seems desirable to give a complete list of the plants which have been recently found growing in Wicken Fen. A * is appended to the names of those which most abound there. The plant which forms the great mass of the herbage is *Cladium Mariscus*, which is still there regarded as a crop, although an uncultivated one.

*Thalictrum flavum.
Ranunculus heterophyllus.
R. Flammula.
R. Lingua.
R. acris.
R. sceleratus.
Caltha palustris.
Nymphæa alba.
Nuphar lutea.
Erysimum cheiranthoides
Armoracia amphibia.
*Viola stagnina.
Lychnis Flos-cuculi.
Sagina nodosa.
Stellaria glauca.
Malachium aquaticum.
Hypericum quadrangulum.
Linum catharticum.
Rhamnus catharticus.
Vicia Cracca.
Lathyrus palustris.
Spiræa Ulmaria.
Potentilla anserina.
Comarum palustre.
Rubus Balfourianus.
Lythrum Salicaria.
Epilobium hirsutum.
Myriophyllum verticillatum.

M. spicatum.
Hippuris vulgaris.
Hydrocotyle vulgaris.
Apium graveolens.
Sium latifolium.
S. angustifolium.
Œnanthe fistulosa.
Œ. Lachenalii
Œ. Phellandrium.
Angelica sylvestris.
*Peucedanum palustre.
Galium uliginosum.
G. palustre.
G. elongatum.
Valeriana sambucifolia.
*V. dioica.
Eupatorium Cannabinum.
Senecio aquaticus.
S. paludosus.
Centaurea nigra.
Carduus palustris.
*C. pratensis.
Thrincia hirta.
Menyanthes trifoliata.
Convolvulus sepium.
Symphytum officinale.
Myosotis palustris.
Scrophularia aquatica.

Pedicularis palustris.
Rhinanthus Crista-galli.
Veronica Anagallis.
Mentha aquatica.
Lycopus europæus.
Scutellaria galericulata.
Utricularia vulgaris.
Hottonia palustris.
Lysimachia vulgaris.
L. nummularia.
Samolus Valerandi.
Plantago lanceolata.
Rumex Hydrolapathum.
Ceratophyllum demersum.
Callitriche verna.
Salix cinerea.
S. Caprea.
S. fusca.
Hydrocharis Morsus-ranæ.
Stratiotes aloides.
*Orchis incarnata.
*Iris Pseud-acorus.
Juncus effusus.
J. obtusiflorus.
J. acutiflorus.
J. lamprocarpus.
J. supinus.
*Luzula multiflora.
Alisma Plantago.
A. ranunculoides.
Sagittaria sagittifolia.
Butomus umbellatus.
Triglochin palustre.
Sparganium ramosum.
S. minimum.
Lemna trisulca.
L. minor.

L. polyrhiza.
L. gibba.
Potamogeton natans.
P. plantagineus.
P. heterophyllus.
P. lucens.
P. pectinatus.
*Schœnus nigricans.
*Cladium Mariscus.
Scirpus cæspitosus.
Carex disticha.
C. panicea.
C. flava.
C. Œderi.
C. fulva.
C. filiformis.
C. glauca.
C. hirta.
C. paludosa.
C. riparia.
Alopecurus geniculatus.
Calamagrostis lanceolata.
Phragmites communis.
*Agrostis canina.
Holcus lanatus.
Arrhenatherum avenaceum.
Molinia cærulea.
Poa trivialis.
Glyceria aquatica.
G. fluitans.
Briza media.
Dactylis glomerata.
Festuca pratensis.
*Lastrea Thelypteris.
Ophioglossum vulgatum.
Chara vulgaris.
C. hispida.

The names of a few plants may be added which either are now, or were formerly, natives of the Fens, although they have not been noticed in Wicken Fen.

Senecio palustris. We learn from Ray and Relhan that this plant was formerly found in several places. It is now believed to be extinct.

Sonchus palustris has not been found for many years.

Utricularia minor. This probably grows in Wicken Fen, for it is found in many parts of the Level.

Populus nigra. Is supposed to have been a native of the Fen country. Large trees of it are now common, but most, if not all, of them have been planted.

Myrica Gale, was formerly abundant in the Fens, as we learn from Ray.

Epipactis palustris. Abundant in the Fens according to Ray. Still found in several places.

Sturmia Loeselii was very plentiful in the years 1835 and 1836 near Reche, but is now extirpated there. It is doubtful if this plant still exists in our Fens.

Potamogeton pusillus, and

P. rufescens, and

Chara flexilis, and

C. tenuissima, are found in other places.

No. IX. List of lost Plants.

The following list contains the names of those plants which, although recorded upon good authority as natives of Cambridgeshire, have not been found there for very many years.

Thlaspi arvense.
Sysymbrium Irio.
Lepidium latifolium.
Drosera intermedia.
D. anglica.
Frankenia lævis.
Dianthus Caryophyllus.
Geranium rotundifolium.

Vicia sylvatica.
Lathyrus Nissolia.
Prunus Cerasus.
Pyrus torminalis.
Sedum Telephium.
S. album.
S. sexangulare.
S. reflexum.

Ribes nigrum.
Cicuta virosa.
Lactuca saligna.
Sonchus palustris.
Gnaphalium luteo-album.
Senecio viscosus.
Asperugo procumbens.
Mentha rotundifolia.
M. pratensis.
Lysimachia nemorum.
Centunculus minimus.
Littorella lacustris.
Chenopodium urbicum.
Beta maritima.
Obione pedunculata.
Polygonum minus.

Aristolochia Clematitis.
Salix undulata.
S. purpurea.
Fritillaria Meleagris.
Ornithogalum pyrenaicum.
Colchicum autumnale.
Ruppia rostellata.
Carex strigosa.
Setaria viridis.
Phleum asperum.
P. arenarium.
Sclerochloa loliacea.
Equisetum hyemale.
E. sylvaticum.
Lastrea Oreopteris.
Polystichum aculeatum.

The following plants are also probably now extirpated. The date of the last certain notice of each plant is added to its name.

Myosurus minimus..............1835.
Hypericum elodes ....1842.
Radiola millegrana.............1821.
Œnanthe silaifolia ....1833.
Caucalis latifolia ....1833.
Centranthus ruber ....1835.
Senecio palustris, before......1850.
Limosella aquatica ....1827.
Veronica spicata................1829.
Urtica pilulifera, before......1838.
Myrica Gale, before...........1850.
Ophrys aranifera ....1837.
Sturmia Loeselii ... .1836.

No. X. Geographical range of Plants.

The physical character of this county is not such as to limit the range of plants, nevertheless there are a few plants which seem to have the limits of their geographical districts within it. In all probability more complete information will prove that they all extend into the neighbouring counties. The following is a list of the plants which, as far as my information extends, have their northern, southern, or western limit in Cambridgeshire.

I. Plants not known certainly to grow to the north of Cambridgeshire.

Bunium Bulbocastanum.
Œnanthe fluviatilis.
Seseli Libanotis.
Caucalis latifolia.
Filago spathulata.
Senecio campestris?
Villarsia nymphæoides.
Melampyrum cristatum.
Orobanche Picridis.
Ajuga Chamæpitys.
Primula elatior.
Statice caspia.
Thesium humifusum.
Euphorbia platyphylla.
Herminium monorchis.
Phleum Boehmeri.

II. Plants not certainly known to grow to the west of Cambridgeshire.

Silene otites.
Medicago sylvestris.
M. minima?
Trifolium ochroleucum.
Primula elatior.
Phleum Boehmeri.
Apera interrupta.

III. Plant which appears to have its southern limit in this county.

Senecio paludosus.

IV. When we compare our Flora with that of Great Britain, as shewn by the arrangement of all the native British species contained in Watson's *Cybele Britannica* (iv. 234—281), it is found that of the 120 commonest species we only want one,

Lychnis diurna.

APPENDIX. 317

1. Of those included between 121 and 174 we want one, namely,

   Digitalis purpurea.

2. We possess all those between 175 and 218.

3. Between 219 and 259 we want

   Chrysosplenium oppositifolium.
   Orobus tuberosus.
   Fumaria capreolata?

4. Between 260 and 305,

   Gentiana campestris.
   Vaccinium myrtillus.
   Festuca duriuscula.
   Lycopodium Selago.

5. Between 306 and 344,

   Eriophorum vaginatum.
   Viola palustris.
   Osmunda regalis.

6. We have all between 345 and 380.

7. Between 381 and 431 we want

   Tragopogon pratensis?
   Silene maritima.
   Ammophila aruudinacea.

8. Between 432 and 471,

   Epilobium angustifolium.
   Scirpus sylvaticus.
   Honkeneja peploides.
   Pilularia globulifera.

9. Between 472 and 509,

   Corydalis claviculata.
   Hieracium murorum.
   Cakile maritima.
   Cerastium tetrandrum.
   Salix Smithiana.

10. Between 510 and 546,

    Cistopteris fragilis.
    Lepidium Smithii.
    Cochlearia officinalis.

Potamogeton polygonifolius.
Carduus tenuiflorus.
Zostera marina.

11. Between 547 and 578,
    Hypericum Androsæmum.
    Polypodium Phægopteris.
    Hypericum dubium.

12. Between 579 and 606,
    Cardamine amara.
    Ceterach officinarum.
    Sagina maritima.
    Salsola Kali.
    Callitriche pedunculata.

13. Between 607 and 632,
    Empetrum nigrum.
    Scutellaria minor.
    Teesdalia nudicaulis.
    Juncus maritimus.
    Sedum anglicum.
    Vicia lathyroides.
    Asplenium marinum.
    Carex extensa.
    Anagallis cærulea?

14. Between 633 and 660,
    Chrysosplenium alternifolium.
    Pyrola minor.
    Lathræa squamaria.
    Pyrus Aria.
    Diplotaxis tenuifolia.

15. Between 661 and 687,
    Polypodium Dryopteris.
    Cotyledon Umbilicus.
    Rubus saxatilis.
    Eryngium maritimum.
    Myosotis repens.

16. And finally, between 688 and 718 we want
    Viola lutea.
    Glaucium luteum.

Lycopodium alpinum.
Allosorus crispus.
Cochlearia danica.
Convolvulus Soldanella.

Mr Watson divides Great Britain into 38 districts, which he calls Sub-provinces. The first group mentioned above consists of the *commonest plants*, or those found in all the 38 Sub-provinces. The succeeding groups, numbered 1, 2, 3, 4, &c. consist of plants whose range extends to 1, 2, 3, &c., fewer Sub-provinces than the thirty-eight.

ALPHABETICAL LIST OF GENERA.

Acer, 44.
ACERACEÆ, 44.
Aceras, 226.
Achillea, 122.
Acorus, 245.
ACOTYLEDONES, 289.
Adoxa, 105.
Ægopodium, 93.
Æthusa, 98.
Agrimonia, 70.
Agrostis, 271.
Aira, 272.
Ajuga, 185.
Alchemilla, 70.
Alisma, 241.
ALISMACEÆ, 241.
Alliaria, 19.
Allium, 235.
Alnus, 218.
Alopecurus, 268.
Alsine, 35
Althæa, 41.
Alyssum, 22.
AMARANTHACEÆ, 194.
Amaranthus, 194.
AMARYLLIDACEÆ, 233.
AMENTIFERÆ, 213.
Anacharis, 222.
Anagallis, 190.
Anchusa, 155.
Anemone, 2.
Angelica, 99.
Antennaria, 127.
Anthemis, 121.
Anthoxanthum, 267.
Anthriscus, 103.
Anthyllus, 60.
Antirrhinum, 163.
Apargia, 139.
Apera, 270.
Apium, 91.
APOCYNACEÆ, 149.

AQUIFOLIACEÆ, 148.
Aquilegia, 9.
Arabis, 17.
ARACEÆ, 245.
ARALIACEÆ, 105.
Arctium, 130.
Arenaria, 36, 302.
Aristolochia, 207.
ARISTOLOCHIACEÆ, 207.
Armeria, 192.
Armoracia, 22.
Arnoseris, 137.
Arrhenatherum, 273.
Artemisia, 124.
Arum, 246.
ASPARAGACEÆ, 233.
Asparagus, 233.
Asperugo, 155.
Asperula, 108.
Asplenium, 293.
Aster, 118.
Astragalus, 60.
Athyrium, 293.
Atriplex, 198.
Atropa, 160.
Avena, 273.

Ballota, 184.
BALSAMINACEÆ, 49.
Barbarea, 16.
Bellis, 119.
BERBERIDACEÆ, 10.
Berberis, 10.
Beta, 197.
Betula, 218.
Bidens, 120.
Blechnum, 294.
Blysmus, 255.
BORAGINACEÆ, 155.
Burago, 155.
Botrychium, 295.
Brachypodium, 285.

Brassica, 20.
Briza, 279.
Bromus, 283.
Bryonia, 84.
Bunium, 94.
Bupleurum, 96.
Butomus, 242.

Calamagrostis, 270.
Calamintha, 177.
CALLITRICHACEÆ, 210.
Callitriche, 210.
Calluna, 147.
Caltha, 8.
Camelina, 23.
Campanula, 145.
CAMPANULACEÆ, 145.
CAPRIFOLIACEÆ, 106.
Capsella, 25.
Cardamine, 17.
Carduus, 134.
Carex, 256.
Carlina, 130.
Carpinus, 219.
Carum, 93.
CARYOPHYLLACEÆ, 31.
Catabrosa, 280.
Caucalis, 101.
CELASTRACEÆ, 50.
Centaurea, 132.
Centranthus, 113.
Centunculus, 191.
Cephalanthera, 230.
Cerastium, 39.
CERATOPHYLLACEÆ, 209.
Ceratophyllum, 209.
Chærophyllum, 103.
Chara, 297.
CHARACEÆ, 297.
Cheiranthus, 15.
Chelidonium, 13.

ALPHABETICAL LIST OF GENERA. 321

CHENOPODIACEÆ, 194.
Chenopodium, 195.
Chlora, 150.
Chrysanthemum, 122.
Cichorium, 137.
Cicuta, 91.
Circæa, 82.
CISTACEÆ, 27.
Cladium, 252.
Clematis, 1.
Cochlearia, 22.
COLCHICACEÆ, 237.
Colchicum, 237.
Comarum, 72.
COMPOSITÆ, 117.
CONIFERÆ, 220.
Conium, 104.
Convallaria, 234.
CONVOLVULACEÆ, 153.
Convolvulus, 153.
Coriandrum, 105.
CORNACEÆ, 105.
Cornus, 105.
Corylus, 219.
CRASSULACEÆ, 87.
Cratægus, 79.
Crepis, 143.
Crocus, 232.
CRUCIFERÆ, 15.
CRYPTOGAMEÆ, 289.
CUCURBITACEÆ, 84.
Cuscuta, 154.
Cynoglossum, 155.
Cynosurus, 280.
CYPERACEÆ, 252.

Dactylis, 280.
Daphne, 206.
Datura, 161.
Daucus, 100.
Delphinium, 10.
Dianthus, 31.
DICOTYLEDONES, 1.
DIOSCOREACEÆ, 221.
Diplotaxis, 21.
DIPSACEÆ, 115.
Dipsacus, 115.
Draba, 22.
Drosera, 29.
DROSERACEÆ, 29.

Echium, 157.
Eleocharis, 253.
ENDOGENÆ, 221.
Endymion, 236.
Epilobium, 81.
Epipactis, 229.

EQUISETACEÆ, 289.
Equisetum, 289.
Erica, 147.
ERICACEÆ, 147.
Erigeron, 118.
Eriophorum, 256.
Erodium, 48.
Erysimum, 19.
Erythræa, 151.
Euonymus, 50.
Eupatorium, 117.
Euphorbia, 207.
EUPHORBIACEÆ, 207.
Euphrasia, 168.
EXOGENÆ, 1.

Fagopyrum, 206.
Fagus, 218.
Festuca, 280.
Filago, 126.
FILICES, 290.
Fœniculum, 98.
Fragaria, 73.
Frankenia, 30.
FRANKENIACEÆ, 30.
Fraxinus, 149.
Fritillaria, 234.
Fumaria, 14.
FUMARIACEÆ, 14.

Galeopsis, 182.
Galium, 109.
Genista, 52.
Gentiana, 152.
GENTIANACEÆ, 150.
GERANIACEÆ, 44.
Geranium, 44.
Geum, 75.
Glaucium, 13.
Glaux, 191.
Glyceria, 277.
Gnaphalium, 127.
GRAMINEÆ, 266.
GROSSULARIACEÆ, 89.
Gymnadenia, 225.

Habenaria, 226.
Hedera, 105.
Helianthemum, 27.
Helleborus, 9.
Helminthia, 140.
HALORAGACEÆ, 83.
Helosciadium, 92.
Heracleum, 100.
Herminium, 228.
Herniaria, 85.
Hieracium, 144.

Hippocrepis, 66.
Hippuris, 84.
Holcus, 272.
Hordeum, 287.
Hottonia, 189.
Humulus, 212.
HYDROCHARIDACEÆ, 222.
Hydrocharis, 222.
Hydrocotyle, 90.
Hyoscyamus, 160.
HYPERICACEÆ, 42.
Hypericum, 42.
Hypochæris, 138.

Iberis, 23.
Ilex, 148.
Impatiens, 49.
Inula, 119.
IRIDACEÆ, 232.
Iris, 232.
Isatis, 25.

Jasione, 145.
JUNCACEÆ, 237.
Juncus, 238.
Juniperus, 220.

Koantia, 116.
Koeleria, 275.

LABIATÆ, 173.
Lactuca, 141.
Lamium, 180.
Lapsana, 137.
Lastrea, 291.
Lathyrus, 64.
LEGUMINOSÆ, 51.
Lemna, 246.
LENTIBULARIACEÆ, 186.
Leontodon, 142.
Leonurus, 182.
Lepidium, 24.
Lepigonum, 85.
Lepturus, 287.
Ligustrum, 149.
LILIACEÆ, 234.
Limosella, 166.
LINACEÆ, 48.
Linaria, 164.
Linum, 48.
Listera, 229.
Lithospermum, 157.
Littorella, 194.
Lolium, 288.
Lonicera, 107.
LORANTHACEÆ, 106.

Lotus, 59.
Luzula, 240.
Lychnis, 33.
LYCOPODIACEÆ, 296.
Lycopodium, 296.
Lycopsis, 156.
Lycopus, 176.
Lysimachia, 189.
Lythrum, 80.

Malachium, 38.
Malaxis, 231.
Malva, 40.
MALVACEÆ, 40.
Marrubium, 184.
Matricaria, 123.
Medicago, 53.
Melampyrum, 167.
Melica, 275.
Melilotus, 55.
Mentha, 173.
Menyanthes, 153.
Mercurialis, 209.
Milium, 269.
Mœhringia, 36.
Mœnchia, 38.
Molinia, 276.
MONOCOTYLEDONES, 221.
Monotropa, 148.
Montia, 84.
Muscari, 236.
Myosotis, 158.
Myosurus, 3.
Myrica, 217.
Myriophyllum, 83.

Narcissus, 233.
Nardus, 269.
Narthecium, 237.
Nasturtium, 16.
Neottia, 229.
Nepeta, 180.
Nuphar, 11.
Nymphæa, 11.
NYMPHÆACEÆ, 11.

Obione, 199.
Œnanthe, 96.
ONAGRACEÆ, 81.
Onobrychis, 66.
Ononis, 53.
Onopordum, 134.
Ophioglossum, 296.
Ophrys, 227.
ORCHIDACEÆ, 223.
Orchis, 223.
Origanum, 176.

Ornithogalum, 234.
Ornithopus, 65.
OROBANCHACEÆ, 161.
Orobanche, 161.
Osmunda, 296.
OXALIDACEÆ, 50.
Oxalis, 50.

Papaver, 11, 300.
PAPAVERACEÆ, 11.
Parietaria, 210.
Paris, 221.
Parnassia, 30.
PARONYCHIACEÆ, 85.
Pastinaca, 100.
Pedicularis, 167.
Peplis, 81.
Petasites, 117.
Petroselinum, 92.
Peucedanum, 99.
Phalaris, 266.
Phleum, 267.
Phragmites, 270.
Picris, 140.
Pimpinella, 94.
Pinguicula, 186.
PLANTAGINACEÆ, 192.
Plantago, 192.
PLUMBAGINACEÆ, 192.
Poa, 276.
Polygala, 30.
POLYGALACEÆ, 30.
POLYGONACEÆ, 200.
Polygonum, 203.
Polypodium, 290.
Polystichum, 293.
Populus, 216.
PORTULACEÆ, 84.
Potamogeton, 247.
POTAMOGETONACEÆ, 247.
Potentilla, 71.
Poterium, 69.
Primula, 188.
PRIMULACEÆ, 188.
Prunella, 179.
Prunus, 66.
Pteris, 295.
Pulicaria, 120.
Pyrus, 79.

Quercus, 219.

Radiola, 49.
RANUNCULACEÆ, 1.
Ranunculus, 3.
Raphanus, 26.

Reseda, 26.
RESEDACEÆ, 26.
RHAMNACEÆ, 50.
Rhamnus, 50.
Rhinanthus, 168.
Rhynchospora, 253.
Ribes, 89.
Rœmeria, 13.
Rosa, 76.
ROSACEÆ, 66.
Rubia, 112.
RUBIACEÆ, 108.
Rubus, 73, 305.
Rumex, 200.
Ruppia, 251.
Ruscus, 234.

Sagina, 33.
Sagittaria, 242.
Salicornia, 197.
Salix, 213.
Salvia, 176.
Sambucus, 106.
Samolus, 191.
Sanguisorba, 69.
Sanicula, 90.
SANTALACEÆ, 206.
Saponaria, 31.
Sarothamnus, 52.
Saxifraga, 89.
SAXIFRAGACEÆ, 89.
Scabiosa, 116.
Scandix, 102.
Schœnus, 252.
Scirpus, 254.
Scleranthus, 86.
Sclerochloa, 278.
Scolopendrium, 294.
Scrophularia, 166.
SCROPHULARIACEÆ, 162.
Scutellaria, 179.
Sedum, 87.
Sempervivum, 88.
Senebiera, 25.
Senecio, 128.
Serrafalcus, 284, 307.
Serratula, 132.
Seseli, 98.
Setaria, 266.
Sherardia, 108.
Silaus, 99.
Silene, 32.
Silybum, 137.
Sinapis, 20.
Sison, 93.
Sisymbrium, 18.
Sium, 95.

Smyrnium, 104.
SOLANACEÆ, 159.
Solanum, 159.
Solidago, 119.
Sonchus, 142.
Sparganium, 245.
Specularia, 147.
Spergula, 86.
Spiræa, 68.
Spiranthes, 228.
Stachys, 183.
Statice, 192.
Stellaria, 37.
Stratiotes, 222.
Sturmia, 231.
Suæda, 194.
Symphytum, 156.

Tamus, 221.
Tanacetum, 125.
Taxus, 220.
Teesdalia, 23.
Teucrium, 185.

Thesium, 206.
Thlaspi, 23.
Thalictrum, 1, 299.
Thrincia, 139.
THYMELACEÆ, 206.
Thymus, 177.
Tilia, 43.
TILIACEÆ, 42.
Torilis, 101.
TRAGOPOGON, 140.
Trifolium, 55.
Triglochin, 243.
TRILLIACEÆ, 221.
Triodia, 275.
Trisetum, 273.
Triticum, 286, 309.
Tussilago, 118.
Typha, 244.
TYPHACEÆ, 244.

Ulex, 51.
ULMACEÆ, 212.
Ulmus, 212.

UMBELLIFERÆ, 90.
Urtica, 211.
URTICACEÆ, 210.
Utricularia, 187.

Vaccinium, 148.
Valeriana, 113.
VALERIANACEÆ, 113.
Valerianella, 114.
Verbascum, 162.
Verbena, 186.
VERBENACEÆ, 186.
Veronica, 169.
Viburnum, 107.
Vicia, 61.
Villarsia, 152.
Vinca, 149.
Viola, 27, 302.
VIOLACEÆ, 27.
Viscum, 106.

Zannichellia, 252.

INDEX OF THE ENGLISH NAMES.

Abele, 216.
Adder's-tongue, 296.
Agrimony, 70.
Alder, 218.
Alehoof, 180.
Alexanders, 104.
Allgood, 197.
Allheal, 184.
Archangel, 181.
Arrowhead, 242.
Ash, 149.
Ashweed, 93.
Aspen, 217.
Asphodel, 237.

Barberry, 10.
Barley-grass, 287.
Basil, 178.
Betony, 166, 183.
Bear's-foot, 9.
Beech, 218.
Beet, 197.
Bell flower, 145.
Bent-grass, 271.
Birch, 218.
Bird Cherry, 67.
Bird's-foot, 65.
Birds-foot trefoil, 59.
Bird's-nest, 100, 148, 229.
Birthwort, 207.
Bindweed, 153, 205.
Bitter-sweet, 159.
Black Alder, 51.
Black Mustard, 20.
Blackthorn, 66.
Bladder Campion, 32.
Bladderwort, 187.
Blinks, 84.
Blue-bottle, 132.
Brakes, 295.
Brank, 206.
Brandy Bottle, 11.

Briar, 78.
Brooklime, 170.
Brook-weed, 191.
Broom, 52.
Broom-rape, 160.
Bryony, 84, 221.
Bog Rush, 252.
Borage, 155.
Buckbean, 153.
Buckthorn, 50.
Buckwheat, 206.
Bugle, 185.
Bugloss, 156.
Bullace, 67.
Bullrush, 254.
Burdock, 131.
Burnet, 69.
Burnet-saxifrage, 94, 95.
Bur-reed, 245.
Butcher's Broom, 234.
Butterbur, 117.
Butterwort, 186.

Calamint, 178.
Caltrops, 249, 251.
Calves-snout, 165.
Campion, 32, 33.
Candytuft, 23.
Caraway, 93.
Carrot, 100.
Catchfly, 32.
Cat-mint, 180.
Cat's-ear, 138.
Cat's-foot, 127.
Cat's-tail grass, 268.
Celandine, 13.
Celery, 91.
Centaury, 151.
Chamomile, 121, 122.
Charlock, 21, 26.
Cheese-rening, 111.
Cherry, 67.

Chervil, 103.
Chickweed, 37, 38.
Chickweed Spurry, 85.
Clary, 176.
Cleavers, 110.
Clover, 54, 55, 57.
Club-moss, 296.
Cock's-foot-grass, 280.
Coleseed, 20.
Coltsfoot, 118.
Columbine, 9.
Comfrey, 156.
Coriander, 105.
Corn Cockle, 33.
Corn-grass, 272.
Corn Rose, 12.
Corn Salad, 114.
Corn Marigold, 123.
Corn Parsley, 102.
Cotton Tree, 107.
Cotton Thistle, 134.
Couch-grass, 286.
Cow-parsnep, 100.
Cowslip, 188.
Cow-wheat, 167.
Crab Tree, 79.
Cranberry, 148.
Crane's-bill, 45, 48.
Crosswort, 109.
Crowfoot, 4, 5, 6.
Cuckoo-flower, 18, 33.
Cuckoo-pint, 246.
Cudweed, 126, 127.
Currants, 89.

Daffodil, 233.
Daisy, 119.
Dandelion, 142.
Danewort, 106.
Darnel, 288.
Dead-Nettle, 181.
Devil's-bit, 116.

INDEX OF ENGLISH NAMES. 325

Dewberry, 75.
Dittander, 24.
Dock, 200, 201, 202.
Dock-cresses, 137.
Dodder, 154.
Dog's Mercury, 209.
Dog Rose, 78.
Dog's-tail-grass, 280.
Dog-wood, 105.
Dove's-foot, 46.
Dropwort, 68, 96.
Duck-weed, 246.
Dutch Clover, 57.
Dwale, 159.
Dyer's-weed, 52.
Dyer's Woad, 25.

Earth Nut, 94.
Elder, 106.
Elecampane, 119.
Elm, 212.
Enchanter's Nightshade, 82.
Everlasting Pea, 65.
Eye-bright, 168.

Fat-hen, 195.
Fellwort, 152.
Fennel, 98.
Fescue-grass, 281, 282.
Feverfew, 123.
Figwort, 166.
Florin-grass, 271.
Five-leaved Grass, 71.
Flag, 232.
Flax, 48.
Fleabane, 118, 120, 129.
Fleabane-Mullet, 129.
Flixweed, 19.
Float-grass, 278.
Flowering Rush, 242.
Fluellin, 164.
Fool's Parsley, 98.
Forget-me-not, 158.
Fox-tail Grass, 268.
Fritillary, 234.
Frogbit, 222.
Frog's Lettuce, 249.
Fumitory, 14.
Furze, 51.

Gale, 217.
Garlic, 235.
Gatter-tree, 105.
Germander, 185.
Gipsy-wort, 176.
Gladdon, 232.

Goat's-beard, 140.
Golden Rod, 119.
Goldilocks, 7.
Gooseberry, 89.
Goose-grass, 110.
Gorse, 51.
Goutweed, 93.
Grass of Parnassus, 30.
Green Saxifrage, 99.
Green-weed, 52.
Gromwell, 157, 158.
Ground Ivy, 180.
Ground Pine, 186.
Groundsel, 128.
Guelder-Rose, 107.

Hair-grass, 275.
Hammersedge, 264.
Hard-fern, 294.
Harebell, 146, 236.
Hare's-ear, 96.
Hare's-foot Trefoil, 56.
Hart's-tongue, 294.
Hawkbit, 139.
Hawthorn, 79.
Hazel, 219.
Heart's-ease, 29.
Heath-grass, 275.
Hedge Mustard, 18.
Hedge Parsley, 101.
Hedgehog-grass, 262.
Hedgehog Parsley, 102.
Hellebore, 9.
Hemlock, 104.
Hemp, wild, 182.
Hemp-Agrimony, 117.
Hemp-Nettle, 183.
Henbane, 159.
Henbit, 180.
Heptree, 78.
Herb Gerard, 93.
Herb Paris, 221.
Herb Truelove, 221.
Herb Robert, 47.
Herb Twopence, 189.
High-taper, 162.
Holly, 148.
Honewort, 92.
Honeysuckle, 108.
Hop, 212.
Hop Trefoil, 58.
Horehound, 176, 184.
Hornbeam, 219.
Horned Pond-weed, 252.
Horned Poppy, 13.
Hornwort, 209.
Horse Radish, 22.

Horseshoe Vetch, 66.
Horse-tail, 289.
Hound's-tongue, 155.
Houseleek, 88.
Hyacinth, 236.

Ironwort, 182.
Ivy, 105.

Juniper, 220.

Kidney Vetch, 60.
Knapweed, 132, 133.
Knawell, 86.
Kneeholm, 234.
Knot-grass, 205.
Knotted Spurrey, 35.

Lady-fern, 293.
Lady's Bedstraw, 111.
Lady's Finger, 60.
Lady's Hair, 279.
Lady's Mantle, 70.
Lady's Smock, 18.
Lady's Tresses, 228.
Lamb's Lettuce, 114.
Lamb's-tongue, 193.
Larkspur, 10.
Lettuce, 141.
Lily of the valley, 234.
Lime, 42.
Ling, 147.
Liquorice, 61.
Livelong, 87.
Loosestrife, 80, 189.
Lousewort, 167.
Lowry, 206.

Madwort, 155.
Maithes, 121.
Mallow, 40, 41.
Maple, 44.
Mare's-tail, 84.
Marjoram, 176.
Marsh Cinque-foil, 72.
Marshlocks, 72.
Marsh Marigold, 8.
Matfellon, 132, 133.
Mat-grass, 269.
Mayweed, 121.
Meadow-grass, 276, 277.
Meadow Pink, 33.
Meadow Rue, 2.
Meadow Saffron, 237.
Meadow Sweet, 68.
Medick, 53, 54.
Melilot, 55.

28

Mercury, 209.
Mignonette, 26.
Milk Thistle, 137.
Milkwort, 30.
Millefoil, 3, 122.
Millet-grass, 269.
Mil Mountain, 49.
Mint, 174.
Mistletoe, 106.
Moneywort, 189, 190.
Moonwort, 295.
Moschatel, 105.
Motherwort, 182.
Mountain Ash, 79.
Mouse-ear, 144.
Mouse-tail, 3.
Mudwort, 166.
Mugweed, 109.
Mullen, 162, 163.
Mustard, 20, 21.

Navew, 20.
Needle Furze, 52.
Neppe, 180.
Nettle, 211.
Nightshade, 158, 159.
Nipplewort, 137.

Oak, 219.
Oat-grass, 273, 274.
Oats, 273.
One-berry, 221.
Orache, 195.
Orchis, 223—228.
Orpine, 87.
Osier, 215.
Ox-eye, 122.
Ox-lip, 188.
Ox-tongne, 140.

Paigle, 188.
Pansies, 29.
Parsley, 92, 98, 102.
Parsley Piert, 70.
Parsnep, 100.
Pasque Flower, 2.
Pearlwort, 33.
Pear Tree, 79.
Pellitory, 210, 211.
Penny Cress, 23.
Penny Royal, 175.
Pennywort, 90.
Peppermint, 174.
Periwinkle, 149, 150.
Petty Whin, 52.
Pig Nut, 94.
Pilewort, 6.

Pimpernel, 190, 191.
Pink, 31, 33.
Pipperidge Bush, 10.
Plantain, 192.
Ploughman's Spikenard, 119.
Plum, 67.
Polypody, 290.
Pond-weed, 247, 251, 252.
Poplar, 216, 217.
Poppy, 12.
Prickmadam, 87.
Prickwood, 50.
Primrose, 188.
Privet, 149.

Quaking-grass, 279.
Quicken Tree, 79.

Ragged Robin, 33.
Ramsons, 236.
Rape, 20.
Raspherry, 73.
Red Bryony, 84.
Red Rattle, 168.
Reed, 270.
Reed-grass, 266, 277.
Reed Mace, 244.
Rest Harrow, 53.
Ribwort, 193.
Rock Rose, 27.
Rowan Tree, 79.
Rue Witlow-grass, 90.
Rush, 238, 252.
Rye grass, 288.

Saffron, 232, 237.
Sage, wood, 184.
Sanicle, 90.
Salfern, 158.
Sallow, 215.
Saltwort, 191, 197.
Sawwort, 132.
Scurvy-grass, 22.
Sea Lavender, 192.
Sea Spurry, 86.
Sedge, 252.
Self-heal, 179.
Service Tree, 80.
Sheep's Scabious, 145.
Shepherd's Needle, 102.
Shepherd's Rod, 115.
Sowbane, 196.
Silver-weed, 71.
Skull-cap, 179.
Sloe, 66.

Smallage, 91.
Snake's-head, 234.
Snake-weed, 203.
Snapdragon, 163.
Sneezewort, 122.
Soapwort, 31.
Soft-grass, 272.
Sorrel, 203.
Sow Thistle, 142, 143.
Spatling Poppy, 32.
Spearwort, 6.
Speedwell, 170.
Spindle Tree, 50.
Spleenwort, 293.
Spurge, 207, 208.
Spurge Laurel, 206.
Spurry, 85, 86.
Squinancy Berries, 89.
Squinancywort, 108.
St John's-wort, 42.
St Peter's-wort, 42, 43.
Star of Bethlehem, 234, 235.
Star Thistle, 133.
Starwort, 118, 210.
Stitchwort, 37.
Strawberry, 72, 73.
Strawberry Trefoil, 58.
Stone Basil, 178.
Stonecrop, 87.
Stonewort, 93.
Sundew, 29.
Succory, 137.
Sweet Briar, 77.
Sweet Flag, 245.
Sycamore, 44.

Tansy, 125.
Tare, 61, 62, 63.
Tare Everlasting, 64.
Tassel Pond-weed, 251.
Teasel, 115.
Tentwort, 294.
Tetterwort, 137.
Thistle, 134—137.
Thorn Apple, 160.
Thorough-wax, 96.
Throatwort, 146.
Thyme, 177, 178.
Tinetare, 61.
Toad-flax, 165, 266.
Toad-grass, 240.
Tormentil, 72.
Traveller's Joy, 1.
Treacle Wormseed, 19.
Trefoil, 54—59.
Tunhoof, 180.

INDEX OF ENGLISH NAMES.

Turnip, 20.
Tway-blade, 229.

Valerian, 113.
Vernal-grass, 267.
Vervain, 186.
Vetch, 63, 64.
Vetchling, 64.
Violet, 27, 28.
Viper's Bugloss, 157.

Wake Robin, 246.
Wall Pepper, 87.
Water Avens, 76.
Water Caltrops, 249, 251.
Water Cress, 16.
Water Crowfoot, 4, 5.
Water Elder, 107.
Water Fennel, 3.
Water Hemlock, 91.
Water Ivy, 246.
Water Lily, 11.

Water Milfoil, 83, 187.
Water Parsnip, 95.
Water Pepper, 205.
Water Plantain, 241.
Water Purslane, 81.
Water Soldier, 222.
Water Thyme, 222.
Water Violet, 189.
Water Yarrow, 83.
Wallflower, 15.
Wall Rue, 294.
Wallwort, 106.
Way-bennet, 287.
Way-bread, 193.
Wayfaring Tree, 107.
Weld, 26.
Whin, 51.
White Bottle, 32.
Whiterot, 90.
Whitethorn, 79.
Whitlow-grass, 22.

Wild Oats, 273.
Wild Tansy, 71.
Wild Williams, 33.
Willow, 213.
Willow-herb, 81.
Wind Flower, 2.
Wind-grass, 270.
Wood Avens, 75.
Woodbine, 108.
Woodruff, 109.
Woodrush, 240.
Wood Sage, 185.
Wood Sorrel, 50.
Wood Waxen, 52.
Wormwood, 124.

Yarrow, 122.
Yellow Flag, 232.
Yellow Rocket, 16.
Yellowwort, 150.
Yew, 220.

THE END.

CAMBRIDGE: PRINTED BY C. J. CLAY, M.A. AT THE UNIVERSITY PRESS.

www.ingramcontent.com/pod-product-compliance
Lightning Source LLC
Chambersburg PA
CBHW022334230426
43664CB00040B/606